P9-ECN-338

Gumhalter

Power Supply Systems in Communications Engineering Part 2

Power Supply Systems in Communications Engineering

Part 2 Equipment Engineering and Planning Instructions

By Hans Gumhalter

Siemens Aktiengesellschaft
John Wiley & Sons

Title of German original edition:
Stromversorgungssysteme der Kommunikationstechnik
Hans Gumhalter
Siemens Aktiengesellschaft 1984
ISBN 3-8009-1413-1

Deutsche Bibliothek Cataloguing in Publiation Data:

Gumhalter, Hans:
Power supply systems in communications engineering/Hans Gumhalter. – Berlin;
München: Siemens-Aktiengesellschaft, [Abt. Verl.];
Chichester; New York; Brisbane; Toronto;
Singapore: Wiley
Dt. Ausg. u.d.T.: Gumhalter, Hans: Stromversorgungssysteme der Kommunikationstechnik

Pt. 2. Equipment engineering and system design.
1985.
ISBN 3 8009 1443 3 (Siemens)
ISBN 0 471 90627 1 (Wiley)

ISBN 3-8009-1413-1 (Siemens Aktienges.)

Library of Congress Cataloging in Publication Data:
(Revised for vol. 2)

Gumhalter, Hans.
Power supply systems in communications engineering.

Bibliography: v. 1, p. 221–223, v. 2, p.
Includes indexes.
Contents: pt. 1. Principles—pt. 2. Equipment engineering and planning instructions.
1. Telecommunication systems—Power supply.
I. Title.
TK5102.5.G85 1984 621.38 83–13666
ISBN 0 471 90290 X (pbk. : v. 1)
ISBN 0 471 90627 1 (pbk : v. 2)

British Library Cataloguing in Publication Data:

Gumhalter, Hans
Power supply systems in communications engineering.
Pt 2 : Equipment engineering and planning instructions
1. Telecommunication—Power supply
I. Title II. Stromversorgungssysteme der Kommunikationstechnik. *English*
621.38 TK5102.5

ISBN 0 471 90627 1

Printed and bound in Great Britain.

This second part of the book, like the recently published first part, is based in its choice of material and manner of presentation upon knowledge and experience gained within the framework of training programmes; it reviews the present state of systems incorporating thyristor- and transistor-controlled converters, as developed by manufacturers and applied by users, and explains the circuit techniques and relationships with the aid of extensive generally-conceived illustrations. The subjects of battery application, earthing and protection techniques are also considered.

Keywords

Central power supply equipments
Thyristor-controlled rectifier equipments
Single-phase equipments
Three-phase equipments
Assemblies
Panels
Compensators
Radio link supplies
Ground communication station supplies
Mains-independent power supplies
Primary power sources
Diesel generators
Solar generators
Wind generators
Hybrid power supplies
Decentralized power supplies
Transistor-controlled rectifiers
Equipments with longitudinal controllers
Equipments with switching controllers
Switching-mode power supplies
D.C./D.C. converters
Inverters
Energy stores
Earthing
Functional and protective earthing
Potential equalization
Protective measures

Foreword

The equipments used for telecommunications power supplies constitute an important part of every communications system. Part 2 of this book, like the recently published Part 1, is based in its choice of material and manner of presentation upon knowledge and experience gained within the framework of pertinent training programmes.

Building upon the foundation of the first part, the present volume reviews the present state of systems incorporating thyristor- and transistor-controlled converters, as developed by manufacturers and applied by users, and explains the circuit techniques and relationships with the aid of extensive generally-conceived illustrations. The book also deals with the subjects of battery application, earthing and protection techniques. Information is given on the planning of battery and earthing installations.

The individual sections are arranged and written in such a way that the book will be equally useful for readers making their first acquaintance with the subject and for those who are directly concerned with planning, installation, commissioning and operation.

Readers who need details of the equipments discussed are referred to the relative circuit principles and to detailed descriptions and operating instructions.

Thanks are due especially to Herr Biagini for his assistance with Chapters 4 and 5.

Munich, November 1984

Siemens Aktiengesellschaft

Contents

1 Central Power Supply Equipments

1.1 Rectifiers with Thyristor Power Sections

At this point a selection of equipment will be described; they all employ the principle of phase-angle control.

1.1.1 Single-phase equipments

1.1.1.1 Type 60 V (48 V)/25 A GR2

Application

The rectifier is suitable for supplying 48 and 60 V communications systems, such as, for example:

▷ PABX systems,
▷ supervisory installations,
▷ clock installations.

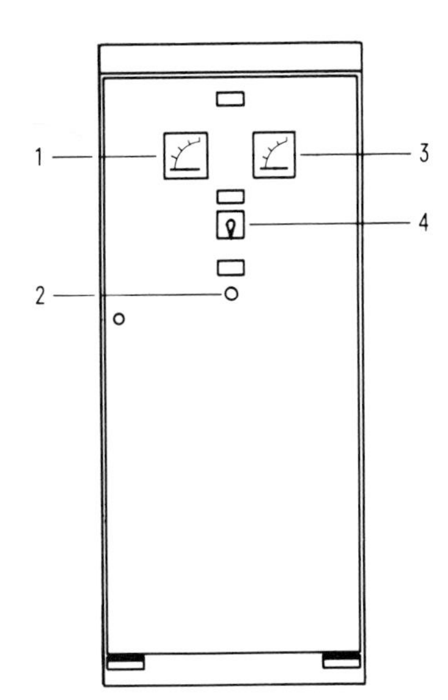

1 Voltmeter
2 Voltmeter measurement switch
3 Ammeter
4 Operating mode switch

Fig. 1.1 Rectifier type 60 V/25 A GR2, exterior

A maximum of three rectifiers can be connected in parallel (Figs 1.1 and 1.2). A current consumption of up to 75 A for the communications system and the battery can thus be provided for.

As well as the 60 V (48 V)/25 A GR2 rectifier (cubicle construction), there are also the technically very similar equipments:

> 60 V (48 V)/6 A GR2 (wall-mounting),
> 60 V (48 V)/12 A GR2 (wall-mounting) and
> 48 V/30 A GR2 (cubicle construction).[1]

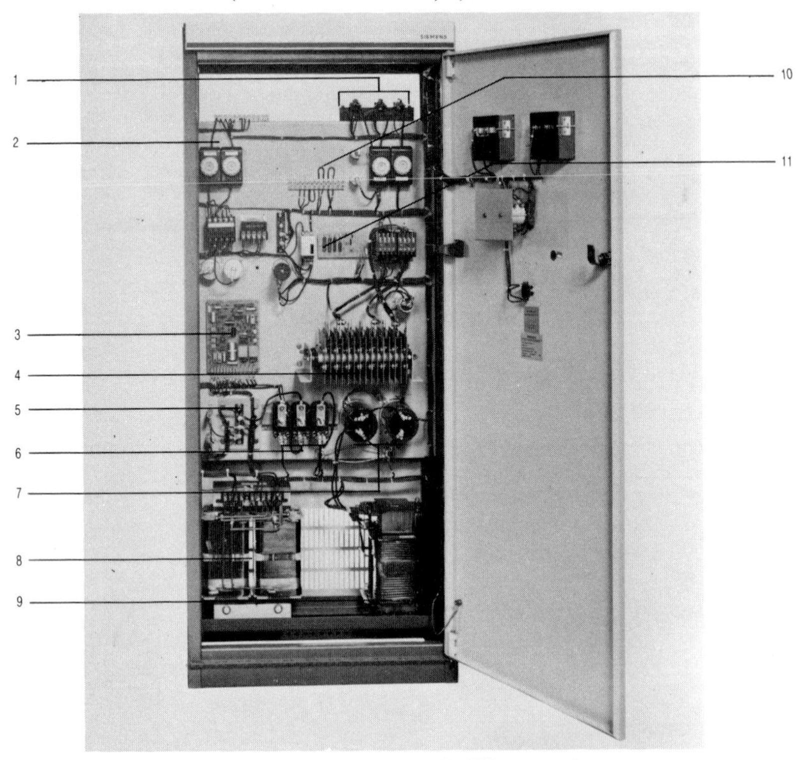

1 D.C. terminals
2 Mains input terminals
3 Closed-loop and open-loop control A1
4 Reducing diodes
5 Thyristors
6 Diodes
7 Filter capacitors
8 Main transformer
9 Filter inductor
10 Mounting position for reducing-diode control A5
11 Voltage limiter A2

Fig. 1.2 Rectifier type 60 V/25 A GR2 with door open

---

[1]) Contains no reducing diodes (counter cells).

Type designations

Rectifier equipments are identified as follows:

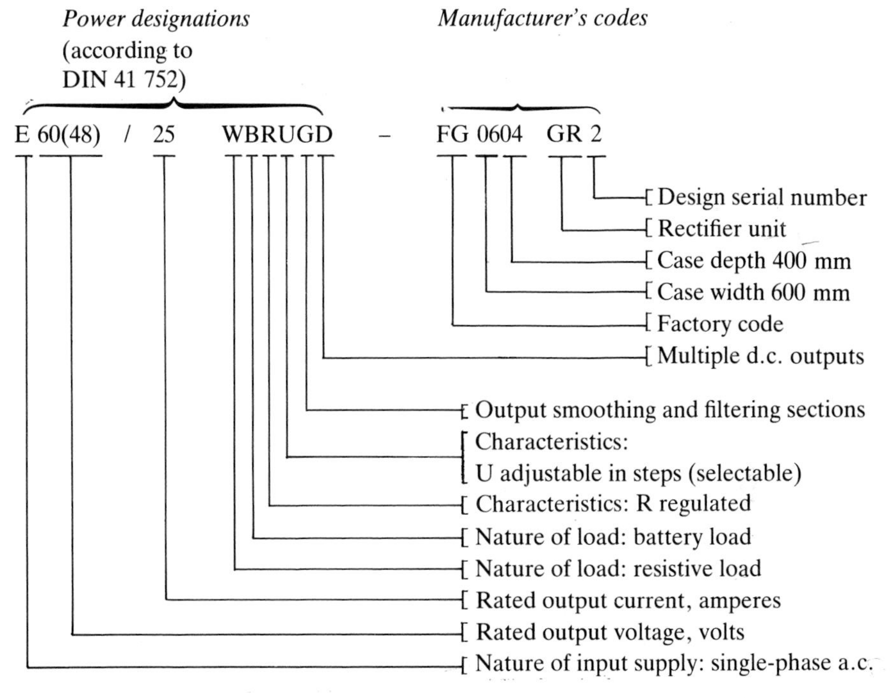

The example given above of a complete type designation applies by analogy to all the other thyristor-controlled rectifiers referred to in this book.

The significance of the codes is as follows:

Nature of input supply

E single-phase a.c.
D three-phase a.c.
G d.c.

Nature of load

W resistive load.

In this case the rectifier can be connected directly to the communications system, without a parallel-connected battery.

B battery load.

The rectifier can be connected to a battery, which is supplied at the required charging voltage.

Characteristics

R rectifiers with characteristics determined by closed-loop control, and so with stabilized characteristics (controlled equipments).

Regardless of the specified supply voltage, supply frequency and load variations the d.c. output voltage is held constant within a tolerance band of, for example, ±0.5%.

U rectifier with characteristic adjustable in steps (selectable).

Smoothing and filtering sections

G rectifier with smoothing or filtering sections.

The filtering sections limit the magnitude of interference voltages to a level that is permissible for the communications system.

Multiple inputs or outputs

No code rectifier with only one input and one output.
D rectifier with several d.c. outputs.

When several outputs are provided, the power code indicates only the rating of the main output.

The code D thus appears in connection with thyristor-controlled rectifiers when several outputs are available. This is the case where a rectifier is used in a parallel mode and provides two different direct voltages simultaneously, one for the communications system and one (higher) for the battery (trickle charging or charging).

The *abbreviation GR* (rectifier unit) may be noted as an example. There are also other abbreviations, e.g.:

GG for rectifiers and
LGG for charging rectifier.

Where significant electrical or constructional variations are introduced in a rectifier, the *design serial number* is correspondingly altered. There are, for example, rectifiers with design serial numbers 1, 2, 3, 6, 10, 11, 12, 31.

Operating modes

The rectifier unit can be used either in the rectifier mode or in the parallel mode with (voltage)-reducing diodes.

Rectifier mode

In the rectifier mode the rectifier supplies the load directly from the supply mains at a constant direct voltage of 51 or 62 V. In the event of a power (mains) failure or a fault in the rectifier the supply to the load is lost.

Parallel mode with reducing diodes

In the parallel mode with reducing diodes (Fig. 1.3) the rectifier, the battery and the load are connected in parallel. The battery is supplied at a trickle-charging voltage of 2.23 V/cell (normal operation). This voltage, which is too high for the load, has to be reduced to 51 or 62 V by reducing diodes, which are inserted into the load circuit by a voltage-responsive means.

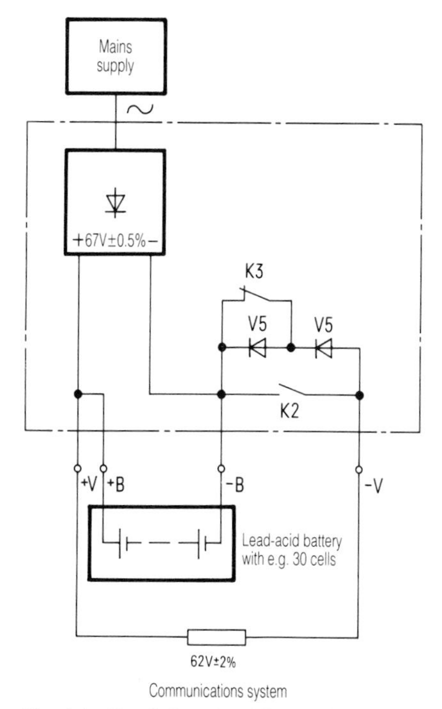

Fig. 1.3 Parallel mode with reducing diodes (normal operation)

To replace the energy taken from the battery during a protracted mains failure, the rectifier is switched *manually*, upon restoration of the supply, to parallel operation at a charging voltage of 2.33 V/cell. The resulting excessive output voltage, from the point of view of the load, is reduced to 51 or 62 V by the voltage-dependent insertion of further reducing diodes in series with the load circuit.

As a special treatment for a new or reconditioned battery it may be necessary to carry out an initial charge (with the load disconnected).

Basic circuit and operation

The basic circuit of the rectifier unit (Fig. 1.4) can be divided into the following functional units and assemblies:

▷ power section,
▷ voltage limiter A2,
▷ reducing-diode control A5,
▷ closed-loop and open-loop control A1.

Power section

The supply voltage reaches the transformer T1 by way of the fuses F1, F2 and the contactor K1. The transformer supplies the semi-controlled single-phase bridge V1, V3, V6; this provides at its output a constant controlled direct voltage, which is smoothed by the filter formed by L1 and C2 (frequency-weighted interference (noise) voltage $\leqq 2$ mV).

The measurement shunt R1 monitors the actual output current I_{Ist} for the purpose of current limiting. The utilization of I_{Ist} is a function of the current regulator (controller) within the closed-loop and open-loop control assembly.

---

① Internal power supply		FPE	Functional and protective earth conductor
② Characteristic changeover		A1	Closed-loop and open-loop control
③ Actual-voltage value adjustment		A2	Voltage limiter
④ Voltage regulator		A5	Reducing-diode control
⑤ Current regulator		V1	Thyristor module
⑥ Active filter (180° phase shift)		T1	Transformer
⑦ 7 kHz clock generator		S1	Switch
⑧ Trigger set			
⑨ Integrated triggering-pulse generator			
⑩ Pulse amplifier			
⑪ Pulse transformers			
⑫ Pulse-enabling, pulse-cancelling and automatic start-up			
⑬ Matching transformer			

Fig. 1.4 Basic circuit of rectifier type 60V (48 V)/25 A GR2

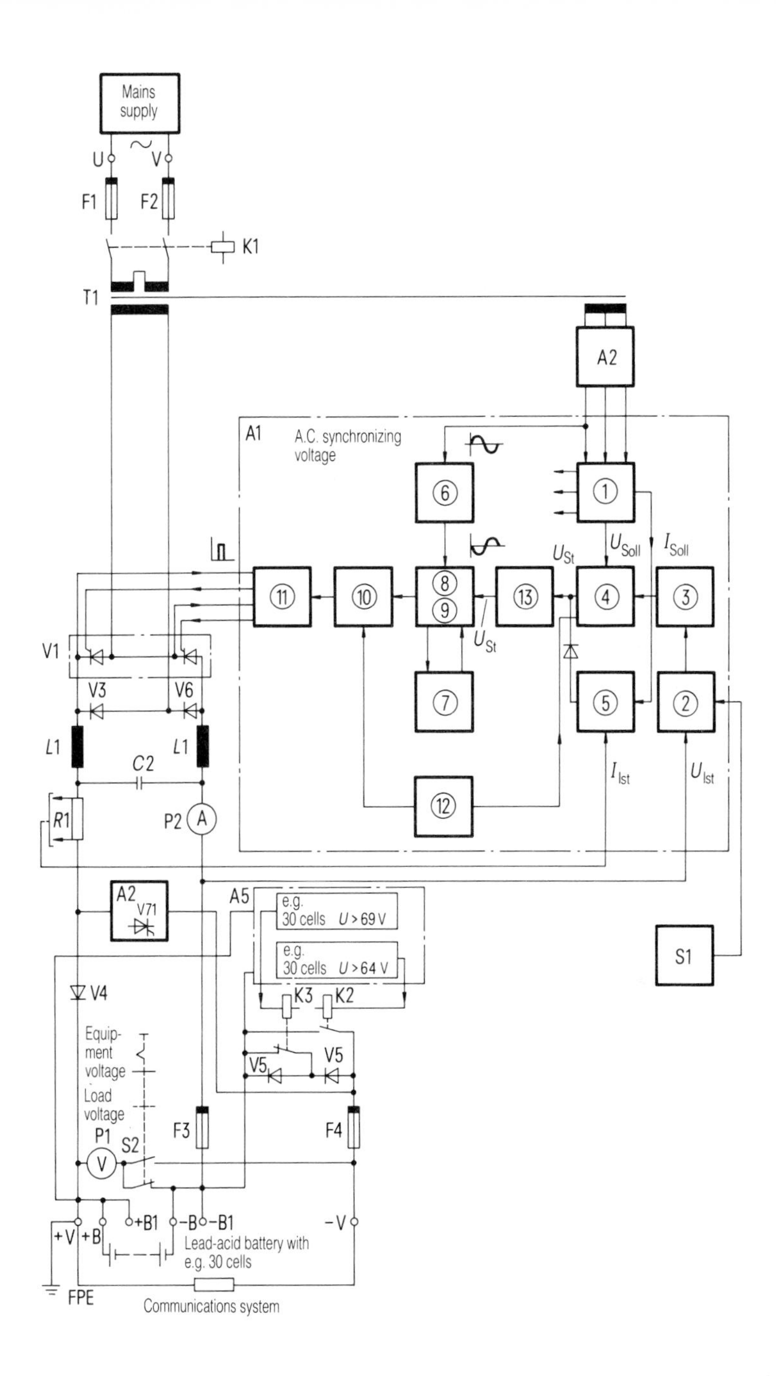

Mains supply
U
V
F1
F2
K1
T1
A2
A1
A.C. synchronizing voltage
V1
V3
V6
L1
L1
C2
R1
P2
A
U_St
U_Soll
I_Soll
U_St
I_Ist
U_Ist
A2
V71
A5
e.g. 30 cells U > 69 V
e.g. 30 cells U > 64 V
S1
V4
K3
K2
Equip-ment voltage
Load voltage
V5
V5
P1
V
S2
F3
F4
+V
+B
+B1
-B
-B1
-V
Lead-acid battery with e.g. 30 cells
FPE
Communications system

The output current of the equipment (load current plus battery-charging current) is measured by the ammeter P2.

Fuse F3 is the equipment fuse; F4 is the load fuse.

The diodes V5 serve as reducing diodes.

The voltmeter P1 enables either the equipment output voltage or the load voltage to be measured according to the position of the switch S2. The difference between the two voltages is the voltage drop across the reducing diodes.

Voltage limiter A2

Protection of the communications system against overvoltage spikes is effected by the voltage limiter A2. In the event of a disturbance, if the present voltage level is exceeded (60 V for a 48 V system or 73 V for a 60 V system), the voltage-limiting thyristor V71 is triggered. This momentarily draws a high current from the rectifier, corresponding to around two or three times the rated current of the equipment. As a result, the d.c. output voltage of the equipment falls. Subsequently the contactor K1 opens in response to an automatic circuit breaker (not shown in the basic circuit diagram) and switches off the rectifier.

The decoupling diode V4 prevents a feedback of current from the battery or a parallel-connected rectifier when the voltage limiter A2 operates.

Assembly A2 also fulfills the function of mains overvoltage limiting (protection of the closed-loop and open-loop control assembly A1).

Reducing-diode control A5

The reducing diodes are inserted in series with the load or bridged in two groups, depending on the voltage, by the reducing-diode control A5 through the bridging contactors K2 and K3. In the rectifier mode the bridging contactor K2 is closed and both the reducing-diode groups are bridged. In this mode of operation the equipment voltage corresponds to the load voltage, so that the voltage does not have to be reduced (51 V ± 0.5% in 48 V systems and 62 V ± 0.5% in 60 V systems).

When the parallel mode at 2.23 V/cell is employed, if a predetermined voltage is exceeded (53 V for 25 cells, 64 V for 30 cells or 65 V for 31 cells) A5 causes the bridging contactor K2 to be de-energized and its contact K2 to open. Contact K3 remains closed and thus *only one* reducing-diode group is inserted.

The equipment output voltage is thereby reduced to the permitted level for the load (51 V ± 2% for 48 V systems and 62 V ± 2% for 60 V systems). If the equipment output voltage drops below a predetermined level (52 V for 25 cells,

63 V for 30 cells or 64 V for 31 cells)—because of a discharged battery, for example—K2 operates and bridges both groups of reducing diodes again.

In the parallel mode at 2.33 V/cell, if a higher predetermined voltage is exceeded (58 V for 25 cells, 69 V for 30 cells or 70 V for 31 cells), in addition to the de-energization of K2, the bridging contactor K3 is energized. Contact K3 is now opened and the second group of reducing diodes is therefore similarly inserted into the load circuit. Both reducing-diode groups are then effective and the equipment output voltage falls to the level permitted for the load (51 V ± 2% for 48 V systems and 62 V ± 2% for 60 V systems). When the voltage drops below a predetermined level (57 V for 25 cells, 68 V for 30 cells or 69 V for 31 cells) the bridging contactor K3 drops out again and one reducing-diode group is bridged.

Closed-loop and open-loop control A1

The closed-loop and open-loop control assembly embraces the functional units represented in Fig. 1.4.

Variation of the output voltage of the voltage and current regulator (control voltage U_{St}) applied to the trigger set (control unit) causes the trigger pulses to be shifted in time relative to the supply voltage. This signifies phase-angle control and determines a mean direct voltage level at the output of the thyristor set. An internal power supply ① is fed from the auxiliary secondary winding of the transformer T1 through the assembly A2 (supply voltage limiting). This provides supplies of ±15 V stabilized and ±24 V unstabilized. It also provides the constant reference values for the voltage and current (U_{Soll} and I_{Soll}). The supply-voltage-limiting functional unit serves to protect the assembly A1 from overvoltage spikes.

The a.c. synchronizing voltage is also taken from the secondary winding of T1 and supplied to the trigger set through an active filter ⑥ (180° phase shift). The coupling of voltage disturbances into the control circuit of the rectifier is thereby largely eliminated. The a.c. synchronizing voltage is used in the trigger set to enable it to produce trigger pulses in synchronism with the supply voltage.

From each half-wave of the a.c. synchronizing voltage a sawtooth pulse is produced in the trigger set ⑧ by an integrated trigger pulse generator ⑨. The sawtooth voltage remains substantially constant and serves as a reference voltage for the variable control voltage U_{St}.

A properly defined (matched) actual voltage value U_{Ist} is supplied through the characteristic changeover ② and actual voltage value adjustment ③ to the voltage regulator ④. This compares U_{Ist} with the constant set-point (reference) voltage U_{Soll} obtained from the power supply. The voltage resulting from this comparison is amplified by the voltage regulator and supplied as a variable control voltage U_{St} via a matching transformer (voltage follower) ⑬ to the trigger set. This phase-shifts the trigger pulses – according to the magnitude of the control voltage U_{St} – and feeds them through pulse amplifiers and pulse transformers to the thyristor set. Figure 1.5 illustrates the shifting of the triggering pulses in dependence upon the control voltage.

Each trigger pulse consists of a pulse chain (7 kHz repetition frequency). The mark–space ratio is 1:1. The pulse transformer may therefore be small. The pulse chain is produced by a 7 kHz clock generator ⑦.

The trigger pulses are amplified by the pulse amplifier ⑩.

Darlington transistors are used as pulse amplifiers. The pulse transformer ⑪ provides electrical isolation between closed-loop and open-loop control A1 and the power section; in addition it transforms the trigger pulses down to a voltage suitable for the thyristors.

The functional unit pulse enabling, pulse cancelling (inhibit) and automatic start-up (soft start) ⑫ is provided in order to avoid an overswing of the equipment output voltage when the rectifier is switched on. First the pulse amplifier must be enabled; then the voltage controller must be constrained to give a high control voltage U_{St} at its output, so that the first trigger pulses occur at the inverter stability limit (trailing-edge triggering threshold) α_W. Only then are the trigger pulses shifted slowly towards the rectifier stability limit (leading-edge trigger threshold) α_G and the equipment output voltage thereby slowly increased ('soft start').

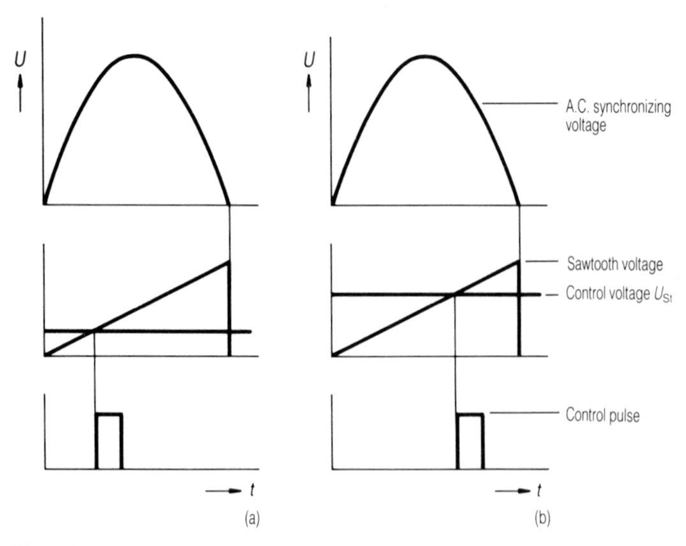

Fig. 1.5
Shifting of triggering pulse as controlled by the relationship of sawtooth voltage to control voltage U_{St}

Table 1.1 Technical data for rectifier type 60 V (48 V)/25 A GR2

Input supply		
Voltage	(V)	220 +10 to –15% or 110 +25 to –15%
Frequency	(Hz)	50 or 60 ±5%
Supply distribution fuse protection	(A)	35 (at 220 V)
Equipment fuse protection	(A)	20 (at 220 V)
Degree of radio interference		N (VDE 0875)

D.C. output

Operating mode or condition	Rated direct voltage (V) (tolerances for 1.5 to 100% I_{Nenn})					
	60 V version			48 V version		
	Equipment voltage for lead-acid battery of		Load voltage	Equipment voltage for lead-acid battery of		Load voltage
	30 cells	31 cells		24 cells	25 cells	
Rectifier operation	62 ± 0.5%	62 ± 0.5%	62 ± 0.5%	51 ± 0.5%	51 ± 0.5%	51 ± 0.5%
Parallel operation/trickle charging (2.23 V/cell)	67 ± 0.5%	69 ± 0.5%	62 ± 2%	53.5 ±0.5%	56 ± 0.5%	51 ± 2%
Parallel operation/charging (2.33 V/cell)	70 ± 0.5%	72.5 ± 0.5%	62 ± 2%	56 ± 0.5%	58.5 ± 0.5%	51 ± 2%
Initial charging	84	84	Communications system disconnected	65	67.5	Communications system disconnected

Rated direct current	(A)	25
Interference voltage	(mV)	≦2 (frequency-weighted with CCITT 'A' filter)
Dimensions ($H \times W \times D$)	(mm)	Cubicle construction: 1400 × 600 × 400

So long as the current set-point (reference) value I_{Soll} is greater than the actual current value I_{Ist} obtained from the measurement shunt $R1$ in the power section, the current regulator (controller) ⑤ has no effect upon the regulation. Only when the preset current-limit value (usually 100% of the rated current I_{Nenn}) is exceeded does the current limiting take effect. Then $I_{Ist} > I_{Soll}$ and there appears at the output of the current regulator a positive voltage, which can be applied to the trigger set through the diode (conducting under this condition). The current regulator turns off the voltage regulator and in the trigger set the trigger pulses are shifted towards the limit α_W. This causes a reduction of the d.c. output voltage and the rectifier is thus protected against overload.

Technical data

The principal technical data for the rectifier type 60 V (48 V)/25A GR2 are listed in Table 1.1.

1.1.2 Three-phase equipments

1.1.2.1 Type 48 V/100 A GR2

The thyristor-controlled rectifier 48 V/100 A GR2 is illustrated in Figs 1.6 and 1.7, which show respectively the exterior and the cubicle with the door opened.

Application

The equipment is intended mainly for supplying large 48 V PABX systems. Three rectifiers can be connected in parallel, so that a current demand of up to 300 A for the communications system and the battery can be met.

The rectifiers 48 V/100 A GR2 and 48 V/40 A GR2 are essentially similar in design (cubicle construction).

Operating modes

The rectifier can be used either in the rectifier mode or in the parallel mode (generally in the standby parallel mode) with reducing diodes.

Rectifier mode

Normal operation The communications system is fed from the supply mains via the rectifier at a direct voltage of 51 V ±0.5%.

Power failure In the event of a mains failure or fault in the rectifier, the supply to the communications system is lost.

Mains return When the mains supply is restored the rectifier switches on again and resumes the supply to the communications system.

Both groups of reducing diodes are permanently bridged under all of the above operating conditions.

Standby parallel mode with reducing diodes

Normal operation The communications system is fed from the mains supply via the rectifier (Fig. 1.8). The lead-acid battery of 25 cells sustains a trickle-charging voltage of about 56 V. This voltage, too high for the communications system, is reduced to 51 V ±2% by means of reducing diodes V3/1 and V4/1 inserted into the load circuit.

Power failure – battery discharge The communications system is fed from the battery. Immediately following a mains failure the reducing-diode groups remain in circuit (as in normal operation). The voltage provided to the communications system during a supply interruption – with the battery discharging – is approximately equal to the battery voltage.

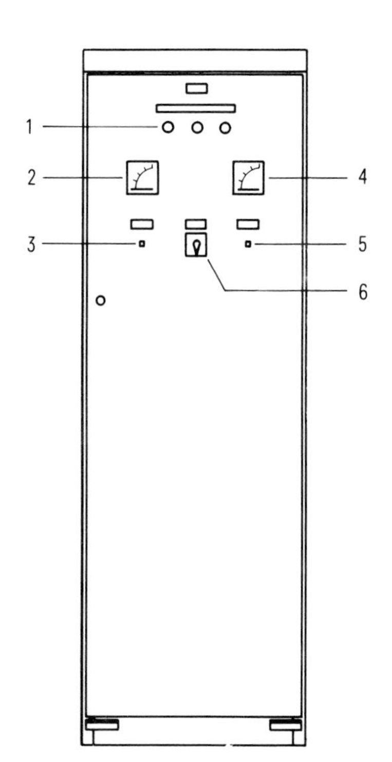

1 Indication lamps
from left to right:
operation/fault/voltage deviation
2 Voltmeter
3 Voltmeter measurement switch
4 Ammeter
5 Ammeter measurement switch
6 Equipment switch (off/operation)

Fig. 1.6 Rectifier type 48 V/100 A GR2, exterior

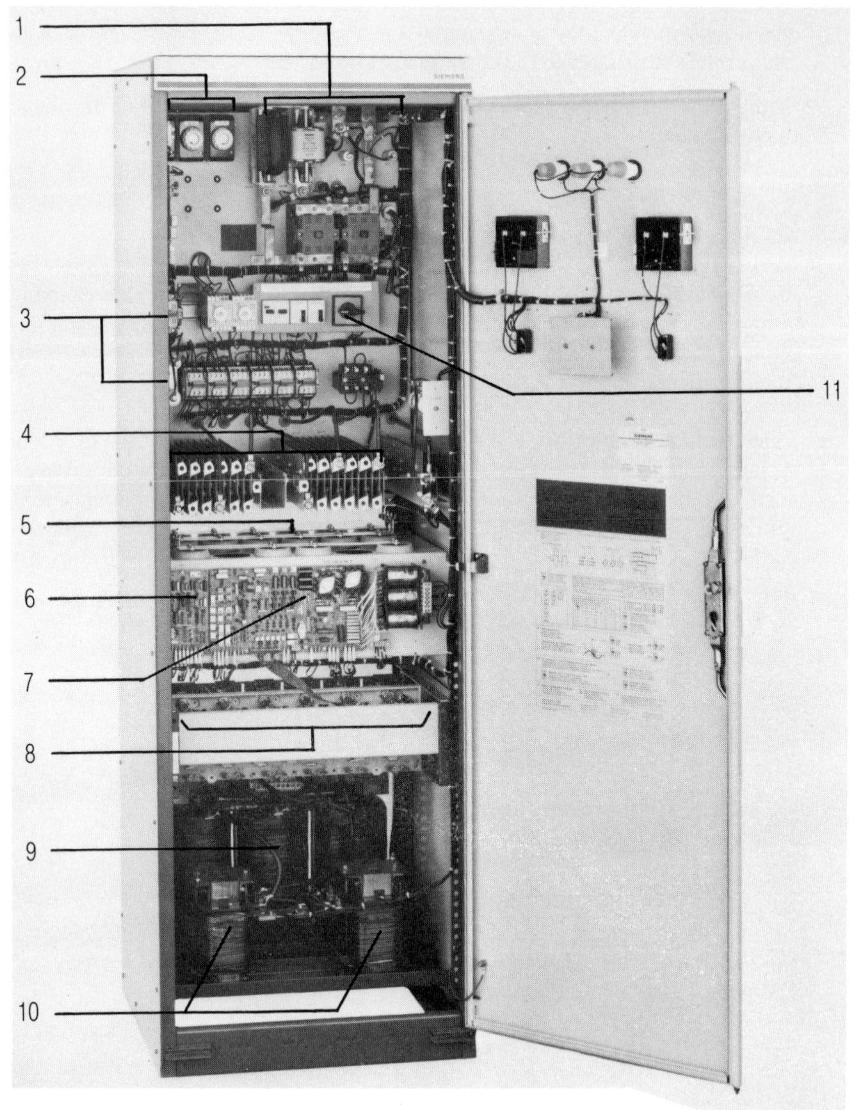

1 D.C. terminals
2 Mains input terminals
3 Relay and contactor control
4 Reducing diodes
5 Filter capacitors
6 Mains monitor A4
7 Closed-loop and open-loop control A3
8 Thyristor set A1
9 Main transformer
10 Filter inductors
11 Operating mode switch

Figure 1.7 Rectifier type 48 V/100 A GR2 with door open

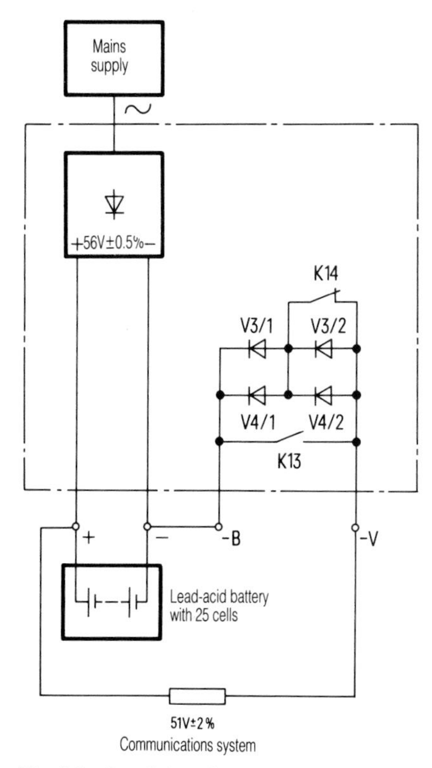

Fig. 1.8 Parallel mode with voltage-reducing diodes (normal operation)

If the voltage drops below 52 V, the bridging-contactor contact K13 closes, so that the battery is connected directly to the communications system.

Mains return The rectifier automatically switches on again and resumes the supply to the communications system.

During the operating time of the 'charging time' time-delay relay (typically a maximum of 24 h) the equipment voltage rises to about 58.5 V. Both groups of reducing diodes (V3/1–V4/1 and V3/2–V4/2) are inserted in series with the load circuit and the voltage supplied to the communications system is reduced to 51 V ±2%.

At the end of the charging period the equipment is automatically switched back to normal operation.

It may be necessary for a new or reconditioned battery to undergo initial charging (a special commissioning charge), with a charging voltage of up to 67.5 V in the case of a lead-acid battery of 25 cells. Since this would result in an excessive load voltage, even with the reducing diodes interposed, the initial charging must be carried out with the communications system disconnected.

Basic circuit and operation

The rectifier equipment comprises the following functional units and assemblies (Fig. 1.9):

- ▷ power section,
- ▷ actual current value matching A10,
- ▷ voltage monitor with test circuit A7 and voltage limiting,
- ▷ reducing-diode control A5,
- ▷ voltage monitor A8,
- ▷ mains monitor A4,
- ▷ closed-loop and open-loop control A3,
- ▷ relay and contactor control.

Power section

The a.c. supply is taken through the fuses F1, F2 and F3 and the main contacts of the supply contactor K1 to the main transformer T1; this serves to separate the communications system 'galvanically' from the mains and to provide for the appropriate d.c. output from the thyristor set (a fully-controlled three-phase bridge). This output is in the form of a constant, controlled direct voltage. A1 fulfils the functions of rectifying and stabilizing the output voltage of the equipment.

By means of the filters L1, L2, C1 and C2 in the d.c. circuit the output is smoothed and the interference voltage is reduced to less than the maximum permissible value of 0.5 mV. In addition, by virtue of its ability to store and release energy, the smoothing filter limits variations in the direct voltage due to load transients to within a certain range.

The actual current value I_{Ist} is determined by means of the measurement shunt R1 for the purpose of current limiting. R1 also serves as a shunt for the ammeter P2 when the switch S37 'current measurement' is turned to the equipment current (load current + battery current) position. If the switch is turned to the load current position the ammeter is connected to the measurement shunt R2 instead of R1.

Fuse F9 is the equipment fuse; F8 is the load fuse.

The voltmeter P1 provides for measurement of the equipment voltage, which corresponds to the battery-charging voltage in the case of parallel operation or of the load voltage only, according to the position of the switch S36, 'voltage measurement'.

The switch 'mode of operation' S31 permits selection of the desired mode and condition of operation. The switch has four positions:

1 Rectifier operation.
2 Parallel operation/automatic.
  This is the normal operating position in battery-supported power supply installations. In this mode (in normal operation) the battery receives a trickle charge at 2.23 V/cell.
3 Parallel operation/charging.
  In this position charging of the battery at 2.33 V/cell may be instituted manually. The rectifier does *not* subsequently switch back automatically to trickle charging.
4 Initial charging.
  Before this switch position can be selected the rectifier must be switched off by means of S32 'equipment'. The initial charging must only be instituted with the communications system disconnected, to avoid applying an excessive voltage to it. The fuse F8 should be removed to this end. The rectifier can then be switched on again with switch S32 and the initial charging of the battery carried out.

Actual current value matching A10

The actual current value matching assembly A10 supplies a voltage proportional to the actual current – about 30 mV, for example, with a current of 100% I_{Nenn} – derived from the measurement shunt R1, to the actual current value amplifier ⑦. The utilization of I_{Ist} is a function of the current regulator (controller) ⑧.

Voltage monitor with test circuit A7 and voltage limiting

The voltage monitor with test circuit A7 is responsible for protecting the communications system against overvoltage spikes (see Fig. 1.13) in conjunction with the voltage limiting. In the event of a disturbance, if the preset maximum voltage value of 57 V is exceeded, the voltage-limiting thyristor V200 is triggered. The rectifier then supplies, for a short period, a high current, corresponding to two or three times the rated current of the equipment.

As a result of the high current the d.c. output voltage of the equipment falls. In addition, the A7 assembly and the closed-loop and open-loop control assembly A3 cause the trigger pulses to thyristors V1 to V6 to be blocked, thus interrupting the further flow of energy from the supply.

The current that flows in consequence of the triggering of thyristor V200 is limited by the resistor $R200$ and the bimetallic relay F200. This protects the thyristor and, if there is a risk of an overload, switches off the rectifier through the fault contactor K80 (relay and contactor control).

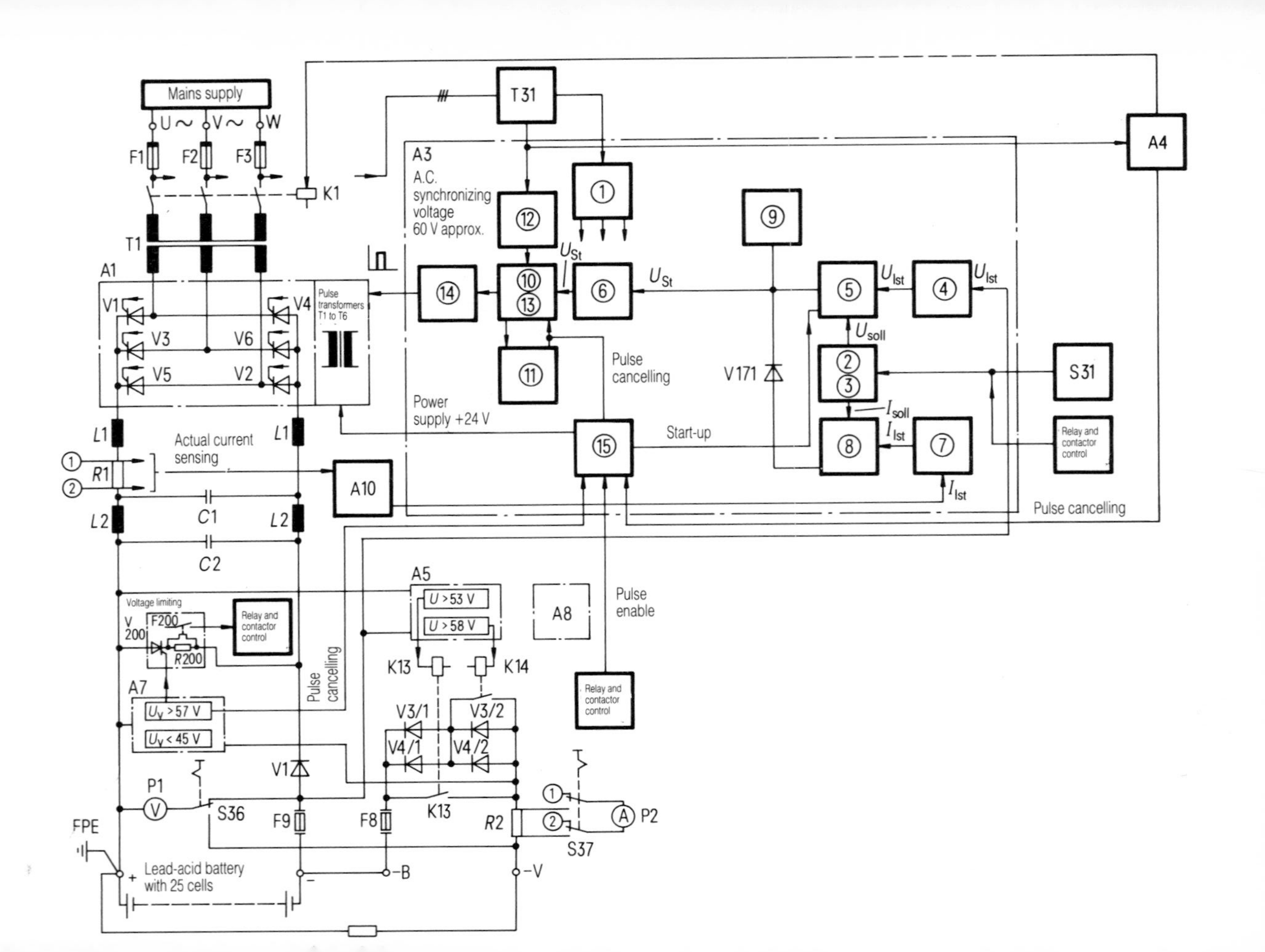
Mains supply
U ~ V ~ W
F1 F2 F3
K1
T1
T 31
A 3
A.C. synchronizing voltage 60 V approx.
A4
A1
V1 V4
V3 V6
V5 V2
Pulse transformers T1 to T6
Power supply +24 V
Pulse cancelling
Start-up
V 171
S 31
Relay and contactor control
Pulse cancelling
L1
Actual current sensing
R1
A10
C1
L2
C2
A5
U > 53 V
U > 58 V
A8
Pulse enable
Voltage limiting
V 200
F200
R200
Relay and contactor control
K13 K14
A7
U_V > 57 V
U_V < 45 V
Pulse cancelling
V3/1 V3/2
V4/1 V4/2
Relay and contactor control
V1
P1
S36
F9
F8
K13
R2
P2
S37
FPE
Lead-acid battery with 25 cells
–B
–V

① Internal power supply
② Set-point device
③ Characteristic control
④ Actual-voltage value matching
⑤ Voltage regulator
⑥ Matching transformer
⑦ Actual-current value amplifier
⑧ Current regulator
⑨ α_W and α_G adjustment
⑩ Trigger set
⑪ 7 kHz clock generator
⑫ 60° phase-shift filter
⑬ Integrated triggering-pulse generator
⑭ Trigger-pulse amplifier
⑮ Pulse enabling, pulse cancelling and automatic start-up

A1 Thyristor set
A3 Closed-loop and open-loop control
A4 Mains monitor
A5 Reducing-diode control
A7 Voltage monitor with test circuit
A8 Voltage monitor (if required: A8 is interchangeable with A5)
A10 Actual-current value matching
S31 Operating-mode switch
T1 Main transformer
T31 Auxiliary transformer
FPE Functional and protective earth conductor

Fig. 1.9 Basic circuit of rectifier type 48 V/100 A GR2

The decoupling diode V1 (power section) prevents a feedback from the battery or a parallel-connected rectifier equipment when the voltage limiting operates.

The A7 assembly also includes an undervoltage monitor. If the load voltage drops below 45 V, an indication is provided. The rectifier is not switched off, however (see Fig. 1.13).

Reducing-diode control A5

So that the rectifier can supply the communications system and the battery simultaneously, in parallel operation the reducing diodes V3/1, V4/1, V3/2 and V4/2 are employed to reduce the equipment output voltage to the level permissible for the communications system.

The reducing diodes are inserted in series with the load or bridged in two groups, depending on the voltage, by the reducing-diode control A5 in conjunction with the bridging contractors K13 and K14. To allow for the two operating modes, the parallel mode at 2.23 V/cell and the parallel mode at 2.33 V/cell, the reducing diodes are introduced in two steps. Both groups consist of a number of diodes.

In the rectifier mode the bridging-contactor contacts K13 and K14 are closed and both reducing-diode groups are bridged. Under this condition the equipment voltage corresponds to the load voltage (51 V) and there is therefore no requirement to reduce the direct voltage.

In the parallel mode at 2.23 V/cell, if the equipment output voltage exceeds 53 V, the bridging contactor K13 is caused to release by the relay K310 on A5 (see Fig. 1.11). With the opening of the K13 contact the reducing diodes V3/1 and V4/1 (the first group) are inserted into the load circuit and the equipment output is thereby reduced from about 56 V (for 25 lead-acid battery cells) by approximately 5 V to the level of 51 V permissible for the load. The bridging contactor K14 remains energized and bridges the second group of reducing diodes V3/2 and V4/2 through its closed contact K14.

If the equipment output voltage drops below 52 V (on battery discharge, for example) the bridging contactor K13 operates again. Contact K13 therefore bridges both groups of reducing diodes (see Fig. 1.12).

In the parallel mode at 2.33 V/cell, with an equipment output voltage of 58 V, the second group of reducing diodes V3/2 and V4/2 is inserted into the load circuit in addition to the first group V3/1 and V4/1. This is brought about by operation of the relay K320 on assembly A5 (see Fig. 1.11). This causes the bridging contactor K14 to release; contact K14 thus opens and all the reducing diodes are then in circuit.

In addition to the voltage drop of about 5 V across the first reducing-diode group V3/1 and V4/1, there is a drop of about 2.5 V across the second group V3/2 and

V4/2, and the equipment output voltage of about 58.5 V (for 25 lead-acid battery cells) is reduced to 51 V for the communications system.

If the equipment voltage falls below 57 V, the relay K320 (on A5) releases again, K14 operates and the second reducing-diode group V3/2 and V4/2 is bridged (see Fig. 1.11).

Voltage monitor A8

The purpose of voltage monitor A8 (fitted only when required) is to enable the load to be switched off if the battery voltage falls below a preset value – below 46 V, for example – through the relay K310 and output terminal 5 on the rectifier.

Mains monitor A4

Mains monitor A4 receives, as a power supply, the a.c. synchronizing voltage from the auxiliary transformer T31 via assembly A3.

Assembly A4 checks the separation of the zero crossings of the three-phase alternating voltage. It is possible to detect not only phase failure but also unbalance of the three-phase system (normal setting: rated supply frequency ±7 Hz). A supply overvoltage and undervoltage monitor is also included (normal setting: rated supply voltage ±15%). In the event of a response in either of these monitoring circuits, the trigger pulses to the thyristors are blocked by a logic signal via assembly A3, while A4 switches off the rectifier with the relay contact K1 via K63 and the supply contactor K1 (see Fig. 1.12).

Closed-loop and open-loop control A3

The closed-loop and open-loop control assembly A3 embraces the functional units shown in Fig. 1.9. It is incorporated into all the three-phase thyristor-controlled rectifiers described in this book.

The trigger set supplies trigger pulses to the thyristor set via pulse amplifiers. The pulse transformers T1 to T6 provide 'galvanic separation' between A3 and the thyristors V1 to V6; the set also transforms the trigger pulses down to a voltage suitable for the thyristors.

In response to the output voltage of the voltage and current regulators (controllers) (control voltage U_{St}) applied to the trigger set, the timing of the trigger pulses is varied. This determines the mean direct voltage at the output of the thyristor set. The a.c. synchronizing voltage of about 60 V is taken from the secondary side of transformer T31 and supplied to the trigger set through a 60° phase-shift filter ⑫, which largely eliminates the coupling of disturbances into the control circuit of the rectifier. The a.c. synchronizing voltage enables the trigger set to deliver trigger pulses to the thyristors in synchronism with the supply voltage. Transformer T31 provides in addition three further supplies, each of about 24 V, for the internal power supply ① of the assembly A3. From these are produced d.c. supply voltages of ±15 V stabilized and +24 V unstabilized.

The mode of operation switch S31 and the characteristic relays in the characteristic control ③ determine the output voltage produced by the rectifier at any particular time. The voltage regulator ⑤ receives the set-point (reference) voltage U_{Soll} from the set-point device ②. U_{Soll}, and thus the output voltage of the equipment, can be adjusted by means of the reference potentiometer in the set-point device.

The actual voltage U_{Ist} is sensed on the negative line at the output of the equipment and fed to the voltage regulator through the actual voltage value matching ④.

The voltage regulator compares U_{Soll} with U_{Ist}, the particular value of U_{Soll} being constant for a given characteristic while U_{Ist} is a variable. The voltage resulting from this comparison of the reference and actual voltages is amplified in the voltage controller and fed via the matching transformer (voltage follower) ⑥ as a control voltage U_{St} to the trigger set ⑩ with an integrated trigger-pulse generator ⑬.

The output voltage limiting in the controller is used to set the stability limits, α_G (leading-edge triggering threshold) and α_W (trailing-edge triggering threshold). This functional unit is referred to as α_W and α_G adjustment ⑨.

The trigger set phase-shifts the trigger pulses, in accordance with the magnitude of the control voltage U_{St}, and feeds them through the trigger pulse amplifier ⑭ and the pulse transformers to the thyristors V1 to V6. The shifting of the trigger pulses in dependence upon the control voltage is explained in connection with Fig. 1.5.

The functional unit pulse enabling pulse cancelling (inhibit), and automatic start-up (soft start) ⑮ in the closed-loop and open-loop control A3, in the pulse enable condition, causes the +24 V power supply to the pulse transformers T1 to T6 to be switched on, thus permitting the trigger pulses to be conveyed to the thyristors. If a fault is detected by the mains monitor A4 or the voltage monitor with test circuit A7, a pulse inhibit signal is transmitted to the assembly A3. When this takes effect, no further trigger pulses can be delivered to the thyristors by the trigger set. The equipment output voltage falls to zero.

The operation of the 7 kHz clock generator, trigger-pulse amplification, the pulse enable, pulse inhibit and soft start unit, the voltage regulator and the current regulator were explained in connection with Fig. 1.4.

Relay and contactor control

At this point the relay and contactor control for the 48 V/100 A GR2 rectifier (Fig. 1.10) will be discussed, as it is representative of all rectifiers with thyristor

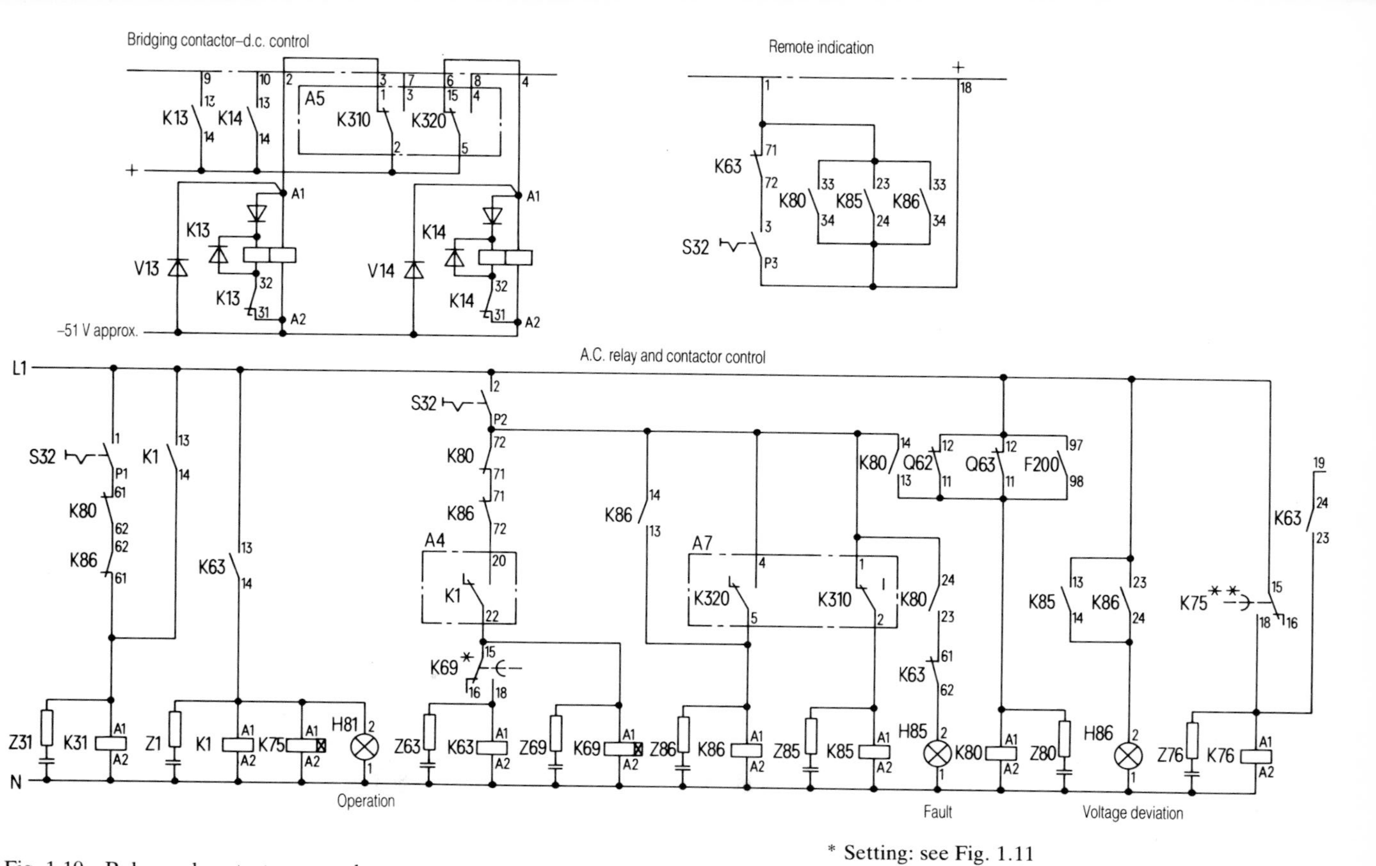

Fig. 1.10 Relay and contactor control

* Setting: see Fig. 1.11

** Setting depends on installation: generally 10 to 24 h

power sections. It can be divided into three parts:

▷ bridging-contactor d.c. control,
▷ remote indications,
▷ a.c. relay and contactor control.

The supply for the bridging-contactor d.c. control is taken from the output of the rectifier unit (about 51 V). The two bridging contactors K13 and K14 are controlled in dependence upon the voltage by the reducing-diode control assembly A5 to switch the reducing diodes in and out of the circuit. Contactor K13 is controlled by relay K310 and contactor K14 by relay K320, both on A5.

The two contacts K13 13/14 and K14 13/14 provide for snychronization of the bridging contactors in the case of operation in parallel with a second or third rectifier. In these circumstances it is desirable that K13 or K14 should operate or release at the same time in all the rectifiers.

The remote indication signals a supply failure, operation of the mains monitor A4 and a fault or voltage deviation.

Via terminal 19, when one of the time-delay relays K75 operates, not only the auxiliary contactor K76 in the first rectifier but also the other K76 contactors (in the second and third equipments) operate. This ensures that all the rectifiers switch back simultaneously from charging at 2.33 V/cell to normal operation with supply to the communications system and trickle-charging of the battery at 2.23 V/cell.

Table 1.2 lists the functions of the relays and contactors.

The functional sequences in the relay and contactor control of a particular equipment are shown in the relay diagrams (Figs 1.11, 1.12 and 1.13).

Technical data

The principal technical data for the rectifier type 48 V/100 A GR2 are listed in Table 1.3.

1.1.2.2 Type 48 V/100 A GR3 (GR31)

Figure 1.14 shows the thyristor-controlled rectifier type 48 V/100 A GR31. It is electrically similar to the 48 V/100 A GR2 equipment, but different in mechanical construction.

Application

The rectifier is intended as a power supply for the telephone systems EMS 600 and 12 000; it can also be used in parallel with a battery.

A maximum of four rectifiers can be operated in parallel. It is thus possible to

Table 1.2 Functions of relays and contactors

Number	Description	Function	Remarks
K1	Mains contactor	Connects rectifier to supply mains	
K13 K14	Bridging contactors	Switch reducing diodes in or out of circuit depending on voltage (controlled by reducing diode control A5)	
K31	Contactor, closed-loop and open-loop control	Switches the supply to transformer T31 to energize closed-loop and open-loop control A3	
K63	Auxiliary mains contactor	Augments the contact availability of K1	
K69	Time-delay relay, switch-on delay	Switches on K1 with a delay	Setting depends upon the operating mode (see Fig. 1.11)
K75	Time-delay relay, charging time	Determines the charging period on restoration of the supply (2.33 V/cell)	At the end of the set time K75 switches the rectifier back to trickle charging (2.23 V/cell)
K76	Auxiliary contactor for K75	Synchronizes the automatic charging when several rectifiers are used in parallel	
K80	Fault contactor	Switches off the rectifier via K63 in the event of a fault	Indication light H85 signals a fault and a remote indication is given.
K85	Contactor, voltage deviation – undervoltage	Operates when the output voltage falls below the value set in voltage monitor A7	Indication light H86 signals voltage deviation and a remote indication is given. The equipment continues to operate.
K86	Contactor, voltage deviation – overvoltage	Operates when the output voltage exceeds the value set in voltage monitor A7	K86 only operates if an overvoltage occurs during the test period (see Fig. 1.13). Indication light H86 signals a voltage deviation and a remote indication is given. K86 switches off the rectifier via K63.

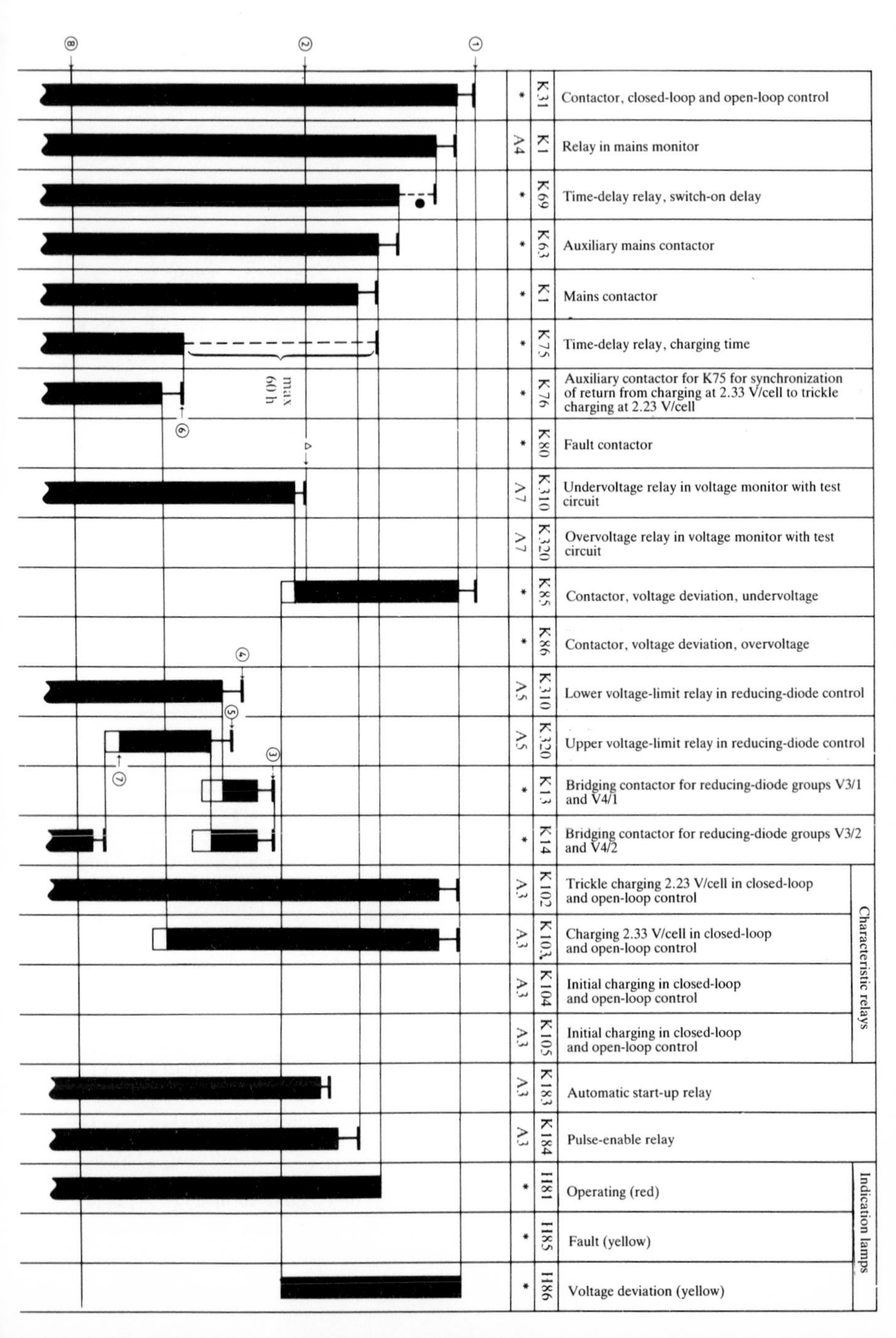
1
2
8
K31
*
Contactor, closed-loop and open-loop control
K1
A4
Relay in mains monitor
K69
*
Time-delay relay, switch-on delay
K63
*
Auxiliary mains contactor
K1
*
Mains contactor
K75
*
Time-delay relay, charging time
max
60 h
K76
*
Auxiliary contactor for K75 for synchronization of return from charging at 2.33 V/cell to trickle charging at 2.23 V/cell
6
Δ
K80
*
Fault contactor
K310
A7
Undervoltage relay in voltage monitor with test circuit
K320
A7
Overvoltage relay in voltage monitor with test circuit
K85
*
Contactor, voltage deviation, undervoltage
K86
*
Contactor, voltage deviation, overvoltage
4
K310
A5
Lower voltage-limit relay in reducing-diode control
5
K320
A5
Upper voltage-limit relay in reducing-diode control
7
3
K13
*
Bridging contactor for reducing-diode groups V3/1 and V4/1
K14
*
Bridging contactor for reducing-diode groups V3/2 and V4/2
Characteristic relays
K102
A3
Trickle charging 2.23 V/cell in closed-loop and open-loop control
K103,
A3
Charging 2.33 V/cell in closed-loop and open-loop control
K104
A3
Initial charging in closed-loop and open-loop control
K105
A3
Initial charging in closed-loop and open-loop control
K183
A3
Automatic start-up relay
K184
A3
Pulse-enable relay
Indication lamps
H81
*
Operating (red)
H85
*
Fault (yellow)
H86
*
Voltage deviation (yellow)

Note: depending upon the voltage, K310 on A5 controls K13 and K320 on A5 controls K14

A Assembly
* Relay and contactor control
● Setting of K69: rectifier mode 3 min, parallel mode 1 min
× Relay K183 de-energized instantaneously by pulse cancelling
△ $U_{Equip} > 46$ V
△△ 3 min disconnect time and 6 min testing time are timed by electronic components on assembly A7
① Supply voltage present: S31 in position 2: parallel operation/trickle charging, automatic operation: S32 at 'On'
② Rectifier delivers output: when K1, K184 and K183 have operated, the battery is charged at 2.33 V/cell during the timing period of time-delay relay K75
③ $U_{Equip} > 51$ V
④ $U_{Equip} > 53$ V
⑤ $U_{Equip} > 58$ V
⑥ Timing period of time-delay relay K75 completed
⑦ $U_{Equip} < 57$ V

Fig. 1.11
Relay diagram—switch-on – charging 2.33 V/cell – reversion to normal operation with supply to communications system and trickle charging of the battery (2.23 V/cell)

Group	Item	Location	Description
	K31	*	Contactor, closed-loop and open-loop control
	K1	A4	Relay in mains monitor
	K69	*	Time-delay relay, switch-on delay
	K63	*	Auxiliary mains contactor
	K1	*	Mains contactor
	K75	*	Time-delay relay, charging time
	K76	*	Auxiliary contactor for K75 for synchronization of return from charging at 2.33 V/cell to trickle charging at 2.23 V/cell
	K80	*	Fault contactor
	K310	A7	Undervoltage relay in voltage monitor with test circuit
	K320	A7	Overvoltage relay in voltage monitor with test circuit
	K85	*	Contactor, voltage deviation, undervoltage
	K86	*	Contactor, voltage deviation, overvoltage
	K310	A5	Lower voltage-limit relay in reducing-diode control
	K320	A5	Upper voltage-limit relay in reducing-diode control
	K13	*	Bridging contactor for reducing-diode groups V3/1 and V4/1
	K14	*	Bridging contactor for reducing-diode groups V3/2 and V4/2
Characteristic relays	K102	A3	Trickle charging 2.23 V/cell in closed-loop and open-loop control
Characteristic relays	K103	A3	Charging 2.33 V/cell in closed-loop and open-loop control
Characteristic relays	K104	A3	Initial charging in closed-loop and open-loop control
Characteristic relays	K105	A3	Initial charging in closed-loop and open-loop control
	K183	A3	Automatic start-up relay
	K184	A3	Pulse-enable relay
Indication lamps	H81	*	Operating (red)
Indication lamps	H85	*	Fault (yellow)
Indication lamps	H86	*	Voltage deviation (yellow)

⑧ ⑨ ⑩ ⑪ ⑫ ⑬ ⑭

⑧ Normal operation with supply to the communications system and trickle-charging of the battery (2.23 V/cell)

⑨ Mains monitor A4 responds (due to phase failure, supply overvoltage or undervoltage or supply unbalance); in addition, triggering-pulse cancelling is effected by the application of the M signal on pin d6 of A4 to pin 22 of A3 – i.e. triggering pulses from A3 to the thyristor set A1 are interrupted

⑩ Battery discharge

⑪ Battery voltage <52 V

⑫ Battery voltage <45 V

⑬ When the mains monitor A4 registers that the supply is again satisfactory, the recifier is automatically switched on again. Pulse enabling follows

⑭ At the end of the timing period of time-delay relay K69, the processes indicated from ① in Fig. 1.11 take place

Fig. 1.12
Relay diagram—normal operation with supply to communications system and trickle charging of the battery – operation of mains monitor A4 – $U_{Batt} < 52$ V – $U_{Batt} < 45$ V – automatic return to operation of the rectifier after resetting of A4.

For key see page 41.

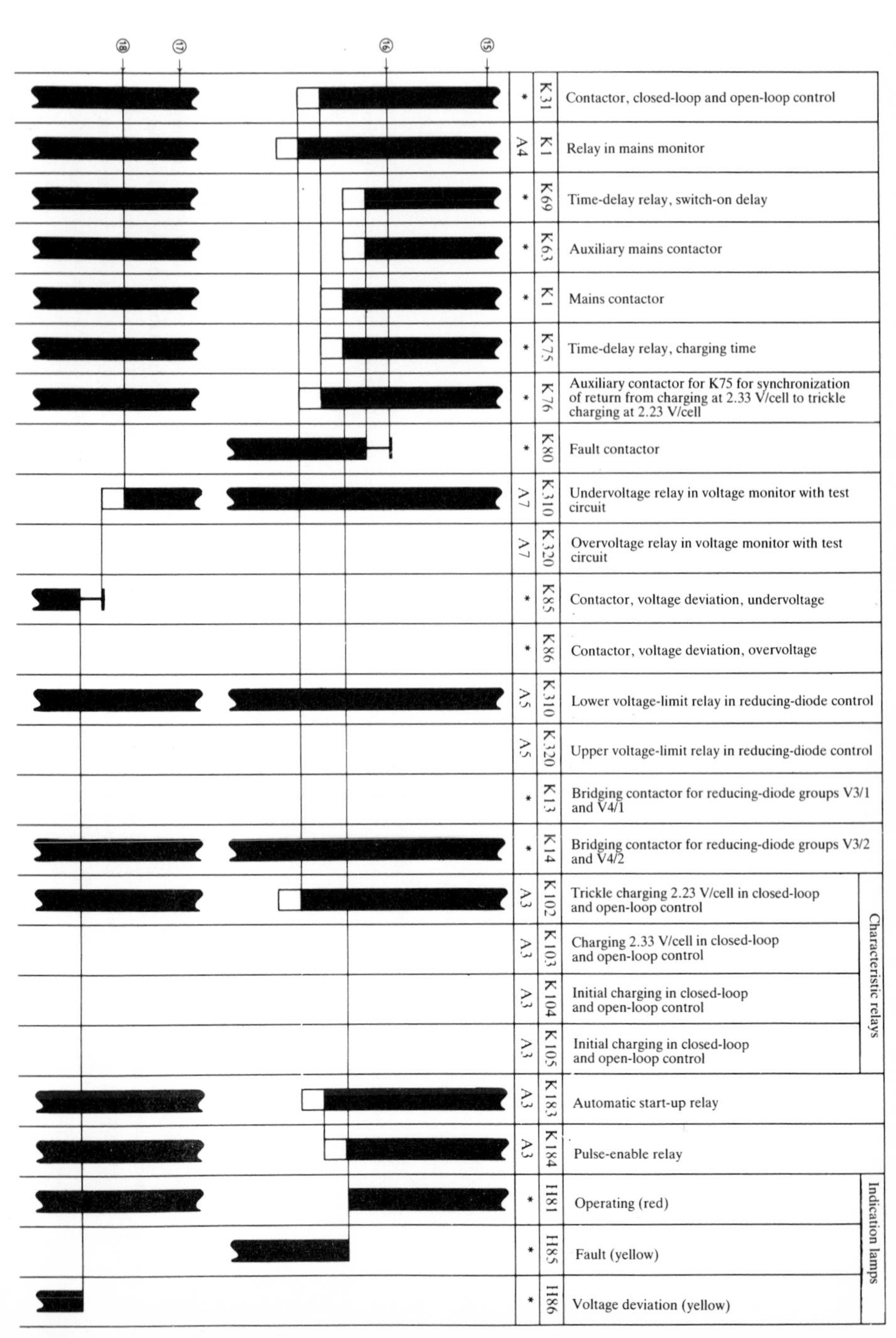
K31 * Contactor, closed-loop and open-loop control
K1 A4 Relay in mains monitor
K69 * Time-delay relay, switch-on delay
K63 * Auxiliary mains contactor
K1 * Mains contactor
K75 * Time-delay relay, charging time
K76 * Auxiliary contactor for K75 for synchronization of return from charging at 2.33 V/cell to trickle charging at 2.23 V/cell
K80 * Fault contactor
K310 A7 Undervoltage relay in voltage monitor with test circuit
K320 A7 Overvoltage relay in voltage monitor with test circuit
K85 * Contactor, voltage deviation, undervoltage
K86 * Contactor, voltage deviation, overvoltage
K310 A5 Lower voltage-limit relay in reducing-diode control
K320 A5 Upper voltage-limit relay in reducing-diode control
K13 * Bridging contactor for reducing-diode groups V3/1 and V4/1
K14 * Bridging contactor for reducing-diode groups V3/2 and V4/2
Characteristic relays
K102 A3 Trickle charging 2.23 V/cell in closed-loop and open-loop control
K103 A3 Charging 2.33 V/cell in closed-loop and open-loop control
K104 A3 Initial charging in closed-loop and open-loop control
K105 A3 Initial charging in closed-loop and open-loop control
K183 A3 Automatic start-up relay
K184 A3 Pulse-enable relay
Indication lamps
H81 * Operating (red)
H85 * Fault (yellow)
H86 * Voltage deviation (yellow)
15
16
17
18

⑮ Normal operation with supply to the communications system and trickle charging of the battery (2.23 V/cell)

⑯ Fault – operation of Q62, Q63, Q100 or F200. The rectifier is switched off and locked out, and the fault is indicated – battery discharge (as in ⑩, ⑪ and ⑫, Fig. 1.12)

⑰ Normal operation with supply to the communications system and trickle charging of the battery (2.23 V/cell)

⑱ Voltage deviation A7, undervoltage – load voltage <45 V. Indication only; rectifier not switched off

Fig. 1.13 Relay diagram – fault and voltage deviation.

For key see page 41.

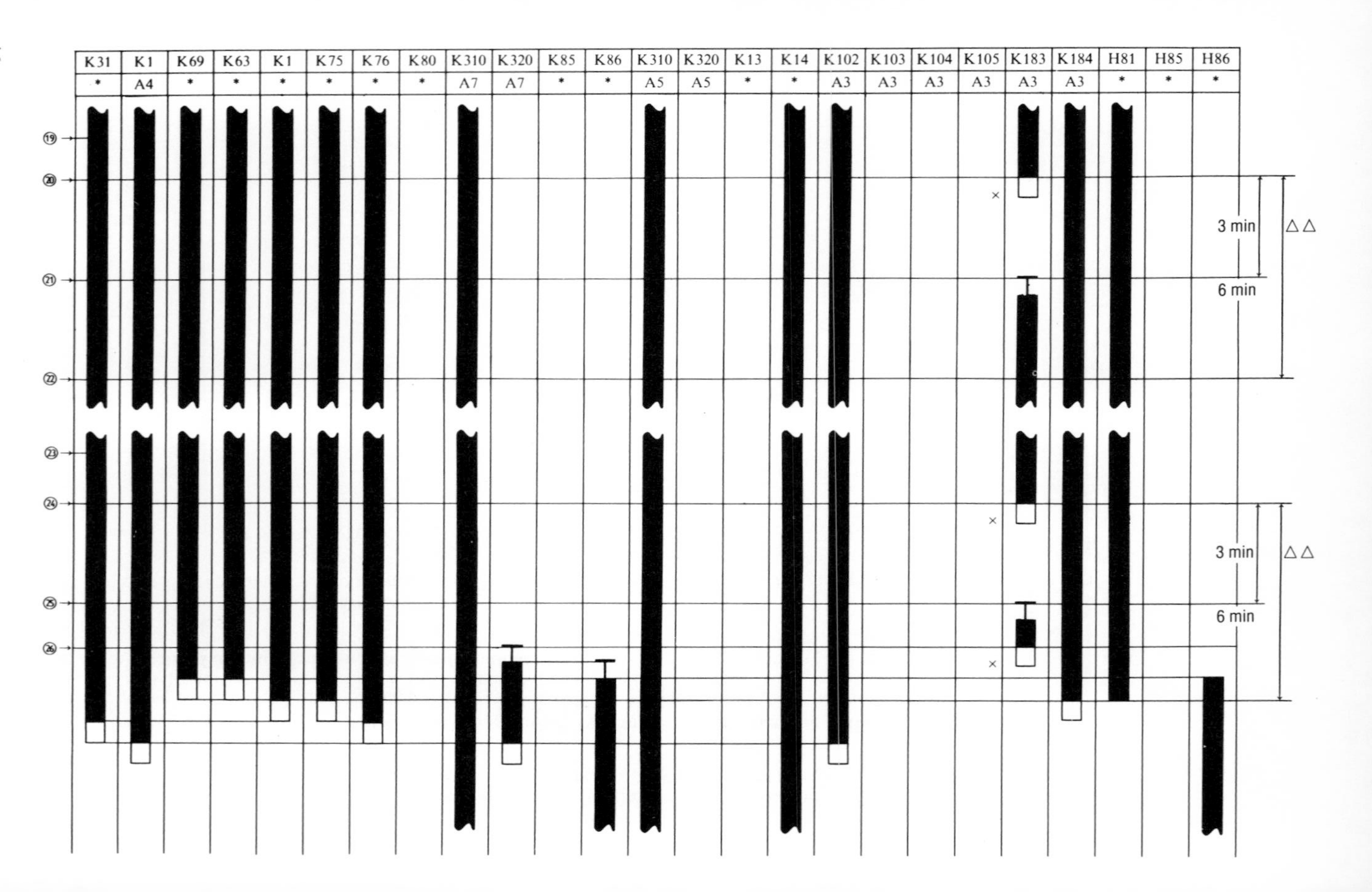
K31 *
K1 A4
K69 *
K63 *
K1 *
K75 *
K76 *
K80 *
K310 A7
K320 A7
K85 *
K86 *
K310 A5
K320 A5
K13 *
K14 *
K102 A3
K103 A3
K104 A3
K105 A3
K183 A3
K184 A3
H81 *
H85 *
H86 *
3 min
6 min
3 min
6 min

⑲ Normal operation with supply to the communications system and trickle-charging of the battery (2.23 V/cell)

⑳ Voltage deviation A7, overvoltage, *first overvoltage*. Load voltage >57 V. Triggering of voltage-limiting thyristor V200 and trigger pulse cancelling

㉑ Trigger pulse enable after disconnect period of 3 min determined by A7

㉒ No second overvoltage within the remainder of the 3 min testing period – normal operation

㉓ Normal operation with supply to the communications system and trickle charging of the battery (2.23 V/cell)

㉔ Voltage deviation A7, overvoltage, *first overvoltage*. Load voltage >57 V. Triggering of voltage-limiting thyristor V200 and trigger pulse cancelling (see ⑳)

㉕ Trigger pulse enable after disconnect period of 3 min determined by A7

㉖ Voltage deviation A7, *second overvoltage* within the remainder of the 3 min testing period. Load voltage again >57 V. Re-triggering of voltage-limiting thyristor V200 and pulse cancelling. Reaction in relay and contactor control and additional indication. Rectifier is locked out. Battery discharge ensues (as at ⑩, ⑪ and ⑫, Fig. 1.12)

Table 1.3 Technical data for rectifier type 48 V/100 A GR2

Input supply			
Voltage	(V)	3 × 380 or 3 × 220 +10 to −15%	
Frequency	(Hz)	50 or 60 ±5%	
Supply distribution fuse protection	(A)	50 (at 380 V)	
Equipment fuse protection	(A)	25 (at 380 V)	
Degree of radio interference		N (VDE 0875)	
D.C. output			
Operating mode or condition		Rated direct voltage (V) (tolerances for 1.5–100% I_{Nenn})	
		Equipment voltage for lead-acid batteries of 25 cells	Load voltage
Rectifier operation		51 ± 0.5%	51 ± 0.5%
Parallel operation/trickle charging (2.23 V/cell)		56 ± 0.5%	51 ± 2%
Parallel operation/charging (2.33 V/cell)		58.5 ± 0.5%	51 ± 2%
Initial charging		67.5	Communications systems disconnected
Rated direct current	(A)	100	
Interference voltage	(mV)	≦0.5 (frequency-weighted with CCITT 'A' filter)	
Dimensions ($H \times W \times D$)	(mm)	Cubicle construction 1800 × 600 × 600	

cater for a current demand for the telephone system and the battery of up to 400 A.

The components of the rectifier equipment are assembled in an 'EMS-system cubicle'. The two versions, GR3 and GR31, are of similar construction. The battery disconnection facility (battery undervoltage monitor and battery disconnection contactor) is included only in the GR31; only one such facility is required for each power supply installation.

Operating modes

The rectifier can be used either in the rectifier mode or in the parallel mode (generally in the standby parallel mode) without voltage-reducing diodes (Fig. 1.15).

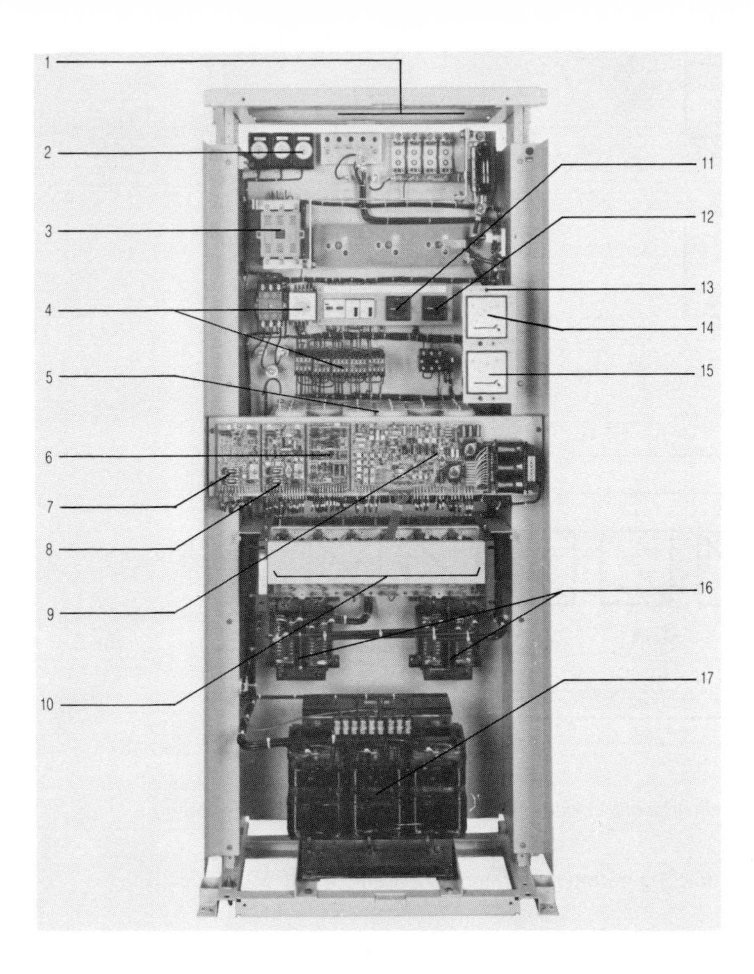

1 D.C. terminals
2 Mains input terminals
3 Battery disconnection contactor
4 Relay and contactor control
5 Filter capacitors
6 Mains monitor A4
7 Voltage monitor A6
8 Voltage monitor with test circuit A7
9 Closed-loop and open-loop control A3
10 Thyristor set A1
11 Operating-mode switch
12 Equipment switch (off/operation)
13 Light-emitting diode – operation indication
14 Voltmeter
15 Ammeter
16 Filter inductors
17 Main transformer

Fig. 1.14 Rectifier type 48 V/100 A GR31, cover plate removed

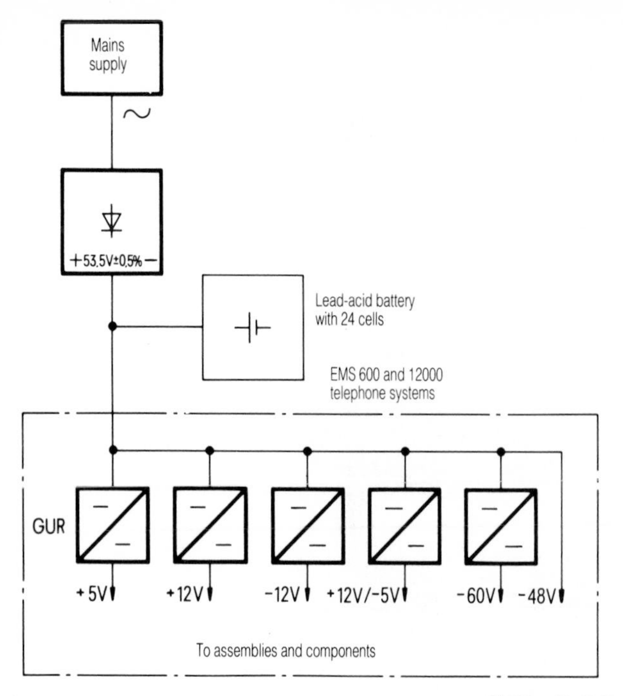

GUR D.C./D.C. converter

Fig. 1.15
Rectifier mode or parallel mode without reducing diodes (normal operation)

As an example, operation in the standby parallel mode will be described.

Standby parallel mode without reducing diodes

Normal operation The telephone system is fed from the supply mains via the rectifier at a direct voltage of 53.5 V ±0.5%. The lead-acid battery (24 cells) is trickle-charged at this voltage.

Power failure – battery discharge The telephone system is supplied from the battery.

Mains return The rectifier switches on again and resumes the supply to the telephone system. The output voltage increases to 56 V ±0.5% for the duration of the operating period of the 'charge time' relay (typically a maximum of 24 h). At the expiry of the timed period the equipment switches back to normal operation.

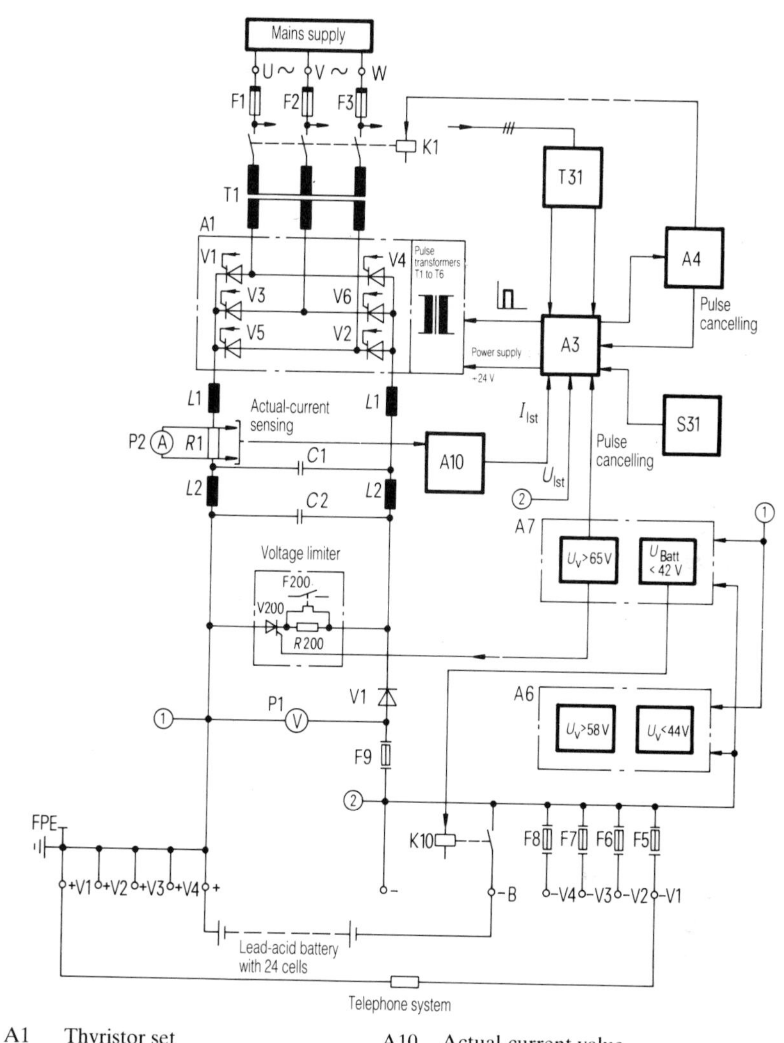

A1	Thyristor set
A3	Closed-loop and open-loop control
A4	Mains monitor
A6	Voltage monitor
A7	Voltage monitor with test circuit
A10	Actual-current value matching
S31	Operating-mode switch
T1	Main transformer
T31	Auxiliary transformer
FPE	Functional and protective earth conductor

Fig. 1.16 Basic circuit of rectifier type 48 V/100 A GR31

Basic circuit and operation

The basic circuit of the equipment is shown in Fig. 1.16. The power section and the assemblies such as A3, A4, A7 and A10 have been described with reference to Fig. 1.9. Relative to that diagram the arrangement shown in Fig. 1.16 embodies only minor differences. Thus there are several positive and negative load connections in Fig. 1.16. This equipment contains no reducing diodes.

Voltage monitor A6

Voltage monitor A6 (constructionally similar to assembly A5 – see Sections 1.1.1.1 and 1.1.2.1) monitors the load voltage. Two similar monitoring circuits respond at the operating limits $U_V < 44$ V and $U_V > 58$ V.

A fall in voltage below the lower limit results simply in a remote indication. In the event of a gradual overstepping of the upper voltage limit the rectifier is switched off by the relay and contactor control in a 'locked-out' condition; i.e. the rectifier can only be put into operation again by moving the equipment switch S32 to the 'off' position and switching on again by hand. An overvoltage also gives rise to a remote signal.

Voltage monitor with test circuit A7

The voltage monitor with test circuit assembly A7 consists of the functional units:

undervoltage monitor and
overvoltage monitor with test circuit (see Section 1.1.2.1).

The undervoltage monitor in the 48 V/100 A GR31 equipment is used to monitor the battery voltage and is set to $U_{Batt} < 42$ V. To prevent the battery from being heavily discharged in the event of a supply failure or a fault in the rectifier, the battery undervoltage monitor causes the battery to be disconnected from the telephone system through the battery disconnection contactor K10.

On restoration of the supply the rectifier switches on again automatically. As the output voltage rises, at $U_A > 43$ V the battery is reconnected in parallel with the telephone system by K10, again controlled by the battery undervoltage monitor.

If the load voltage rises rapidly to >65 V, as a result of a fault in the control circuit of the rectifier, for example, the overvoltage monitor with test circuit responds. The pulse inhibition is then actuated and the voltage-limiting thyristor V200 is triggered (see Section 1.1.2.1).

Technical data

The principal technical data for the rectifier type 48 V/100 A GR3 (GR31) are listed below (Table 1.4).

Table 1.4 Technical data for rectifier type 48 V/100 A GR3 (GR31)

Input supply		
Voltage	(V)	3 × 346/380/400/415 + 10 to −15% * 3 × 200/208/220/240 + 10 to −15% **
Frequency	(Hz)	50 or 60 ±5%
Supply distribution fuse protection	(A)	50
Equipment fuse protection	(A)	25 * 35 **
Degree of radio interference		N (VDE 0875)
D.C. output		
Operating mode or condition		Rated direct voltage (V) (tolerances for 1.5–100% I_{Nenn}). Equipment voltage for lead–acid battery of 24 cells ≙ load voltage (except for initial charging with telephone system disconnected)
Rectifier operation		51 ± 0.5%
Parallel operation/trickle charging (2.23 V/cell)		53.5 ± 0.5%
Parallel operation/charging (2.33 V/cell)		56 ± 0.5%
Initial charging		65 (telephone system disconnected)
Rated direct current	(A)	100
Interference voltage	(mV)	≦ 0.5 (frequency weighted with CCITT 'A' filter)
Dimensions ($H \times W \times D$)	(mm)	EMS system cubicle 1885 × 840 (770 in suite) × 450

1.1.2.3 Type 60 V (48 V)/100 A GR11

The thyristor-controlled rectifier type 60 V(48 V)/100 A GR11 (Fig. 1.17) is mechanically identical and electrically similar to the 48 V/100 A GR2 equipment described in Section 1.1.2.1. At this point, therefore, only the differences will be touched on (see Fig. 1.6).

Application

The rectifier is suitable for supplying 48 V or 60 V communications systems.

Three rectifiers can be connected in parallel, so that a current demand for the communications system and the battery of up to 300 A can be catered for.

The 60 V (48 V)/100 A GR11 and the 60 V (48 V)/50 A GR11 equipments are essentially similar in construction (cubicle construction).

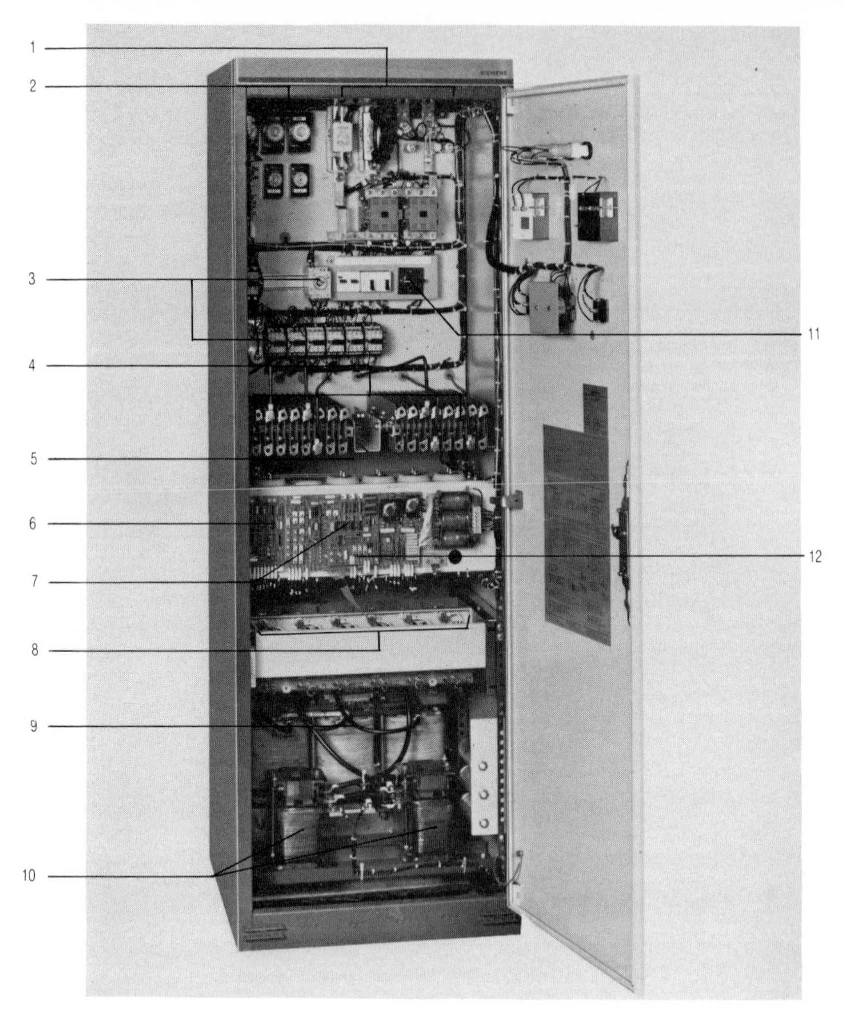

1 D.C. terminals
2 Mains input terminals
3 Relay and contactor control
4 Reducing diodes
5 Filter capacitors
6 Mains monitor A4
7 Closed-loop and open-loop control A3
8 Thyristor set A1
9 Main transformer
10 Filter inductors
11 Operating-mode switch
12 Current balancing and current monitoring A31

Fig. 1.17 Rectifier type 60 V/100 A GR11 with door open

Operating modes

The same operating configurations are possible as described in Section 1.1.2.1 – namely, rectifier mode or standby parallel mode with reducing diodes (see Fig. 1.8).

Basic circuit and operation

The basic circuit of the rectifier type 60 V (48 V)/100 A GR11 is shown in Fig. 1.18.

The power section and the assemblies such as A3, A4, A5, A7 and A10 have been described with reference to Fig. 1.9.

The filter components at the output of the rectifier are designed to keep the interference voltage below 2 mV.

A power-factor correction capacitor (C3) is included to improve cos ϕ.

In addition, various small loads, such as emergency lighting, can be connected.

The *voltage monitor* assembly A6 for the monitoring of the load voltage is fitted as normal equipment. It is identical in construction to the reducing-diode control assembly A5. The set points of A5 – for different numbers of battery cells – were discussed in Section 1.1.1.1.

In place of assembly A6, the *voltage monitor with test circuit* A7 (see Sections 1.1.2.1 and 1.1.2.2) can be fitted.

The *current balancing and current monitoring* assembly A31 included in GR11 is plugged into the closed-loop and open-loop control assembly A3, and in this equipment performs the function of current balancing (see Section 1.1.2.4) when two or three rectifiers are connected in parallel.

Technical data

The principal technical data for the rectifier type 60 V (48 V)/100 A GR11 are listed in Table 1.5.

1.1.2.4 Type 60 V/100 A GR10

The thyristor-controlled rectifier type 60 V/100 A GR10 is illustrated in Figs 1.19 and 1.20.

Application

The rectifier is intended to supply 60 V communications systems – public telephone and transmission systems, for example.

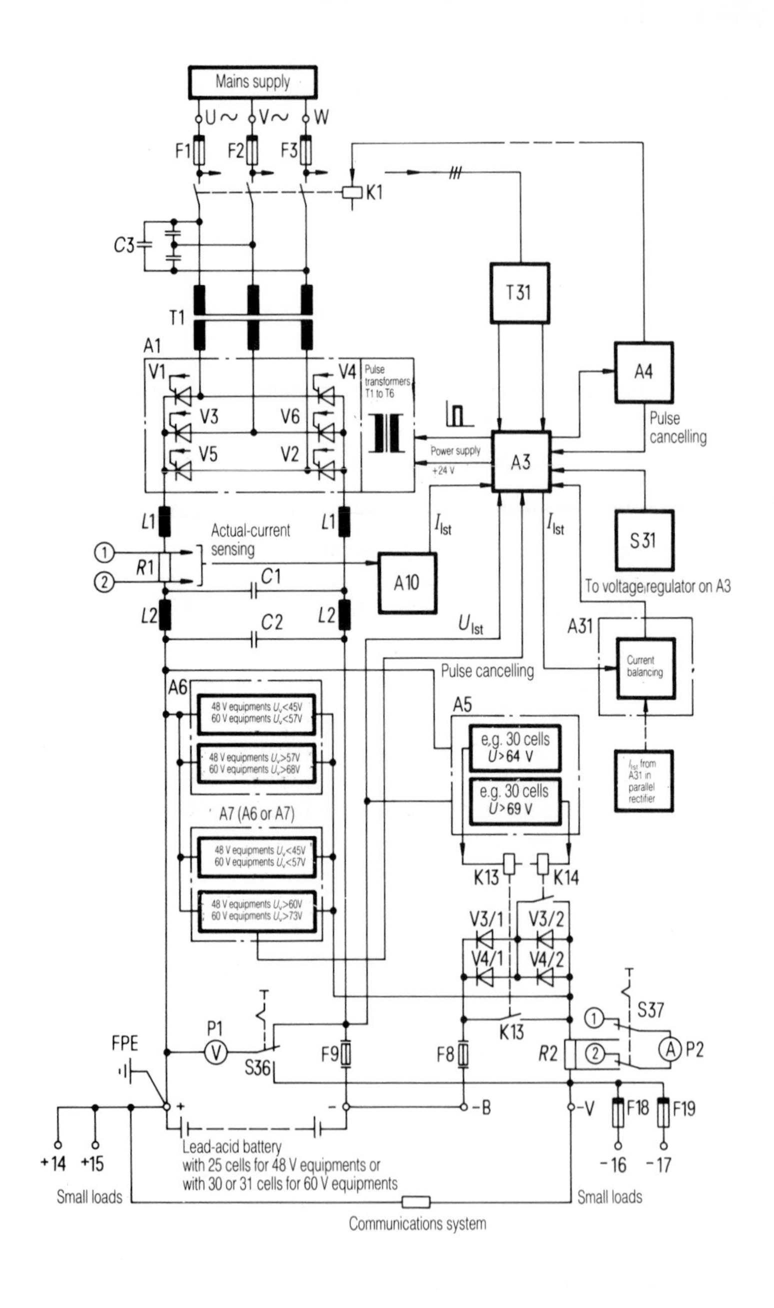

Mains supply
U~
V~
W
F1
F2
F3
K1
C3
T31
T1
A1
V1
V4
Pulse transformers T1 to T6
V3
V6
V5
V2
A4
Pulse cancelling
A3
Power supply
+24 V
L1
L1
I_{Ist}
I_{Ist}
S31
Actual-current sensing
R1
C1
A10
To voltage regulator on A3
L2
C2
L2
U_{Ist}
A31
Pulse cancelling
Current balancing
A6
48 V equipments U_v<45V
60 V equipments U_v<57V
A5
48 V equipments U_v>57V
60 V equipments U_v>68V
e.g. 30 cells
U>64 V
e.g. 30 cells
U>69 V
I_{Ist} from A31 in parallel rectifier
A7 (A6 or A7)
48 V equipments U_v<45V
60 V equipments U_v<57V
K13
K14
48 V equipments U_v>60V
60 V equipments U_v>73V
V3/1
V3/2
V4/1
V4/2
S37
P1
K13
FPE
V
S36
F9
F8
R2
A
P2
-B
-V
F18
F19
Lead-acid battery
with 25 cells for 48 V equipments or
with 30 or 31 cells for 60 V equipments
+14
+15
-16
-17
Small loads
Small loads
Communications system

The equipment can be used as the power supply for the communications system and/or for the trickle-charging and charging of the lead-acid battery.

Eight rectifiers can be connected in parallel, so that the current requirements of the communications system can be catered for up to 800 A. In addition, one or two rectifiers can be employed for the purpose of charging the battery. It is also possible to operate equipments with different rated currents in parallel.

The 60 V/100 A GR10, 60 V/50 A GR10 and 60 V/25 A GR10 equipments are basically of the same construction (cubicle construction).

The battery switching panel with control unit BS (BFS) which is required with the rectifiers 100 A, 50 A and 25 A (two to four equipments) contains the control and monitoring arrangements necessary for operation. In the single 25 A equipment the control system is built in, so no BS is required.

Operating modes

The 25 A equipment is used in the parallel (standby parallel) mode with voltage-reducing diodes. In this mode it is also possible to use the 50 A rectifier in conjunction with a BS 100 A. It is further possible to use the 100 A and 50 A rectifiers with other battery switching panels with control units in the modes:

- ▷ standby parallel mode without reducing diodes,
- ▷ changover mode with battery tap,
- ▷ combined standby parallel and changeover mode.

The operating sequences in the three latter modes will now be illustrated by examples.

Standby parallel mode without reducing diodes (lead-acid battery of 30 cells)

Normal operation The rectifiers supply a voltage of 67 V ±0.5% to the communications system and the parallel-connected battery (trickle-charging voltage 2.23 V/cell) (Fig. 1.21).

A1	Thyristor set	A31	Current balancing and current monitoring
A3	Closed-loop and open-loop control	S31	Operating-mode switch
A4	Mains monitor	T1	Main transformer
A5	Reducing-diode control	T31	Auxiliary transformer
A6	Voltage monitor	FPE	Functional and protective earth conductor
A7	Voltage monitor with test circuit		
A10	Actual-current value matching		

Fig. 1.18
Basic circuit of rectifier type 60 V (48 V)/100 A GR11

Table 1.5 Technical data for rectifier type 60 V (48 V)/100 A GR11

Input supply		
Voltage	(V)	3 × 380 or 3 × 220 +10 to –15%
Frequency	(Hz)	50 or 60 ±5%
Supply distribution fuse protection	(A)	50 (at 380 V)
Equipment fuse protection	(A)	25 (at 380 V)
Degree of radio interference		N (VDE 0875)

D.C. output

Operating mode or condition	Rated direct voltage (V) (tolerances for 1.5–100% I_{Nenn})				
	60 V version			48 V version	
	Equipment voltage for lead–acid battery of		Load voltage	Equipment voltage for lead–acid battery of	Load voltage
	30 cells	31 cells		25 cells	
Rectifier operation	62 ±0.5%	62 ±0.5%	62 ±0.5%	51 ±0.5%	51 ±0.5%
Parallel operation/trickle charging (2.23 V/cell)	67 ±0.5%	69 ± 0.5%	62 ±2%	56 ±0.5%	51 ±2%
Parallel operation/charging (2.33 V/cell)	70 ±0.5%	72.5 ±0.5%	62 ±2%	58.5 ±0.5%	51 ±2%
Initial charging	84	84	Communications system disconnected	67.5	Communications system disconnected

Rated direct current	(A)	100
Interference voltage	(mV)	≦2 (frequency-weighted with CCITT 'A' filter)
Dimensions ($H \times W \times D$)	(mm)	Cubicle construction 1800 × 600 × 600

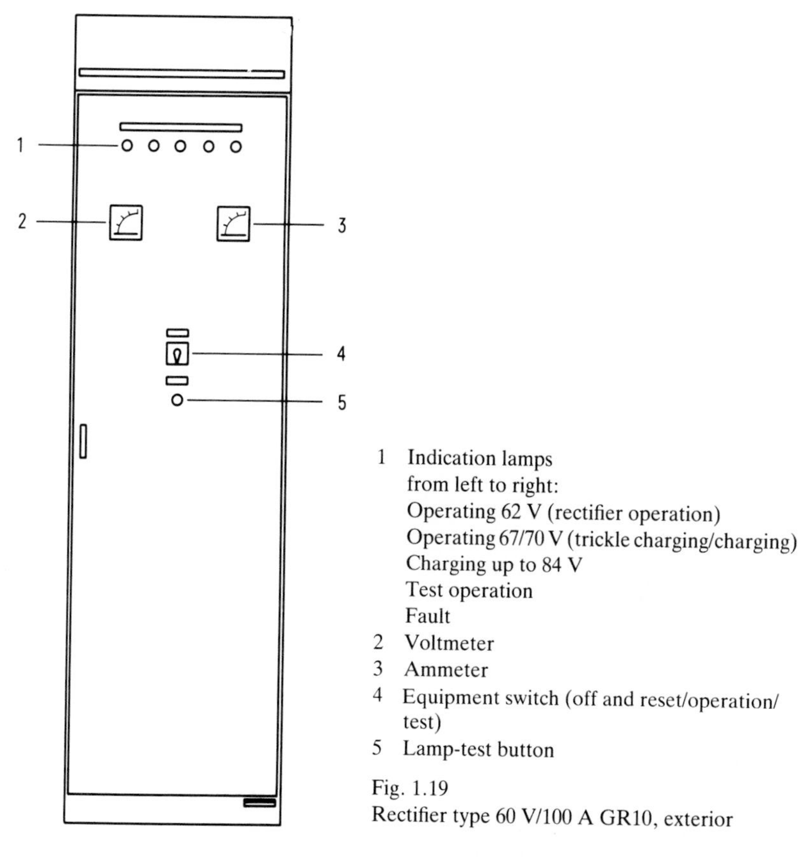

1 Indication lamps
from left to right:
Operating 62 V (rectifier operation)
Operating 67/70 V (trickle charging/charging)
Charging up to 84 V
Test operation
Fault
2 Voltmeter
3 Ammeter
4 Equipment switch (off and reset/operation/test)
5 Lamp-test button

Fig. 1.19
Rectifier type 60 V/100 A GR10, exterior

Power failure The voltage supplied to the communications system corresponds to the battery voltage (battery discharge).

Mains return after an interruption <2 min. All rectifiers switch on with an output voltage of 67 V (normal operation).

Mains return after an interruption >2 min. All rectifiers switch on with an output voltage of 70 V ±0.5% (2.33 V/cell); at the end of the charging period (typically adjustable up to 24 h) the output voltage of the rectifiers is switched back to 67 V and normal operation is resumed.

Changeover mode with battery tap (lead-acid battery of 31 cells)

Normal operation With supply voltage available the operating rectifier (rectifier 1) supplies the communications system at 62 V ±0.5% while the charging

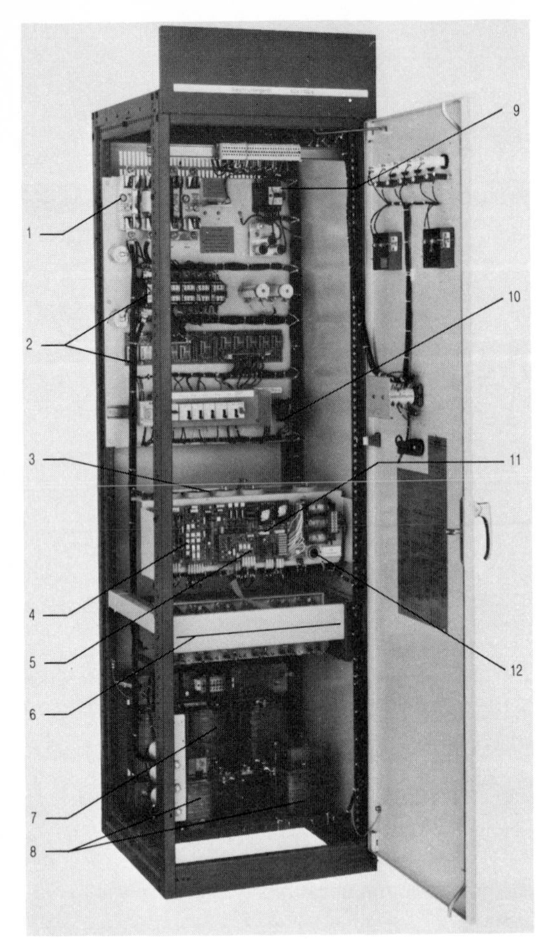

1 D.C. terminals
2 Relay and contactor control
3 Filter capacitors
4 Mains monitor A4
5 Current balancing and current monitoring A31
6 Thyristor set A1
7 Main transformer
8 Filter inductors
9 Mains input terminals
10 Operating-mode switch
11 Closed-loop and open-loop control A3
12 Manual start button

Fig. 1.20
Rectifier type 60 V/100 A GR10 with door open

rectifier (rectifier 2) supplies 69 V ±0.5% to the battery (trickle-charging voltage 2.23 V/cell) (Fig. 1.22).

Power failure The communications system is supplied without interruption, until the contact of the battery discharge contactor K11 closes, through the tapping diode V7 from 26 cells of the battery (battery tap).

When K11 closes the load is supplied from 31 battery cells (battery discharge).

The supply voltage to the load corresponds to the battery voltage.

Mains return – parallel operation The operating rectifier switches on with an output voltage of 63.5 V ±0.5% and the charging rectifier at 63 V ±0.5%. K11

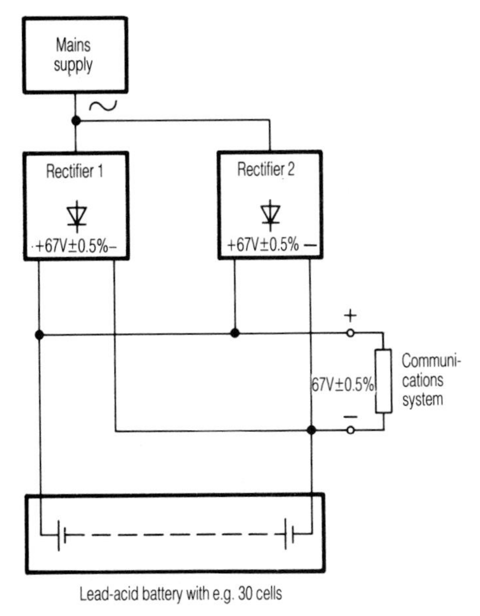

Fig. 1.21 Parallel mode without reducing diodes (normal operation)

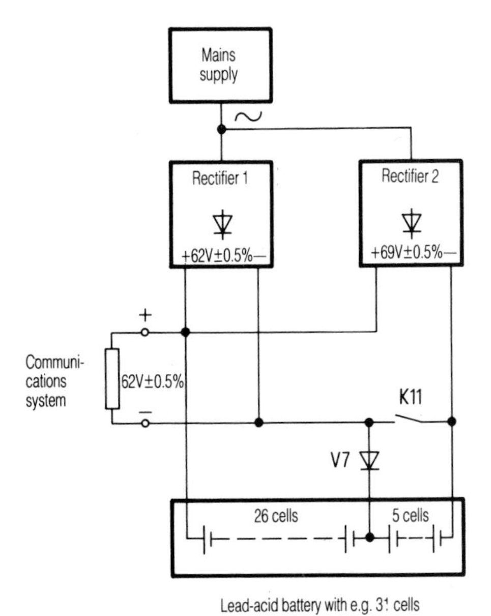

Fig. 1.22 Changeover mode with battery tap (normal operation)

remains closed and consequently all the rectifiers, the communications system and the battery are in parallel; this condition is maintained for an adjustable period (generally 30 min to 3 h).

The rectifiers are thus switched on at a higher voltage (63.5 V and 63 V), in order to ensure an adequate supply of power.

Initially the operating rectifier alone supplies the communications system and the battery. The charging rectifier assumes a proportion of the supply only when the operating rectifier is loaded to its rated current and the voltage falls in consequence below 63 V.

Since there is no current balancing between the operating rectifiers and the charging rectifiers, it would be possible during parallel operation, if all the rectifiers were set to the same voltage, for the operating rectifiers to supply no power. The voltage difference of 0.5 V ensures the preferential loading of the operating rectifiers in parallel operation. This avoids a sudden loading of the operating rectifiers at the changeover from parallel operation to charging, which could give rise to an unacceptable voltage deviation and the switching-off of the operating rectifier.

In the event of any necessary change in the utilization of the equipment – a regrouping, for example, in which an operating rectifier becomes a charging rectifier – the different characteristic settings must be borne in mind.

Normal operation – trickle charging – following a supply interruption <2 min and the termination of parallel operation The battery-discharge contactor contact K11 opens and the output voltage of the charging rectifier is switched to 69 V. Normal operation is reestablised.

Charging after a supply interruption >2 min and the termination of parallel operation The battery-discharge contactor contact K11 opens and the output voltage of the charging rectifier is raised to 72.5 V ±0.5%, so that the charging of the battery is initiated (2.33 V/cell). The communications system receives a supply at 62 V, as in normal operation.

At the end of the charging period (typically adjustable up to 24 h) the output voltage of the charging rectifier reverts to 69 V and normal operation is restored.

Combined standby parallel and changeover mode (lead-acid battery of 30 cells and compensator AGE)

Normal operation With the mains supply available rectifier 1 is arranged to supply the communications system (load 1, also referred to as the 62 V load) at a voltage of 62 V ±0.5% via the AGE (compensator), which is short-circuited under this condition. At the same time rectifier 2 supplies the communications system (load 2, also referred to as the 67 V load) and the parallel-connected battery (on trickle charge at 2.23 V/cell) at a voltage of 67 V ±0.5%) (Fig. 1.23).

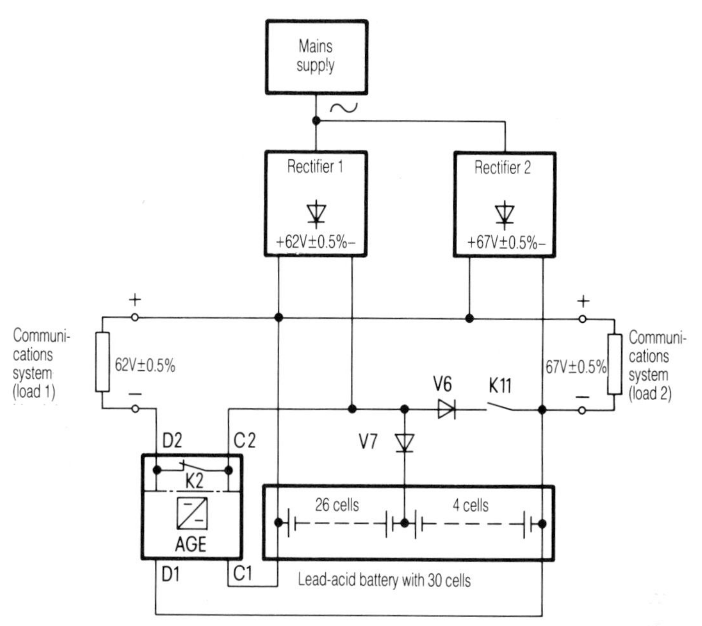

Fig. 1.23
Combined standby parallel and changeover mode (normal operation)

Power failure The communications system (load 1) is supplied without interruption through the tapping diode V7 from 26 cells of the battery until the contact of the battery-discharge contactor K11 closes (battery tap). The voltage on the communications system (load 2) corresponds to the battery voltage (battery discharge). When K11 closes load 1 is supplied through the decoupling diode V6 from 30 battery cells (battery discharge). If in the course of the discharge the battery voltage drops below 62 V, the bridging contactor K2 opens; the AGE produces a boost voltage varying between 0 and 7 V and thereby holds the voltage on the communications system (load 1) constant at 61 V ±1%.

Mains return – parallel mode All rectifiers switch on with an output voltage at 62 V ± 0.5%. K11 remains closed, and consequently all the rectifiers, all the loads and the battery are in parallel. This mode of operation is maintained for an adjustable period (generally 30 min to 3 h). Rectifier 1 supplies load 1. The compensator is switched off and short-circuited about 5 min after the restoration of the supply. Rectifier 2 supplies load 2 and the battery. The decoupling diode V6 prevents charging of the battery and the feeding of load 2 by rectifier 1.

Normal operation – trickle charging after a supply interruption <2 min and the termination of parallel operation At the end of the parallel-operation period the battery-discharge contactor contact K11 opens and the output voltage of rectifier 2 is switched to 67 V. Normal operation is reestablished.

Charging after a supply interruption >2 min and the termination of parallel operation The battery-discharge contactor contact K11 opens and the output voltage of rectifier 2 is raised to 70 V ± 0.5%, initiating the charging of the battery (2.33 V/cell). This voltage is also applied to load 2. Load 1 receives a supply at 62 V, as in parallel or normal operation.

At the end of the charging period (typically adjustable up to 24 h) the output voltage of rectifier 2 is switched back to 67 V (normal operation).

Special features

The following special features apply to all the operating modes described.

In order to avoid the sudden loading of the standby power supply system (combustion engine generating set), and also of the supply mains when the supply is restored, when the rectifiers are switched on, the rectifiers switch on with a time separation (cut-in delay) of 5 s (*staggered switch-on*).

If it is not permissible for all the rectifiers to be switched on during operation with the standby system – the power from the generator not being sufficient for all the rectifiers – a *conditional switch-on* can be established. If the supply-type switch S34 (a.c. power source) is set to mains/generator a rectifier can be switched onto either the mains supply or the standby generator; with the switch in the mains position it can be fed only from the mains.

For the *initial charge*, a lead-acid battery of 30 or 31 cells can be charged by one of the rectifiers to 84 V. The battery concerned has to be switched manually to the initial charge by means of the battery switch (on the battery switching panel). The rectifier selected for the charging is switched to the appropriate output with switch S31, mode of operation, in position 3 (charging to 84 V).

The prescribed charging current is set in the rectifier and is held constant during the whole of the charging period (*I*-charging). The voltage rises in the process to 2.6–2.8 V/cell at the end of charge.

Basic circuit and operation

The basic circuit of the rectifier unit is shown in Fig. 1.24. Assemblies A3, A4 and A10 have been described at length with reference to Fig. 1.9. Explanation at this point is therefore limited to differences in the power section and assembly A31.

Power section

Depending on the type of installation, the rectifier is connected via the corresponding output to the negative load busbar (supply load 1), the negative battery

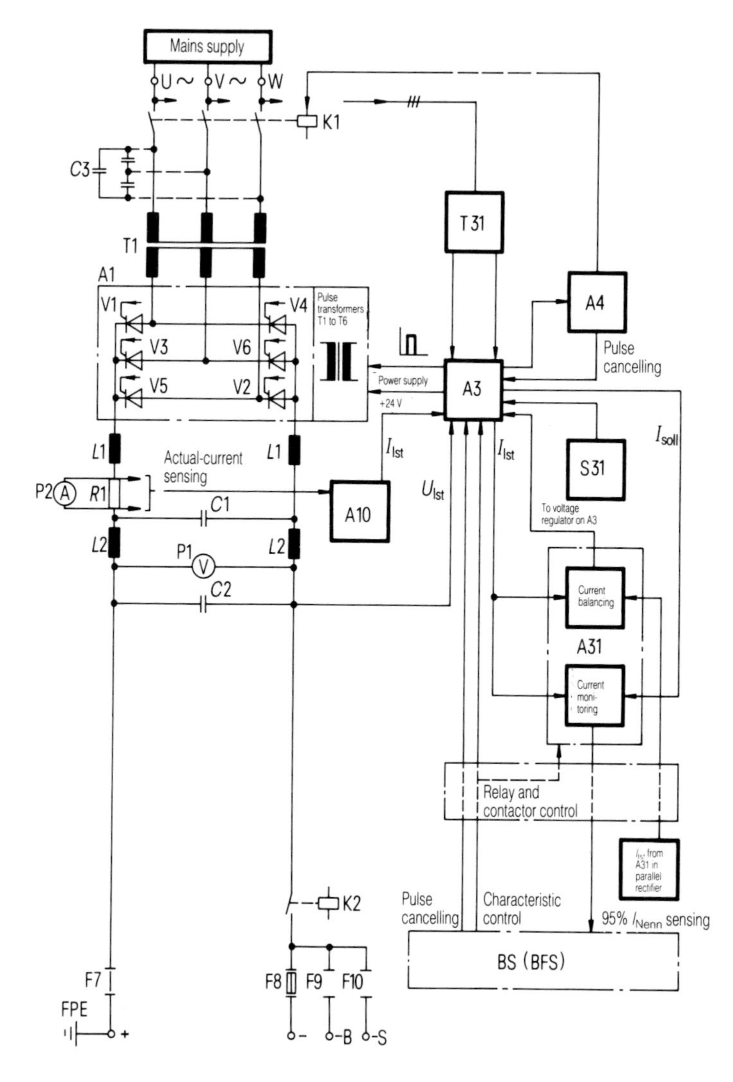

A1	Thyristor set	A31	Current balancing and current monitoring
A3	Closed-loop and open-loop control	S31	Operating-mode switch
A4	Mains monitor	T1	Main transformer
A10	Actual-current value matching	T31	Auxiliary transformer
BS	(BFS) Battery switching panel with control unit		

Fig. 1.24 Basic circuit of rectifier type 60 V/100 A GR10

busbar –B (trickle charging/charging, supply load 2) or the negative special charging busbar –S (initial charging). The required output is selected by moving the output fuses F8, F9 or F10 to the appropriate fuse-holders and the operating mode by means of the switch S31.

S31 determines the operating mode and thus the characteristic of the rectifier. Switch positions:

1 62 V operation (rectifier mode),
2 67 V operation with a 30-cell battery or 69 V with a 31-cell battery (trickle charging 2.23 V/cell, charging 2.33 V/cell) and
3 charging to 84 V

If required, a power-factor correction capacitor (C3) can be incorporated.

A filter circuit on the d.c. output (L1, C1, L2, C2) limits the (frequency-weighted) interference voltage to a level ≦1.8 mV.

The negative terminal of the rectifier is switched onto the predetermined busbar by the output contactor K2.

The voltage monitors and overvoltage limiters are not built into the rectifier unit but are accommodated centrally in the battery switching panel with control unit BS (BFS). If, as an example, the overvoltage limiter in the BS responds, not only is an alarm given but the triggering pulses from the closed-loop and open-loop control A3 are also blocked.

The following control commands (characteristic control) can be passed from the BS to the closed-loop and open-loop control assembly A3 via the relay and contactor control of the rectifier:

- ▷ trickle charging 2.23 V/cell,
- ▷ charging 2.33 V/cell,
- ▷ parallel operation (without voltage rise, e.g. after restoration of the supply),
- ▷ parallel operation (with a voltage rise during standby generator operation).

The control command 'characteristic slope with standby generator operation (max. ±3%)' can be passed from the BS to the current-balancing and current-monitoring assembly A31.

Current-balancing and current-monitoring assembly A31

The A31 assembly provides for:

- ▷ current balancing,
- ▷ current monitoring (95% circuit) and
- ▷ characteristic slope with standby generator operation (max. ±3%).

Current balancing (load sharing) When several rectifiers are used together, it is advantageous to apportion the current (load) approximately equally between all the equipments operating with the same characteristic (e.g. 62 V equipments). The heating of all the equipments is thereby kept substantially the same.

A disparity of ±20% of the rated current between the rectifiers is acceptable. In practice a considerably closer tolerance is obtained.

The constant adjustment of the output voltager is effected by the current-balancing functional unit in conjunction with assembly A3. The actual current value I_{Ist} is taken from the measurement shunt $R1$ and passed to the assembly A3 via the actual current value matching assembly A10; from there it passes via the actual current value amplifier to the current-balancing functional unit of the A31 assembly. Here the value of I_{Ist} registered in the particular rectifier equipment is compared with the average of the actual current values from all the equipments in the common current-balancing system.

The result is passed as a control signal to the voltage regulator (assembly A3).

If I_{Ist} is equal to the average value the voltage regulator is not affected; i.e. A31 provides neither an additional reference voltage value nor an additional actual voltage value in the voltage regulator.

If I_{Ist} is larger than the average, the rectifier is supplying too much power. Now the current balancing supplies a negative additional actual voltage value to the voltage regulator. As a result the output voltage of the equipment is slightly lowered by the voltage regulator and the loading of the equipment thereby reduced. The process operates in a similar way in the opposite direction if I_{Ist} is less than the average; i.e. the voltage regulator now receives a positive additional reference voltage value.

Current monitoring If the set rated current of a rectifier I_{Nenn} (the setting is normally 100%) is exceeded, the current limiting takes effect and the equipment output voltage is controlled downwards.

To avoid current limiting in the operating rectifiers (62 V equipments), parallel operation is instituted, by means of the current-monitoring function, when the current exceeds 95% of the rated value.

The current monitoring compares the actual current value with the set-point (reference) current value. If the actual current exceeds 95% of the set rated current, corresponding information is transmitted via the relay and contactor control to the 95% I_{Nenn} register in the BS and stored.

If *all* the 62V rectifiers are loaded in excess of 95% I_{Nenn} the power supply system is switched by the BS to parallel operation; i.e. all the other equipments that are available in the power supply system (charging rectifiers, for example), and ultimately the battery, are brought into service to supply the communications system.

As the load current subsequently falls again, when it drops below 85% of the rated current the equipment is switched back to normal operation.

Characteristic slope with standby generator operation With the standby generator supply the output voltages of all rectifiers are automatically raised. At the same time the slope of the rectifier characteristic can be made greater than normal (±0.5%) as a means of stabilizing the regulation process, up to a maximum of ±3%. This can be adjusted on assembly A31. A steeper slope reduces oscillations between the standby power supply system and the rectifier.

On no load the output voltage rises to 62 V + 3% and on load it drops to 62 V – 3% (at rated current). To achieve this the actual current value I_{Ist} is passed to the voltage regulator (assembly A3) through a contact (assembly A31) as a negative additional actual voltage value.

Technical data

The principal technical data for the rectifier type 60 V/100 A GR10 are listed in Table 1.6.

Table 1.6 Technical data for rectifier type 60 V/100 A GR10

Input supply			
Voltage	(V)	3 × 380 +10 to –15%	
Frequency	(Hz)	50 ±5%	
Supply distribution fuse protection	(A)	25	
Degree of radio interference		N (VDE 0875)	
D.C. output			
Operating mode or condition		Rated direct voltage (V) (tolerances for 1.5–100% I_{Nenn}). Equipment voltage for lead–acid batteries of	
		30 cells	31 cells
Supply to load–rectifier operation		62 ±0.5%	62 ±0.5%
Standby generator operation		62 ±3%	
Parallel operation			63.5 ±0.5%
Trickle charging (2.23 V/cell)		67 ±0.5%	69 ±0.5%
Charging (2.33 V/cell)		70 ±0.5%	72.5 ±0.5%
Initial charging		84	84
Rated direct current	(A)	100	
Interference voltage	(mV)	≦1.8 (frequency weighted with CCITT 'A' filter)	
Dimensions ($H \times W \times D$)	(mm)	Cubicle construction 2000 × 600 × 600; height with headboard: 2150	

1.1.2.5 Type 60 V (48 V)/200 A GR10

Figures 1.25, 1.26 and 1.27 illustrate the thyristor-controlled rectifier type 60 V/ 200 A GR10.

Application

The rectifiers 60 V/200 A GR10 and 48 V/200 A GR10 (48 V/200 A GR10N) are intended for supplying 60 and 48 V communications systems – e.g. public telephone and transmission systems.

Eight rectifiers can be connected in parallel, so that a current demand for the communications system of up to 1600 A can be catered for. In addition, one or two equipments can be employed for charging the battery. A battery switching panel with control unit BS (BFS) is required for these equipments (see Section 1.1.2.4).

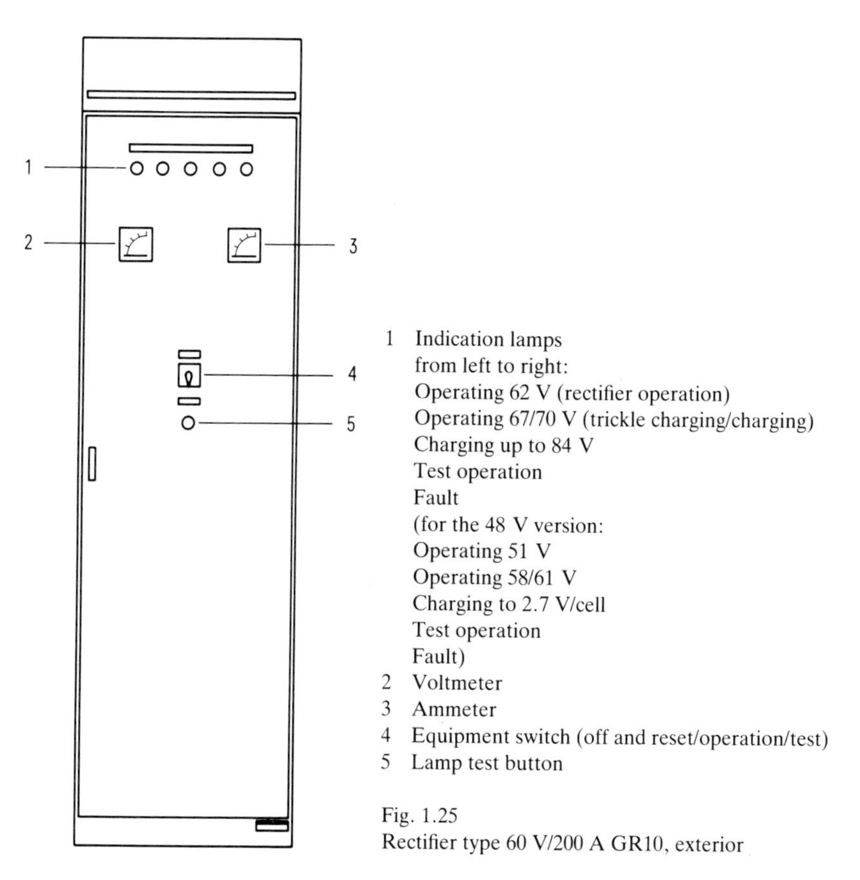

1 Indication lamps
from left to right:
Operating 62 V (rectifier operation)
Operating 67/70 V (trickle charging/charging)
Charging up to 84 V
Test operation
Fault
(for the 48 V version:
Operating 51 V
Operating 58/61 V
Charging to 2.7 V/cell
Test operation
Fault)
2 Voltmeter
3 Ammeter
4 Equipment switch (off and reset/operation/test)
5 Lamp test button

Fig. 1.25
Rectifier type 60 V/200 A GR10, exterior

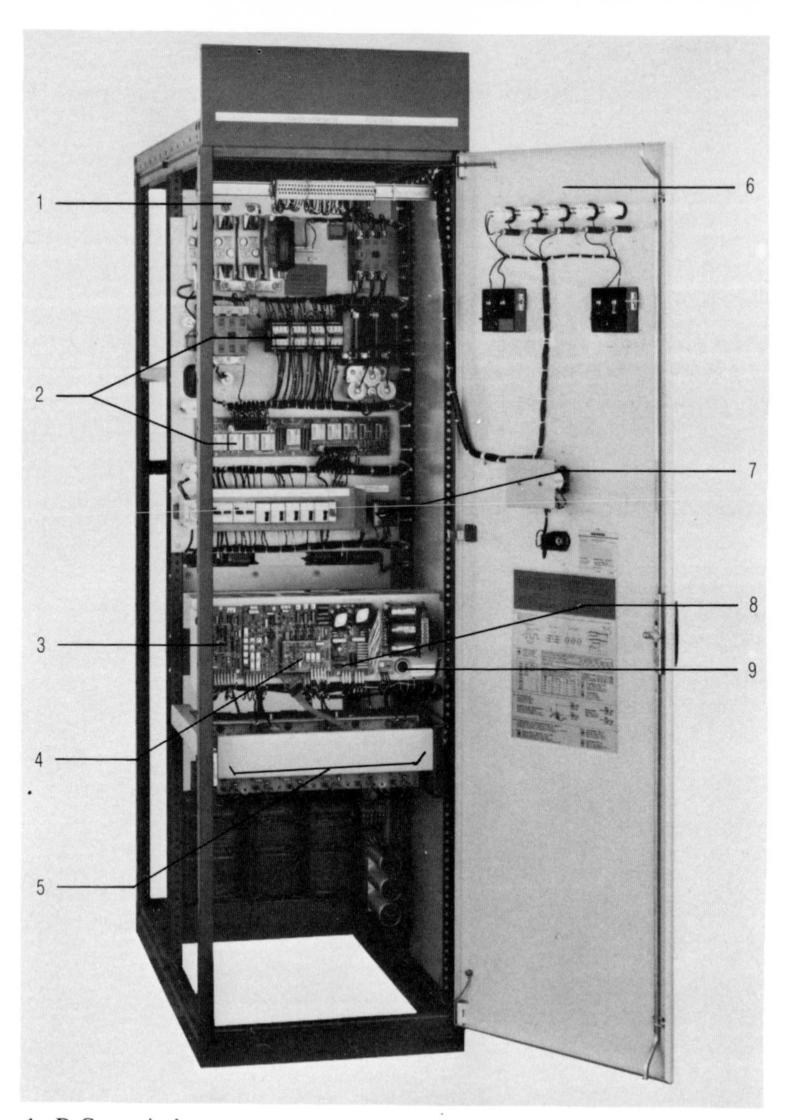

1 D.C. terminals
2 Relay and contactor control
3 Mains monitor A4
4 Current balancing and current monitoring A31
5 Thyristor set A1
6 Mains input terminals
7 Operating-mode switch
8 Closed-loop and open-loop control A3
9 Manual start button

Fig. 1.26 Rectifier type 60 V/200 A GR10 with door open

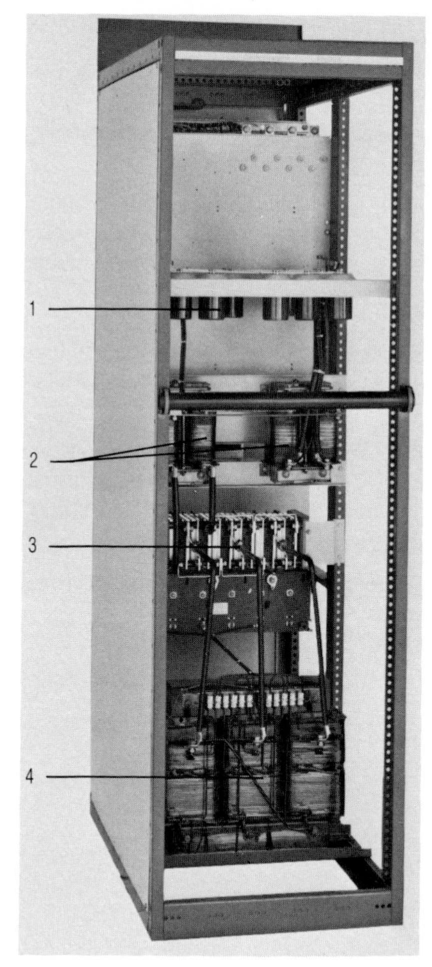

1 Filter capacitors
2 Filter inductors
3 Diode rectifier set A2
4 Main transformer

Fig. 1.27
Rectifier type 60 V/200 A GR10, back view

Operating modes

The rectifier unit can be used in the following modes:

▷ standby parallel mode without reducing diodes,
▷ changeover mode with battery tap,
▷ combined standby parallel and changeover mode.

The operating sequences are set out in Section 1.1.2.4. Further explanation is given here of the changeover mode with battery tap and the operating sequences entailed in the use of the rectifier in a 48 V power supply system (lead-acid battery of 26 cells) and 'voltage-dependent' control of the battery-discharge contactor K11.

Changeover mode with battery tap, with voltage-dependent control of the battery-discharge contactor K11 (lead-acid battery of 26 cells)

Normal operation With the mains supply available the operating rectifier (rectifier 1) supplies the communications system at a voltage of 51 V ± 0.5%. The charging rectifier (rectifier 2) supplies the battery at 58 V ± 0.5% (trickle charging 2.23 V/cell) (Fig. 1.28).

Power failure The communications system is supplied without interruption through the tapping diode V7 from 22 cells of the battery until the battery-discharge contactor contact K11 closes in response to the voltage (battery tap).

If the voltage of the 22 battery cells falls below 46 V, assembly A35 (voltage-dependent control of K11) causes the battery-discharge contactor to operate and all 26 cells of the battery are connected to the communications system (battery discharge).

During the supply interruption the voltage supplied to the communications system corresponds to the battery voltage.

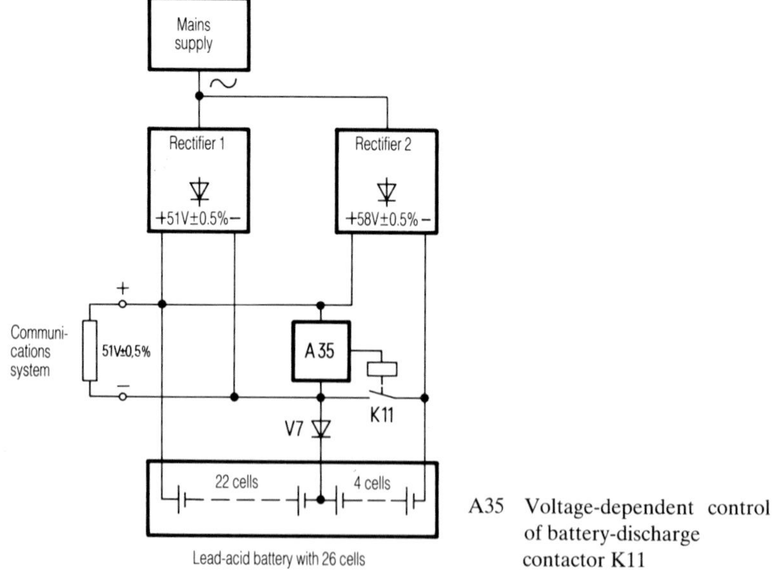

Fig. 1.28 Changeover mode with battery tap (normal operation)

Mains return – parallel operation Rectifier 1 switches on with an output voltage of 52.5 V ± 0.5%, rectifier 2 at 52 V ± 0.5%. K11 remains closed and consequently all the rectifiers, the communications system and the battery are in parallel.

This operating condition is maintained for an adjustable period (generally 30 min to 3 h).

The rectifiers are thus switched to a higher voltage (52.5 and 52 V) in order to ensure a supply of power (see Section 1.1.2.4).

Normal operation – trickle charging after a supply interruption <3 min and termination of parallel operation The battery-discharge contactor contact K11 opens. Charging of the battery at 2.33 V/cell is not necessary. The trickle-charging voltage of 2.23 V/cell is sufficient in this case to replace the energy consumed. Rectifier 2 is therefore switched to 58 V (trickle charging) and rectifier 1 to 51 V. Normal operation is reestablished.

Charging after a supply interruption >3 min and termination of parallel operation The battery-discharge contactor contact K11 opens. Rectifier 2 is switched for an adjustable period to 61 V ± 0.5% (charging at 2.33 V/cell).

To prevent the battery from assuming a proportion of the supply to the communications system (22 × 2.33 V = 51.26 V), rectifier 1 is kept at 52.5 V during the charging period.

At the end of the charging period (adjustable, for example, up to 24 h) rectifier 1 is switched back to 51 V and rectifier 2 to 58 V. Normal operation is thus restored.

Basic circuit and operation

The basic circuit of the rectifier is shown in Fig. 1.29.

Assemblies A3, A4, A10 and A31 have previously been described with reference to Figs 1.9 and 1.24.

Power section

The power section of the equipment embodies several differences in comparison with the rectifiers discussed hitherto.

The thyristor set A1 is disposed on the primary side of the main transformer T1 and arranged as a fully controlled three-phase a.c. controller circuit. This circuit is intended to maintain a constant alternating voltage in dependence upon trigger pulses, supplied to A1 by A3, which are phase-shifted relative to the applied alternating voltage (function of final control element).

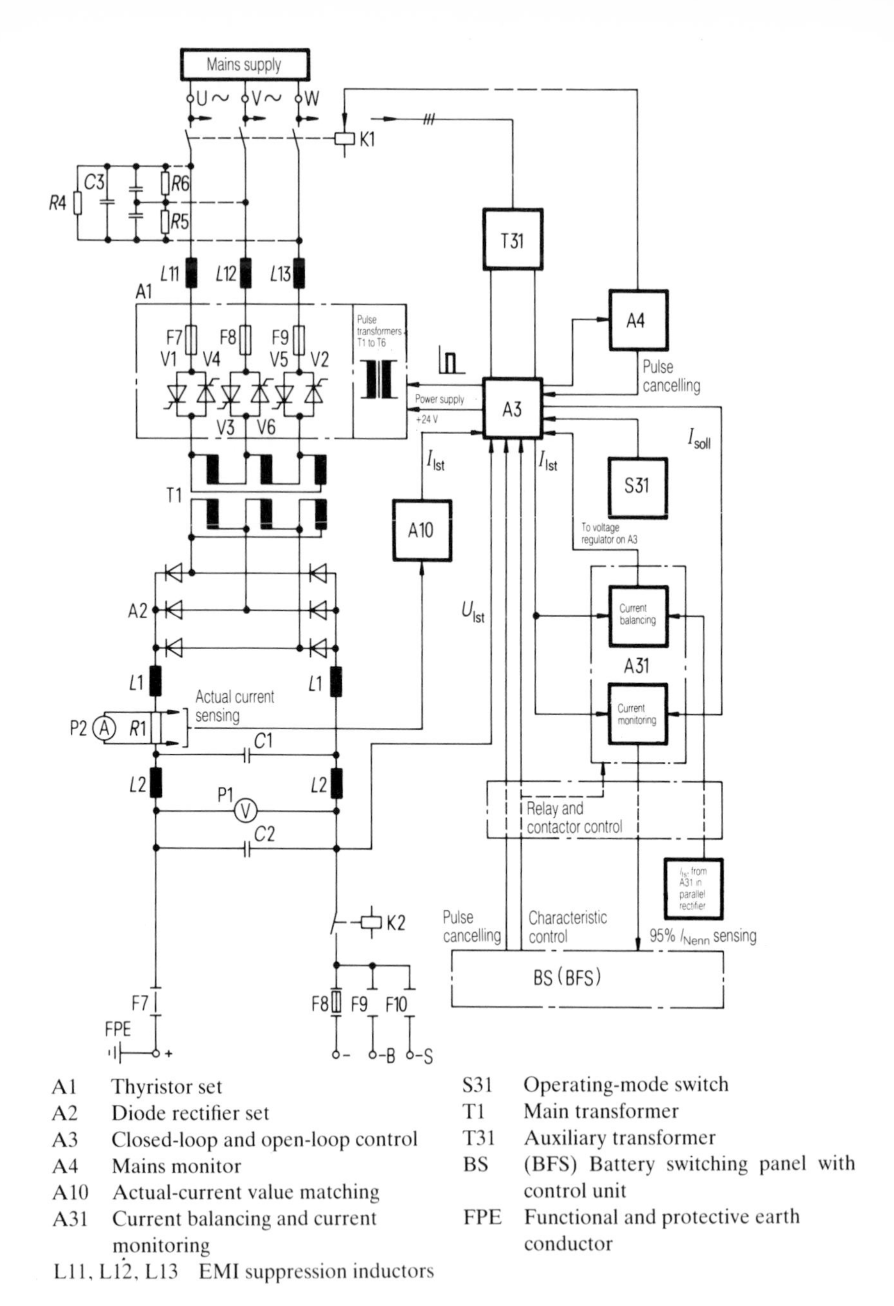

A1 Thyristor set
A2 Diode rectifier set
A3 Closed-loop and open-loop control
A4 Mains monitor
A10 Actual-current value matching
A31 Current balancing and current monitoring
L11, L12, L13 EMI suppression inductors
S31 Operating-mode switch
T1 Main transformer
T31 Auxiliary transformer
BS (BFS) Battery switching panel with control unit
FPE Functional and protective earth conductor

Fig. 1.29 Basic circuit of rectifier type 60 V (48 V)/200 A GR10

Table 1.7 Technical data for rectifier type 60 V (48 V)/200 A GR10

Input supply					
Voltage	(V)	3 × 380 +10 to –15% (connection to other supplies possible)			
Frequency	(Hz)	50 or 60 ±5%			
Fuse protection in mains distribution switchboard or mains switch panel	(A)	50			
Degree of radio interference		N (VDE 0875)			
D.C. output					
Operating mode or condition		Rated direct voltage (V) (tolerances for 1.5–100% I_{Nenn})			
		60 V version Equipment voltage for lead–acid batteries of		48 V version Equipment voltage for lead–acid batteries of	
		30 cells	31 cells	25 cells	26 cells
Supply to load – rectifier operation		62 ±0.5%	62 ±0.5%	51 ±0.5%	51 ±0.5%
Standby generator operation		62 ±3%			
Parallel operation			63.5 ±0.5%	52.5 ±0.5%	52.5 ±0.5%
Trickle charging (2.23 V/cell)		67 ±0.5%	69 ±0.5%	56 ±0.5%	58 ±0.5%
Charging (2.33 V/cell)		70 ±0.5%	72.5 ±0.5%	58.5 ±0.5%	61 ±0.5%
Initial charging		84	84	67.5	70.5
Rated direct current	(A)	200			
Interference voltage	(mV)	≦1.8 (frequency-weighted with CCITT 'A' filter)			
Dimensions ($H \times W \times D$)	(mm)	Cubicle construction: front with door, back open: 2000 × 600 × 1200; height with headboard: 2150			

The fuses F7, F8 and F9 have very high-speed rupturing characteristics, and protect the thyristors of the thyristor set A1.

After down-transformation (T1) the regulated alternating voltage is rectified by the diode rectifier set A2 (three-phase bridge connection).

The output circuit from A2 is essentially similar to the arrangement in Fig. 1.24.

The electromagnetic interference (EMI) suppression inductors L11, L12 and L13 (degree of radio interference N) are included in high-power equipments in addition to EMI suppression capacitors (not shown).

For particular communications systems a voltage-limiter unit must be added (see Sections 1.1.1.1, 1.1.2.1 and 1.1.2.2).

The voltage-limiter unit incorporated into the 48 V/200 A GR10N equipment consists of the voltage-limiter thyristor V200 and the thyristor control assembly A200. On the occurrence of an overvoltage on the load side the thyristor V200 is triggered by the A200 unit and the rectifier is loaded to two or three times the rated current; the voltage is thus prevented from rising further. In addition, as in the rectifiers 48 V/40 and 100 A GR2 and 48 V/100 A GR3 (GR31), provision is made for a bimetallic relay F200 to monitor the limiting circuit, and also a decoupling diode V2.

Technical data

The principal technical data for the rectifier type 60 V (48 V)/200 A GR10 are listed in Table 1.7.

1.1.2.6 Type 60 V (48 V)/1000 A GR10

Figures 1.30 to 1.32 illustrate the thyristor-controlled rectifier type 60 V/1000 A GR10.

Application

The information previously given in Sections 1.1.2.4 and 1.1.2.5 in regard to application is relevent.

A maximum of ten equipments can be connected in parallel, so that a current demand for the communications system of up to 10 000 A can be supplied. In addition, one or two rectifiers can be employed for charging the battery.

As well as the 60 V (48 V)/1000 A GR10 rectifier there is also the variant 60 V (48 V)/500 A GR10. The two equipments are essentially similar in construction (cubicle construction).

In power supply installations for current levels up to 2000 A, a battery switching panel with a control unit BS (BFS) is necessary, and for current levels above

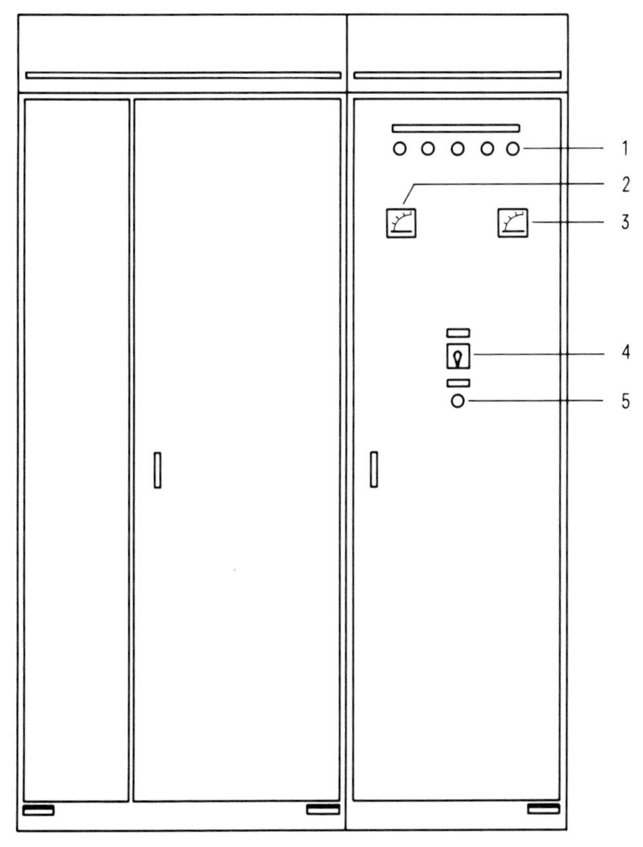

1 Indication lamps
from left to right:
Operating 62 V (rectifier operation)
Operating 67/70 V (trickle charging/charging)
Charging up to 84 V
Test operation
Fault

(for the 48 V version:
Operating 51 V
Operating 58/61 V
Charging to 2.7 V/cell
Test operation
Fault)

2 Voltmeter
3 Ammeter
4 Equipment switch
(off and reset/operation/test)
5 Lamp test button

Fig. 1.30 Rectifier type 60 V/1000 A GR10, exterior

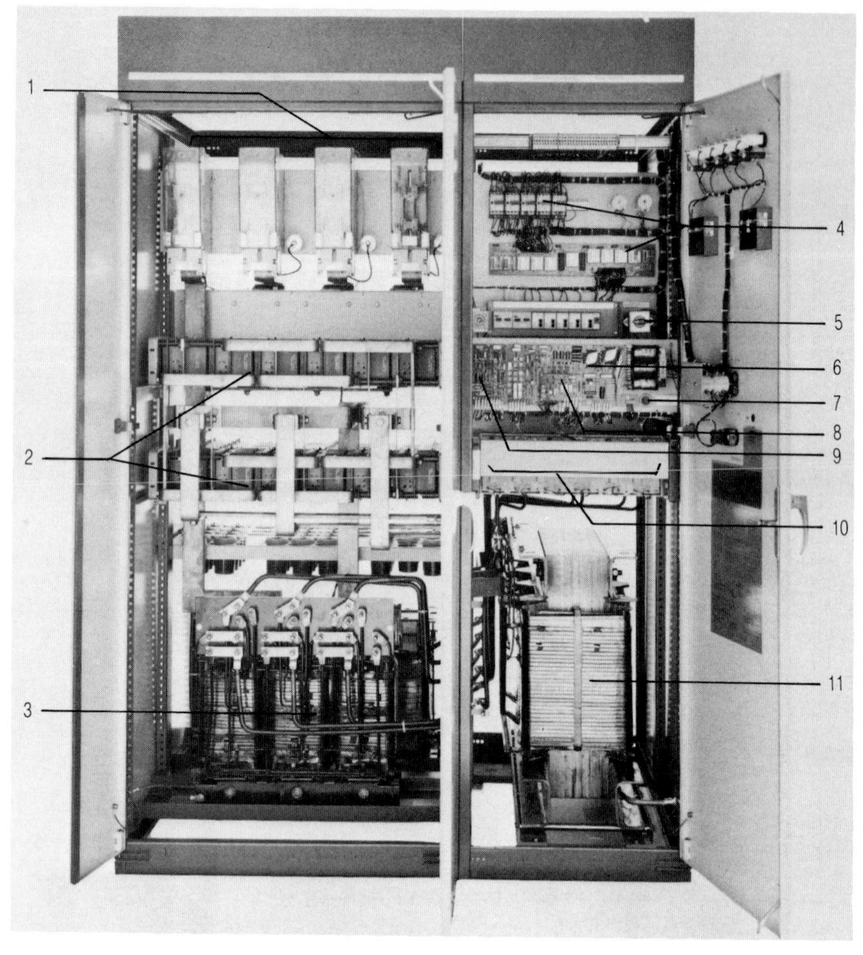

1 D.C. terminals
2 Diode rectifier set A2/1 and A2/2
3 Phase-shifting transformer
4 Relay and contactor control
5 Operating-mode switch
6 Closed-loop and open-loop control A3
7 Manual start button
8 Current balancing and current monitoring A31
9 Mains monitor A4
10 Thyristor set A1
11 Main transformer

Fig. 1.31 Rectifier type 60 V/1000 A GR10 with doors open

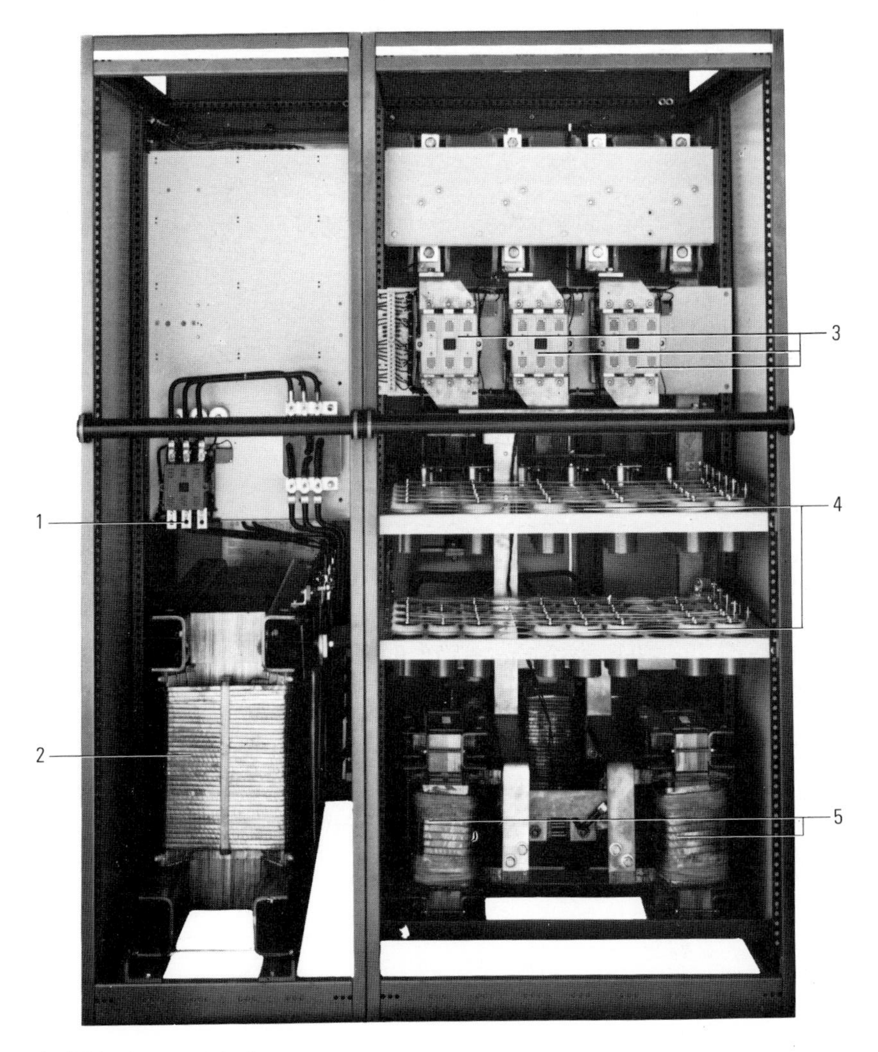

1 Mains input terminals
2 Main transformer
3 Output contactors
4 Filter capacitors
5 Filter inductors

Fig. 1.32 Rectifier type 60 V/1000 A GR10, back view

2000 A a separate battery switching panel BF is required for each battery, together with a control panel SF (one for each installation).

BS (BFS) or SF assumes the control and monitoring functions necessary for operation. One mains switch panel NF is also required for each installation.

Operating modes

The operating modes and sequences described in Sections 1.1.2.4 and 1.1.2.5 are applicable (see Figs 1.21 to 1.23 and 1.28).

Basic circuit and operation

The basic circuit of the rectifier type 1000 A GR10 is shown in Fig. 1.33.

The A3, A4, A10 and A31 assemblies have been described with reference to Figs 1.9 and 1.24.

Power section

The power section of the 500 and 1000 A GR10 equipments is similar in construction to that of the 60 V (48 V)/200 A GR10 rectifier unit (see Fig. 1.29).

A special feature of the 500 and 1000 A GR10 rectifiers in comparison to the 200 A is the phase-shifting transformer T2; this reduces the supply reactions when equipments are operated in pairs. T2 can also be used to match the equipment to the supply voltage: e.g. 3 × 346 V, 3 × 380 V, 3 × 400 V or 3 × 415 V. This transformer is described in detail in Part 1 of the book, Chapter 5.

In the 500 and 1000 A rectifiers it is only necessary to operate switch S31 to change the operating mode, since a separate output contactor is provided for each output (K2, K3 and K4).

A1 Thyristor set
A2 Diode rectifier set
A3 Closed-loop and open-loop control
A4 Mains monitor
A10 Actual-current value matching
A31 Current balancing and current monitoring
L11 EMI suppression inductor
S31 Operating-mode switch
T1 Main transformer
T2 Phase-shifting transformer
T31 Auxiliary transformer
BS (BFS) Battery switching panel with control unit
SF Control panel
FPE Functional and protective earth conductor

Fig. 1.33
Basic circuit of rectifier type 60 V (48 V)/1000 A GR10

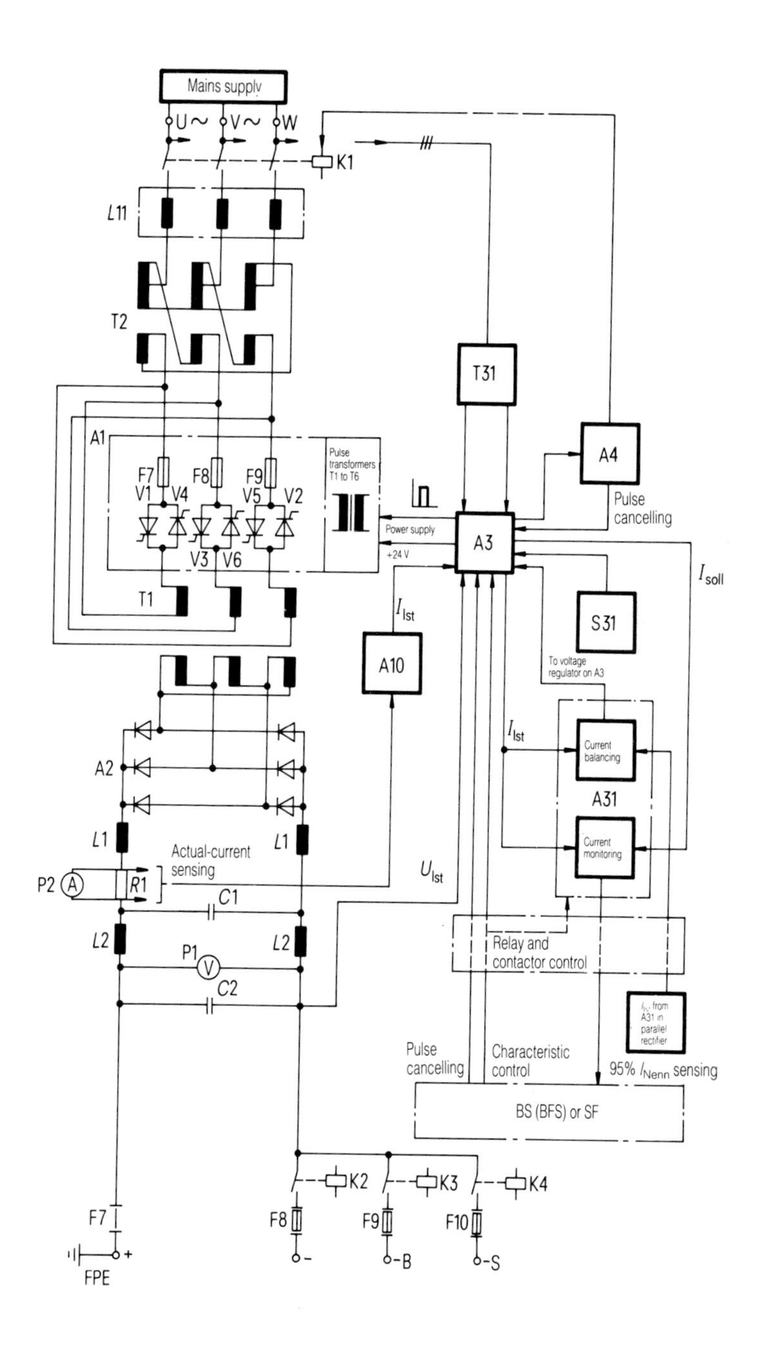

Mains supply
U~
V~
W
K1
L11
T2
A1
F7
F8
F9
V1
V4
V5
V2
V3
V6
Pulse transformers T1 to T6
Power supply
+24 V
T1
T31
A4
A3
Pulse cancelling
I_{soll}
I_{Ist}
A10
S31
To voltage regulator on A3
A2
I_{Ist}
Current balancing
A31
Current monitoring
L1
L1
Actual-current sensing
P2
R1
C1
U_{Ist}
L2
L2
P1
C2
Relay and contactor control
Pulse cancelling
Characteristic control
95% I_{Nenn} sensing
BS (BFS) or SF
K2
K3
K4
F7
F8
F9
F10
FPE
-B
-S

Table 1.8 Technical data for rectifier type 60 V (48 V)/1000 A GR10

Input supply					
Voltage	(V)	3 × 380 +10 to −15% (connection to other supplies possible)			
Frequency	(Hz)	50 or 60 ±5%			
Fuse protection in mains switch panel	(A)	200			
Degree of radio interference		N (VDE 0875)			
D.C. output					
Operating mode or condition		Rated direct voltage (V) (tolerances for 1.5–100% I_{Nenn})			
		60 V version Equipment voltage for lead–acid batteries of		48 V version Equipment voltage for lead–acid batteries of	
		30 cells	31 cells	25 cells	26 cells
Supply to load – rectifier operation		62 ±0.5%	62 ±0.5%	51 ±0.5%	51 ±0.5%
Standby generator operation		62 ±3%			
Parallel operation			63.5 ±0.5%	52.5 ±0.5%	52.5 ±0.5%
Trickle charging (2.23 V/cell)		67 ±0.5%	69 ±0.5%	56 ±0.5%	58 ±0.5%
Charging (2.33 V/cell)		70 ±0.5%	72.5 ±0.5%	58.5 ±0.5%	61 ±0.5%
Initial charging		84	84	67.5	70.5
Rated direct current	(A)	1000			
Interference voltage	(mV)	≦1.8 (frequency-weighted with CCITT 'A' filter)			
Dimensions ($H \times W \times D$)	(mm)	Cubicle construction: front with door, back open: cubicle 1: 2000 × 900 × 1200; cubicle 2: 2000 × 600 × 1200; height with headboard: 2150			

Technical data

The principal technical data for the rectifier type 60 V (48 V)/1000 A GR10 are listed in Table 1.8.

1.1.3 Assemblies

1.1.3.1 Closed-loop and open-loop control A1 and A3

Closed-loop and open-loop control A1

The closed-loop and open-loop control assembly A1 (Figs 1.34 and 1.35) is used for all single-phase rectifiers (see Section 1.1.1).

The functions of the various functional units and their interrelationships have been described previously with reference to Fig. 1.4.

① *Internal power supply*

Transformer T1 supplies an alternating voltage to the bridge V200 via assembly A2. This corresponds to a direct voltage of ±24 V. The stabilizing circuit V201 to V203 produces a constant supply voltage of ±15 V.

The voltage regulator is supplied with a constant set-point (reference) voltage of +6.2 V through V204.

Figure 1.34 Closed-loop and open-loop control A1

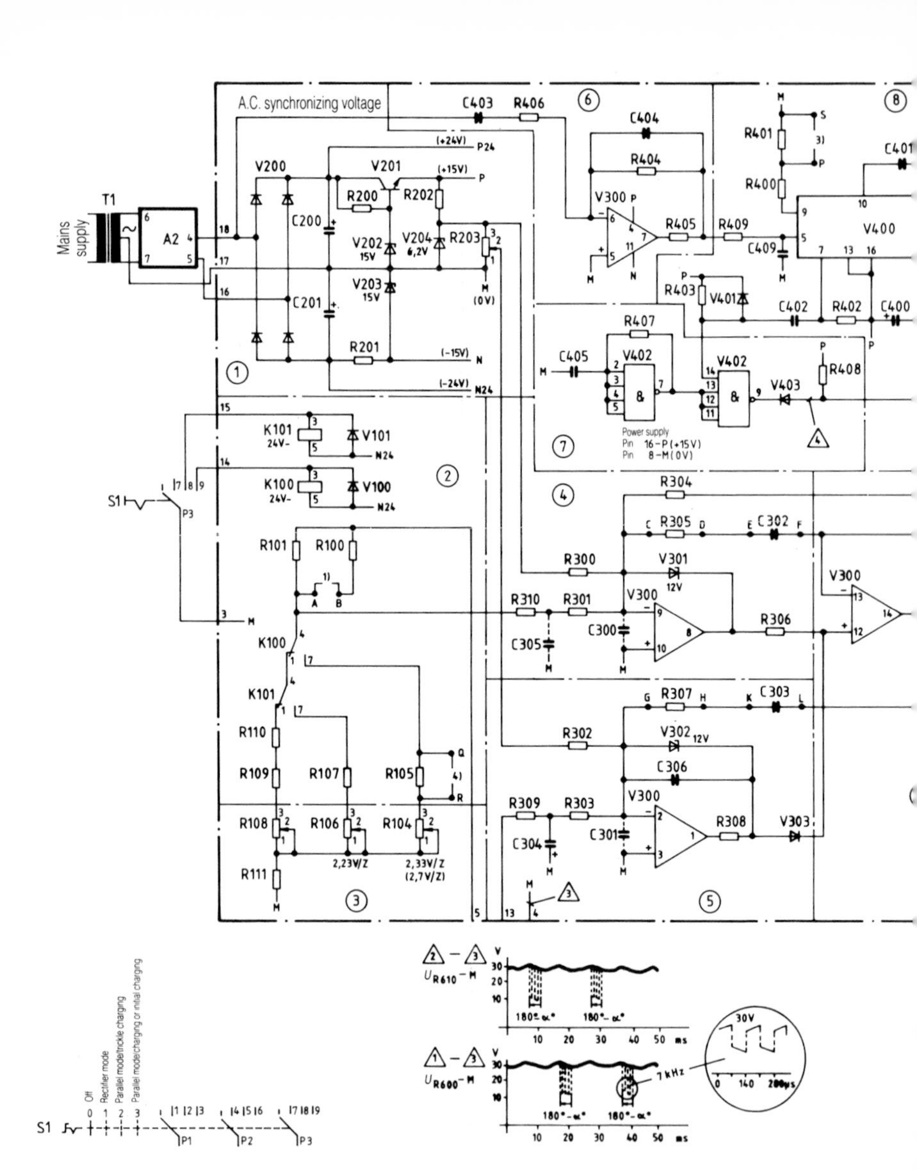

Link variations
1) Link soldered in for 48 V equipment
2) Link N–M may be removed for test purposes
3) Link soldered in for 60 Hz operation
4) Link soldered in for initial charging
△ Oscilloscope test point

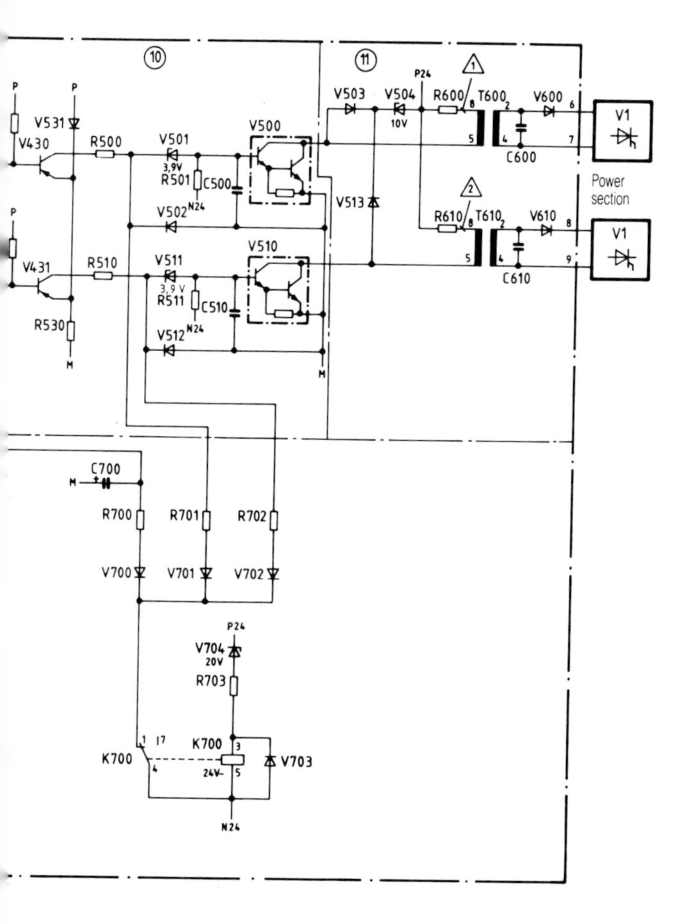

① Internal power supply
② Characteristic changeover
③ Actual-voltage value adjustment
④ Voltage regulator
⑤ Current regulator
⑥ Active filter (180° phase shift)
⑦ 7 kHz clock generator
⑧ Trigger set
⑨ Integrated triggering-pulse generator
⑩ Pulse amplifier
⑪ Pulse transformers
⑫ Pulse enabling, pulse cancelling and automatic start-up
⑬ Matching transformer
A2 Voltage limiter (see Fig. 1.57)
T1 Transformer

Fig. 1.35 Closed-loop and open-loop control A1

The current value (reference) set point, and thereby the current limiting point, is determined by R203.

② *Characteristic changeover and* ③ *actual voltage value adjustment*

The actual voltage value (negative) from the power section is fed to assembly A1 (on pin 5).

Depending on the operating mode (the position of switch S1) this voltage is presented as a 'matched' actual voltage value, via the relay contacts K100 and K101 and the selected voltage divider, to the voltage regulator. The relative switch and relay positions, together with the associated voltage values and adjusting potentiometers, are given in Table 1.9.

To adapt the A1 assembly for use in a 48 V rectifier unit, it is necessary to solder in the link A–B.

④ *Voltage regulator*

The operational amplifier V300, with connections 8, 9 and 10, forms, in conjunction with resistor *R*305 and capacitor C302, a voltage regulator with 'PI' characteristics. The actual voltage value applied through R310 and R301 is compared with the set-point reference voltage applied through R300 at the inverting input 9 of the operational amplifier. The following impedance-matching transformer (voltage follower) ⑬ V300 (connections 12, 13, 14) serves as a high-impedance coupling from the voltage regulator, and also the current regulator,

Table 1.9 Operating modes and direct voltages

Operating mode	Actual voltage value adjustment potentiometer	Switch position S1	Relay K101	Relay K100	Direct voltage (V)			
					48 V version		60 V version	
Rectifier operation	R108	1	4/1	4/1	51		62	
Parallel operation/ trickle charging (2.23 V/cell)	R106	2	4/7	4/1	53.5	56	67	69
Parallel operation/ charging (2.33 V/cell)	R104	3	4/1	4/7	56	58.5	70	72.5
Initial charging	Insert link Q–R	3	4/1	4/7	65	67.5	84	84
Number of lead–acid battery cells					24	25	30	31

to the trigger set ⑧. This prevents the trigger set from affecting the voltage and current regulators. At output 14 this operational amplifier produces a direct voltage varying, according to the control deviation, between 0 and +10 V. This control voltage U_{St} is the input voltage for the trigger set.

A fall in the output voltage of the equipment implies a decrease in actual voltage value. The potential of the comparison point, input 9 of V300, therefore shifts to a positive value. The inverted signal appears at output 8, and reduces the positive output voltage of the regulator. A lower control voltage causes the trigger set to increase the output voltage of the rectifier.

A rise in the output voltage shifts the voltage at the comparison point of the reference and actual values into the negative region, so that (through inversion) the control voltage becomes more positive. The trigger set then reduces the rectifier output voltage.

The zener diode V301 limits the output voltage of the voltage regulator and thereby limits the control range (stability limits) of the trigger set ($\alpha_G \mathrel{\hat{=}} 0$ V, $\alpha_W \mathrel{\hat{=}} 10$ V approximately). The control range is defined by the rectifier limit (leading-edge triggering threshold) α_G (the highest possible output voltage) and the inverter limit (trailing-edge triggering threshold) α_W (the lowest possible output voltage).

⑤ *Current regulator*

The operational amplifier V300 (connections 1, 2 and 3), with *R*307 and C303, forms a PI controller for current regulation. At the inverting input 2 the actual current value derived from the measurement shunt *R*1 in the power section (see Fig. 1.4) and fed through pin 13 on A1 is compared with the adjustable reference obtained from the +6.2 V reference supply. If the reference level set on *R*203 is exceeded, the voltage regulator is turned off by a positive output voltage from the current regulator (gating through the diode V303) and reduces the equipment output voltage through the voltage follower V300 (12, 13 and 14) and the trigger set (onset of current limit).

The output voltage of the current regulator is limited to about +10 V by the zener diode V302.

⑥ *Active filter (180° phase shift)*

To ensure that the triggering pulses are produced in the correct time range for each thyristor, when positive voltage is applied to it the trigger set must be synchronized to this voltage. The a.c. synchronizing voltage is supplied to the trigger set by transformer T1 via an active filter, V300 (5, 6 and 7), C403 and *R*406. The active filter introduces a phase shift of 180°. By this means transient disturbances are prevented from reaching the trigger set.

⑦ *7 kHz clock generator*

The 7 kHz clock generator consists of an integrated circuit (two NAND gates V402) with RC coupling (*R*407 and C405).

The trigger pulses are converted in this circuit into pulse chains. Figure 1.36 shows the form of the output voltage of the 7 kHz clock generator, as seen at the anode of V403.

The operation of the clock generator will be described from the switching-on point (Fig. 1.35).

At the instant of switching on there is no charge on C405. The parallel-connected inputs 2, 3, 4 and 5 of the NAND gate V402 are thus in the low state; by virtue of the parallel connections of the inputs, this gate functions as a NOT gate. When the inputs 2 to 5 are low, the output 7 is high.

The capacitor C405 is now charged through the feedback resistor *R*407 (to about +7.5 V).

The voltage at the inputs 2 to 5 is now received by the NOT gate as a high signal. The output 7 therefore switches to the low state. The capacitor then discharges through *R*407. When the capacitor has discharged to less than about +4.5 V, the inputs 2 to 5 of V402 see a low signal again, and output 7 reverts to the high state. This causes the capacitor C405 to be recharged through *R*407 and the same process is repeated. Terminal 14 of V402 receives a high signal through *R*403. The NAND gate V402 (9) consequently functions similarly as a NOT gate. The 7 kHz clock signal is applied to connection 6 of V400, so that the trigger pulses appearing at connections 2 and 4 are correspondingly chopped.

The triggering pulse is initiated at the point of intersection of the saw-tooth voltage and the control voltage U_{St} (instant t_1, Fig. 1.37a).

Since the clock generator oscillates freely, the 7 kHz clock signal can assume any phase position. An unsatisfactory situation is illustrated in Fig. 1.37(b), where the low state at V400/6 suppresses the beginning of the trigger-pulse chain. The trigger pulses at the outputs of V400 begin without a synchronizing pulse at the

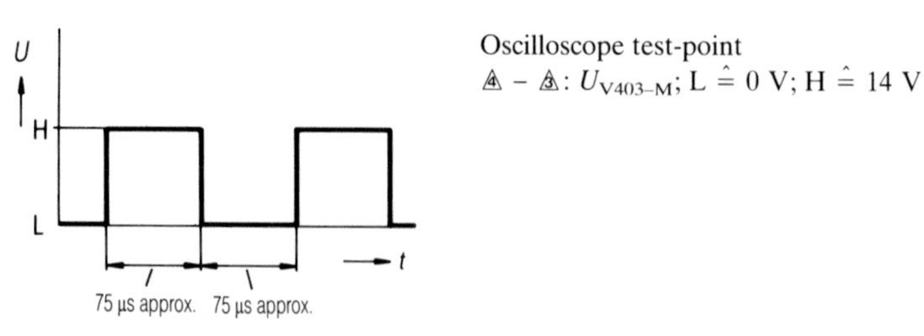

Fig. 1.36 Output voltage of 7 kHz clock generator

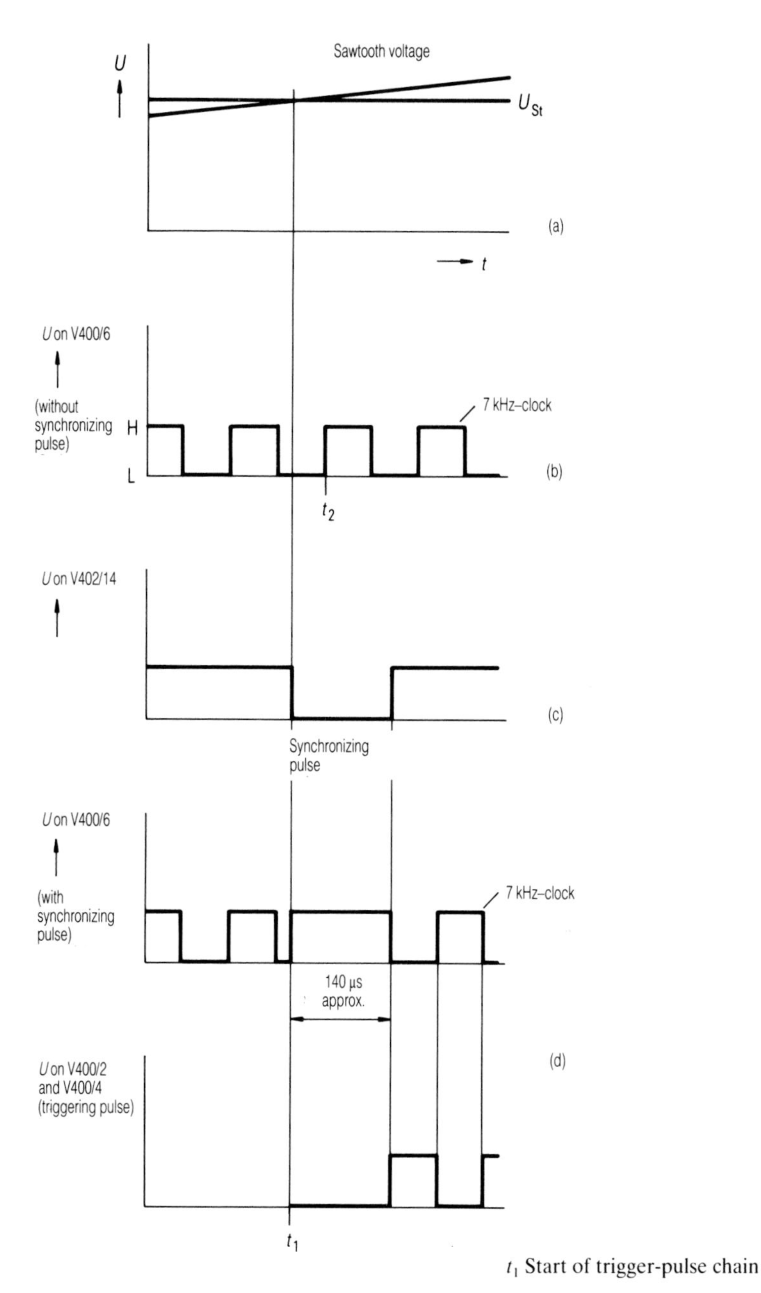

t_1 Start of trigger-pulse chain

Fig. 1.37 Edge synchronization

instant t_2. The thyristor is triggered later than it should be according to the command from closed-loop and open-loop control A1.

To avoid this delay, a low pulse is produced in the trigger set at connection V400/7 at the beginning of a trigger pulse (edge synchronization). The capacitor C402 is now charged through *R*403. A low signal is thus produced at V402/14 (Fig. 1.37(c). With V402/14 low, V402/9, and consequently V400/6, are high, irrespective of other high inputs.

It will be seen from Fig. 1.37(d) that the synchronizing pulse enables the trigger pulse to begin at the 'right' moment (instant t_1). It will also be observed that the first pulse of the chain is longer than the others. Capacitor C402 is discharged after the edge synchronization.

Input 14 of V402 is again high, and the switching pattern is therefore reproduced at output 9 (see Fig. 1.36).

⑧ *Trigger set*

The trigger set consists principally of the integrated trigger-pulse generator V400; at outputs 2 and 4 it produces two trigger pulses mutually displaced by 180° for the two thyristors V1. At each zero-crossing of the synchronizing voltage U_{Syn} a 200 μs pulse is generated in V400, which terminates the charging of the capacitor C401 and brings about a rapid discharge within the 200 μs period. Therefore a 'sawtooth' voltage arises with a frequency of 100 Hz. Whenever the instaneous value of the sawtooth voltage exceeds the control voltage U_{St}, V400 produces a trigger pulse. The pulse width is set at 180°-α by the M potential at input 12 of V400.

⑨ *Integrated trigger-pulse generator*

The purpose of the integrated trigger-pulse generator is to generate trigger pulses for the thyristors. It is produced as a monolithic analogue integrated circuit in a 16-pin dual-in-line (DIL) plastic housing. The trigger-pulse generator is used in the closed-loop and open-loop control assemblies of *all* thyristor-controlled rectifiers.

As an example, the widely used type TCA 780D and its pin layout will be taken as a basis for description (Fig. 1.38).

The trigger-pulse generator performs the functions of:

▷ pulse generation,
▷ pulse cancelling,
▷ pulse phase shifting,
▷ mains synchronization.

Special features of the trigger-pulse generator are:

▷ wide supply voltage range (8–18 V),
▷ low current consumption (5 mA),

- ▷ LSL (slow fault-free logie) compatible,
- ▷ two outputs for 55 mA triggering current,
- ▷ two inverse function outputs,
- ▷ increased trigger-pulse width possible with external capacitor
- ▷ reliable detection of zero-crossings,
- ▷ protection circuit against triggering failure,
- ▷ three switching circuits for six-pulse operation (e.g. in conjunction with the closed-loop and open-loop control assembly A3 for rectifiers with three-phase input).

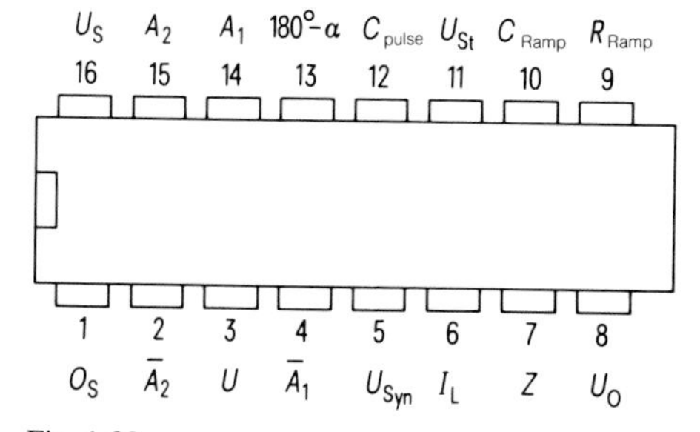

Fig. 1.38
Integrated trigger-pulse generator TCA780 D (SO273B) with pin layout (viewed from above)

Pin 1	Reference potential input M = 0 V (low)
Pin 2	Inverted trigger-pulse output, supplying low trigger pulse
Pin 3	Not used
Pin 4	Inverted trigger-pulse output, supplying low trigger pulse
Pin 5	A.C. synchronizing voltage input U_{Syn}. The trigger pulses at all trigger-pulse outputs are synchronized with respect to the zero-crossings of the a.c. synchronizing voltage. A sawtooth pulse is generated in phase with each half-cycle of the a.c. synchronizing voltage.
Pin 6	When a low signal (M = 0 V) is applied to the pulse-inhibit input (pin 6), all output pulses from the trigger-pulse generator are blocked except those on pin 7. Pulse chains are supplied to pin 6 by the 7 kHz clock generator.
Pin 7	Used for edge synchronization for the 7 kHz clock generator
Pin 8	Used for synchronization in three-phase operation
Pin 9	Resistance connected to pin 9 enables the charging current I_C of the capacitor connected to pin 10 to be adjusted
Pin 10	Connecting point for the sawtooth capacitor
Pin 11	Input for control voltage U_{St}
Pin 12	The application of M to pin 12 results in long trigger pulses (180°-α) at outputs 2 and 4
Pin 13	Supply voltage P (+15 V)
Pin 14[1])	Trigger-pulse output, supplying high trigger pulse
Pin 15[1])	Trigger-pulse output, supplying high trigger pulse
Pin 16	Supply voltage P (+15 V)

[1]) Not used in later assemblies

Figure 1.39 shows the *pulse diagram* of the trigger-pulse generator. The effect of the 7 kHz clock generator connected to pin 6 is not shown in the pulse diagram.

The trigger-pulse generator contains the following *functional units* (Fig. 1.40):

▷ stabilized internal power supply +3.1 V,
▷ synchronizer/zero-voltage detector,
▷ synchronizer latch,
▷ sawtooth generator (ramp generator),
▷ sawtooth reset detector (discharge monitor),
▷ comparison stage (control comparator),
▷ pulse generation,
▷ output stage.

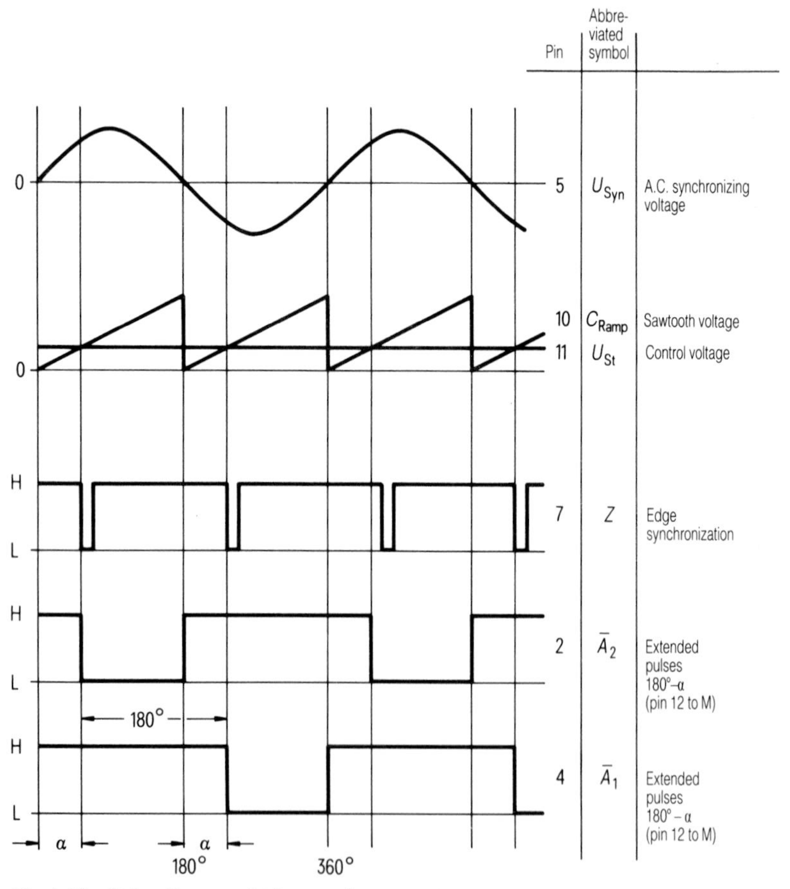

Fig. 1.39 Pulse diagram of trigger-pulse generator

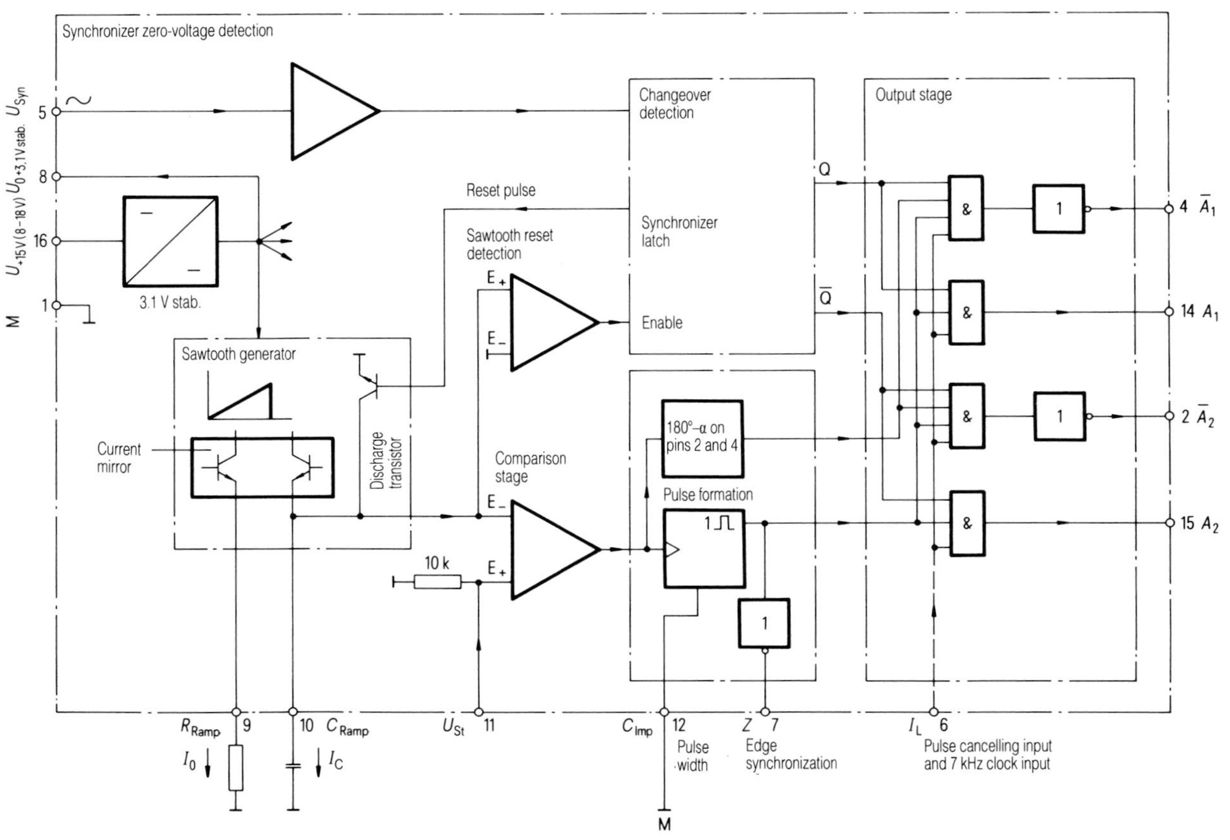

Fig. 1.40 Block diagram of trigger-pulse generator TCA 780 D

The response threshold of the *synchronizer/zero-voltage detector* is in the region of 30 mV and 1.5 μA. At every transition from a positive to a negative half-cycle (and vice versa) of the a.c. synchronizing voltage a change takes place in the output signal – i.e. from high to low or from low to high.

This signal change is registered in the *synchronizer latch* and permits a chosen rate-of-change of the synchronizing voltage at the zero-crossing. It controls the sawtooth generator and the output stage.

The *sawtooth generator* comprises in essence a controllable constant-current source, a current mirror and a discharge transistor. By means of the constant-current source the 'sawtooth capacitor' connected to pin 10 is charged linearly.

The charging current I_C of the capacitor, by virtue of the current mirror, is proportional to a control current I_O, which can be determined by a fixed resistor or a potentiometer connected to pin 9.

At each synchronizing half-cycle a sawtooth pulse is generated for comparison with the control voltage. This is achieved as follows: as the a.c. synchronizing voltage begins to increase, the sawtooth capacitor is charged. The synchronizer latch is now blocked and is prevented from producing a clock signal. At the next zero-crossing the signal level from the synchronizer/zero-voltage detector changes. The synchronizer latch thereupon transmits a pulse to the discharge transistor and causes it to conduct.

When the sawtooth capacitor is discharged, the synchronizer latch is enabled by the *sawtooth reset detector (discharge monitor)*.

The comparison stage (control comparator) compares the sawtooth voltage with the control voltage U_{St}.

In each half-cycle of the synchronizing voltage U_{Syn} the output of the control comparator switches to the opposite state when the sawtooth voltage and the control voltage are equal. This triggers a monostable stage and a trigger pulse is produced.

The *pulse generation* controls the formation of the triggering pulses. The signals from the pulse generation and the sychronizer latch are processed in the *output stage*, which contains the output logic.

⑩ *Pulse amplifier*

The low trigger pulses at connections 2 and 4 of V400 (Fig. 1.35) are inverted by transistors V431 and V430, and subsequently inverted again and amplified by the Darlington transistors V510 and V500.

⑪ *Pulse transformer*

The low trigger pulses (M) at the output pins of the closed-loop and open-loop control assembly A1 are conveyed to the thyristor set in the power section via the pulse transformers T610 and T600. At the switching-on of the trigger pulse a current flows in the primary winding which produces a pulse on the secondary side with a magnitude of about 2 V, which triggers the thyristor through the gate-cathode circuit.

⑫ *Pulse enabling, pulse cancelling (inhibit) and automatic start-up (soft start)*

Provision must be made to ensure that when the equipment is switched on the output voltage does not overshoot.

The enable relay K700 is initially de-energized. The capacitor C700 is now charged (to N24). Up to this point the trigger pulses are suppressed. When K700 operates the pulses are enabled and C700 is discharged by the voltage regulator. The effect of this is that with the diminishing capacitor charge the control voltage U_{St} decreases, so that the trigger pulses are slowly shifted away from the stability limit α_W towards α_G. The output voltage consequently rises slowly.

⑬ *Matching transformer*

The function of the matching transformer (voltage follower) has been explained in connection with the voltage regulator ④.

Closed-loop and open-loop control A3

The closed-loop and open-loop control assembly A3 (Figs. 1.41, 1.42 and 1.43) is used in all rectifiers with three-phase input supplies (see Section 1.1.2).

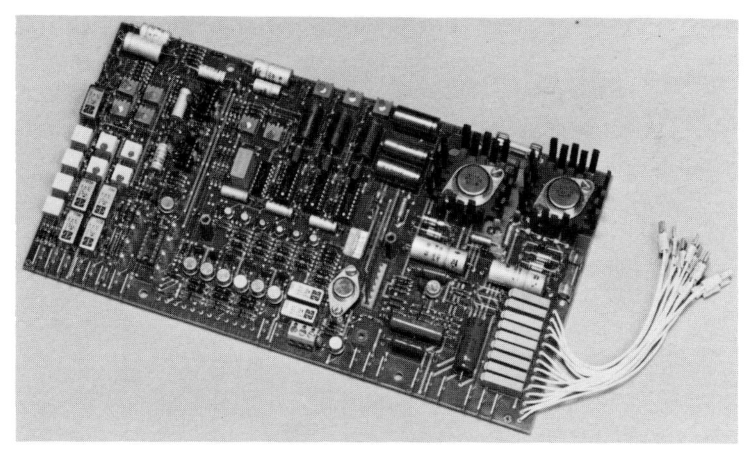

Fig. 1.41 Closed-loop and open-loop control A3

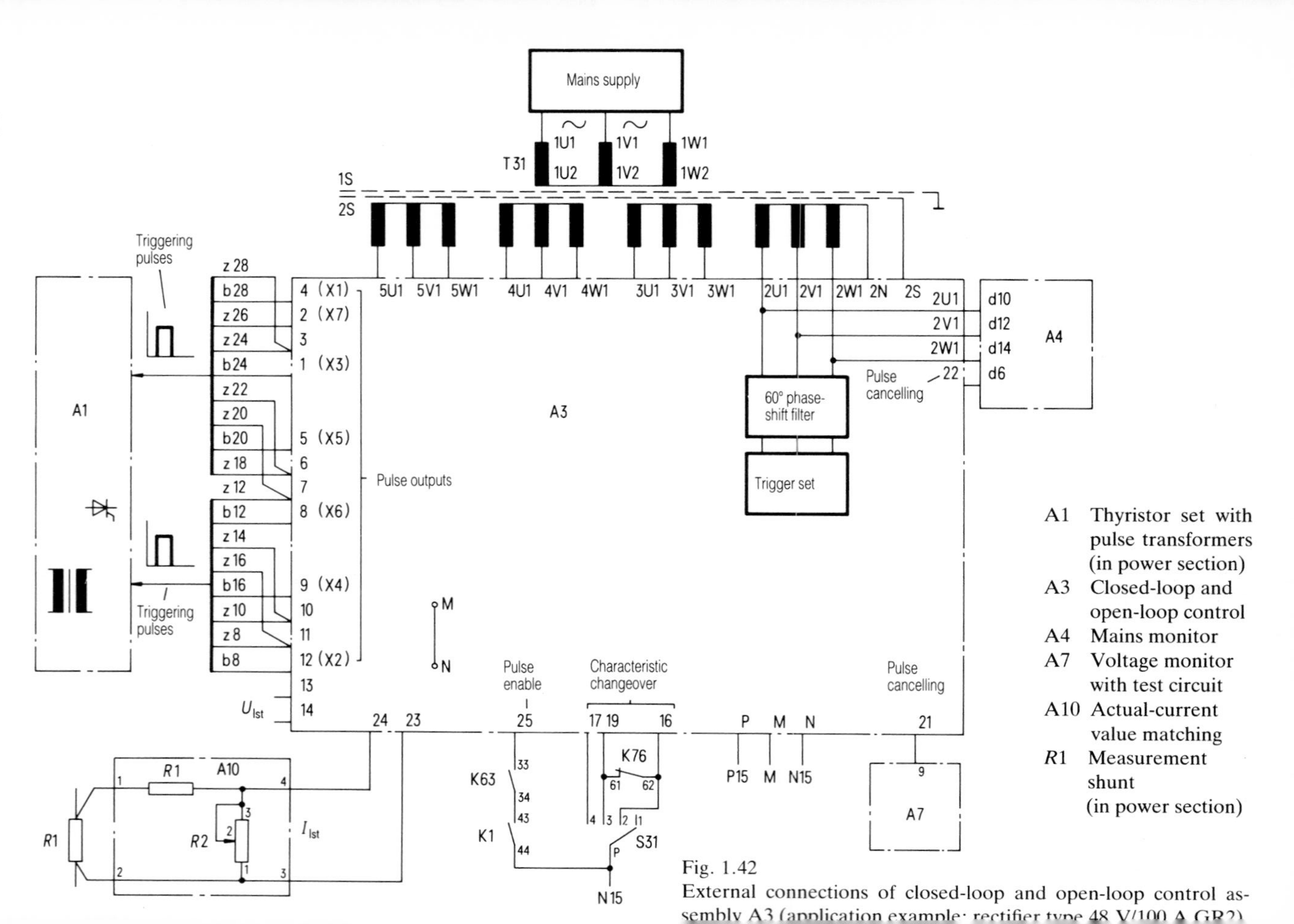

A1 Thyristor set with pulse transformers (in power section)
A3 Closed-loop and open-loop control
A4 Mains monitor
A7 Voltage monitor with test circuit
A10 Actual-current value matching
*R*1 Measurement shunt (in power section)

Fig. 1.42
External connections of closed-loop and open-loop control assembly A3 (application example: rectifier type 48 V/100 A GR2)

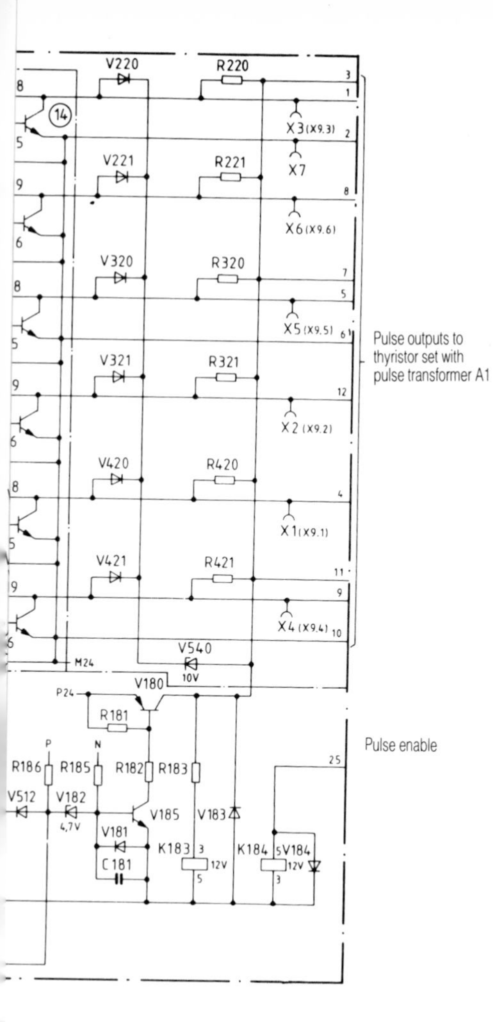

Pulse enable

1. Internal power supply
2. Set-point device
3. Characteristic control
4. Actual-voltage value matching
5. Voltage regulator
6. Matching transformer
7. Actual-current value amplifier
8. Current regulator
9. α_W and α_G adjustment
10. Trigger set
11. 7 kHz clock generator
12. 60° phase-shift filter
13. Integrated trigger-pulse generator
14. Trigger-pulse amplifier
15. Pulse enabling, pulse cancelling and automatic start-up

T31 Auxiliary transformer

Fig. 1.43
Closed-loop and open-loop control A3

at 50% I_{Nenn}		
	60 - V - System	
26C	30C	31C
51V	62V	62V
52,5V	62V	63,5V
58V	67V	69V
61V	70V	72,5V

R106 Current limiting for initial charging

R168 Current matching

R116 Voltage matching

	Voltage regulator	Current regulator
ant	R134	R164
	R141	R171

Current limiting
- 100 % I_N ≙ NE ●— * —● NA
- 80 % I_N ≙ NE ●— * —● NB
- 60 % I_N ≙ NE ●— * —● NC
- 40 % I_N ≙ NE ●— * —● ND

R148 Inverter stability limit α_W

R149 Rectifier stability limit α_G

X8 wrt X7(M) Clock generator

X1 to X6 wt X7(M) Trigger pulses

Figure 1.42 illustrates as an *application example* the use of the closed-loop and open-loop control A3 in the 48 V/100 A GR2 rectifier (see Section 1.1.2.1). This shows the pin connections between A3 and the thyristor set A1, the A4, A7 and A10 assemblies and the relay and contactor control. Also indicated on the A3 assembly is the applied link M–N. This modifies the assembly for a 48 V rectifier. For a 60 V equipment the link has to be removed. This provides a coarse matching of the actual voltage value U_{Ist}.

The power supply for assembly A3 is obtained from transformer T31.

The a.c. synchronizing voltages are taken from the secondary windings to pins 2U1, 2V1 and 2W1 and fed through the 60° phase-shift filter to the trigger set and to the similarly numbered pins 2U1, 2V1 and 2W1 which are connected to the mains monitor A4 (pins d10, d12 and d14). This enables the mains monitor to respond to supply faults.

In the event of a fault the pulse-inhibit command is fed, *inter alia*, from pin d6 (on A4) to pin 22 (on A3).

In the case of an overvoltage on the d.c. side, the voltage monitor with test circuit A7 responds and transmits the pulse-inhibit command to the assembly A3 (from pin 9/A7 to pin 21/A3).

The internal stabilized power supply taken from pin P appears at point P15 (+15 V), and similarly M at point M. The internal stabilized power supply from pin N appears at points N15 (–15 V).

Contacts in the relay and contactor control, and also switch S31 (mode of operation), are connected to A3 at pins 25, 17, 19 and 16. On pulse enable N15 (–15 V) is connected to pin 25 though the closed contacts K1 and K63.

With switch S31 in position 4 (initial charging) point N15 is switched to pin 17. On A3, via pin 17, the two characteristic relays K104 and K105 are energized (see Fig. 1.43), and these switch the rectifier characteristic to the initial charging setting.

If switch S31 is turned to position 3, the parallel mode is instituted, with constant charging at 2.33 V/cell, the characteristic relay K103 on A3 having operated as a result of the connection of point N15 to pin 19. Normal operation with trickle charging at 2.23 V/cell is *not* obtained at this switch position.

Position 2 of switch S31 provides parallel mode/automatic operation.

If contactor K76 is in the released position (automatic charging at 2.33 V/cell) S31 connects N15 to both pins 19 *and* 16. The characteristic relays K102 *and* K103 on A3 (see Fig. 1.43) are energized. Operation of K76 terminates the automatic charging. The contact K76 opens, thereby disconnecting N15 from pin 19. The characteristic relay K103 releases, while K102, connected to pin 16, remains in the operated position. The characteristic is now switched to normal operation (parallel operation with trickle charging at 2.23 V/cell).

With switch S31 in position 1 none of the pins 16, 17 and 19 on A3 is supplied from N15. This means that all the characteristic relays are released (rectifier mode characteristic). This switch position is only selected when the rectifier is used without a battery, since it produces the lowest direct output voltage (see pages 38–47).

The actual current value I_{Ist} is fed to pins 23 and 24 of A3 from the measurement shunt R1 in the output circuit via the actual current value matching assembly A10. Potentiometer $R2$ is adjusted on test in the factory so that with 100% I_{Nenn} at the output of the equipment A10 produces 30 mV between pins 23 and 24 on A3.

The actual voltage value U_{Ist} taken from the equipment output is fed to terminals 13 and 14 of A3.

Trigger pulses are supplied to the thyristor set A1 from pins 1 to 12 on A3.

The trigger pulses can be observed with an oscilloscope at points X1 to X6 (see Fig. 1.46). These are arranged as follows, the measurement being made in each case with respect to test socket X7 (reference point M $\hat{=}$ 0 V):

test socket X1 – trigger pulse for thyristor V1
test socket X2 – trigger pulse for thyrixtor V2
test socket X3 – trigger pulse for thyristor V3
test socket X4 – trigger pulse for thyristor V4
test socket X5 – trigger pulse for thyristor V5
test socket X6 – trigger pulse for thyristor V6

The purposes of the functional units of the closed-loop and open-loop control A3 and their principal interrelationships have already been described with reference to Fig. 1.9.

The circuit diagram is also divided into the corresponding blocks (Fig. 1.43).

① *Internal power supply*

Direct voltages of 15 V (stablized) and 24 V (unstabilized) are produced in the A3 assembly. The a.c. supplies for these are provided by transformer T31, as is also the synchronizing voltage for the trigger set and the mains monitor.

Three of the isolated three-phase a.c. voltages are rectified by three three-phase bridge rectifiers (miniature rectifier sets V601 to V609). Three direct voltages from about 24 to 30 V are thus produced. One of these d.c. supplies is smoothed by capacitor C603 and used to supply the pulse transformers T1 to T6 on the thyristor set A1 (P24 $\hat{=}$ +24 V, unstabilized). The other two d.c. supplies are stablized at 15 V by two integrated voltage regulators N601 and N602 in the negative lines.

The ±15 V supplies are used for the electronics and the relays.

The P24 (+24 V) supply is measured with respect to M24 (0 V), and the P (+15 V) and N (–15 V) supplies with respect to M (also 0 V).

The connections referred to above can in addition be loaded by external assemblies with currents of 200 mA (±15 V) and 250 mA (+24 V) respectively.

The capacitors C606 and C609 smooth the inputs to the voltage regulators. For the protection of the two voltage regulators N601 and N602 against voltage spikes, reverse voltages and overload, there are provided:

- ▷ suppressor diodes V610, V611, V612, V614, V615 and V616,
- ▷ diodes V618 and V619 and
- ▷ fuses F601 and F602.

The suppressor diodes V612, V616, V614 and V615 protect the voltage regulators against overvoltages from the mains supply.

The permissible range of d.c. input voltage for the voltage regulators N601 and N602 is from 18 to 40 V.

The fuses F601 and F602 blow in the event of protracted overvoltages at the input or short-circuits at the output of the voltage regulators. The diodes V618 and V619 provide reverse voltage protection.

The capacitors C601, C602, C604 and C605 represent output filtering of the stablized supplies.

② *Set-point device (reference value selector),* ③ *characteristic control and* ④ *actual voltage value matching*

The set-point device provides the various voltage and current set-point (reference) values for normal operation and for the initial charging.

As previously explained with reference to Fig. 1.42, the characteristic relays in the characteristic control, which effect the changes in the voltage reference, are controlled from the relay and contactor control, via pins 16, 17, 18 and 19, and also by the operating mode switch S31, so that the appropriate reference levels for the voltage regulator are selected.

To hold the equipment output voltage constant the voltage regulator has to be provided with the correct reference voltage.

Because of the various operating modes of the equipment, such as the rectifier mode, parallel mode/automatic, parallel mode/charging and initial charging, different reference values are required. These are derived by means of voltage dividers from the stabilized voltage of +15 V with respect to M3 obtained from the power supply.

In the rectifier mode all the characteristic relays are de-energized, and the load voltage – i.e. the equipment output voltage of, for example, 51 V – can be adjusted by means of the potentiometer *R*107 in the voltage divider *R*102/*R*107/*R*111.

In the parallel mode at 2.23 V/cell (trickle charging/automatic), relay K102 is operated, switching the voltage divider *R*104/*R*109/*R*113 in parallel with the voltage divider for the rectifier mode. A battery voltage of 2.23 V/cell can be set on potentiometer *R*109.

For the parallel mode at 2.33 V/cell (charging), operation of relay K103 connects the voltage divider *R*105/*R*110/*R*114 in parallel with the voltage divider for the rectifier mode. The equipment output voltage of 2.33 V/cell can be adjusted by means of potentiometer *R*110.

When the rectifier switches to charging at 2.33 V/cell in the course of the automatic operating sequence, relays K102 *and* K103 are operated. In the subsequent reversion from charging at 2.33 V/cell to trickle charging at 2.23 V/cell, relay K103 releases while K102 remains energized.

If the operating mode switch S31 is set to position 4 (initial charging), the two characteristic relays K104 and K105 are operated.

Contact K104 changes its position to 4/7, with the result that the maximum voltage reference value P (+15 V), for the maximum available output voltage, is applied to the input of the voltage controller via the link AD–AC and the contact K104.

The voltage for the initial charging can be adjusted by means of potentiometer *R*116 in the actual voltage value matching functional unit. It should be noted that all the voltage reference potentiometers in the set-point device (*R*107, *R*108, *R*109 and *R*110) should be readjusted afterwards.

The actual voltage value (negative) U_{Ist} is taken to pin 13 on A3. As a reference point, pin 14 is connected to the 'plus' output of the equipment and to M3 (0 V). In 48 V equipments U_{Ist} is applied to the input of the voltage controller via the inserted link N–M, *R*117, *R*116, V124 and *R*124. In 60 V equipment the link N–M is omitted, and resistor *R*118 is then included in the actual voltage value circuit (coarse matching of U_{Ist}).

The characteristic relay K101 is only used when the closed-loop and open-loop control assembly A3 is installed in GR10 rectifiers. In a 48 V system, for example, it enables the voltage of the operating rectifiers to be raised to a higher level in parallel operation and in charging at 2.33 V/cell than in normal operation. This level is set with potentiometer *R*108.

For the initial charging, among other things, the characteristic relay K105 is operated, and its contact changed to the 4/7 position. The current reference value I_{Soll} for current limiting during the initial charging can now be adjusted by means of potentiometer *R*106. The voltage divider *R*100/*R*101/*R*106 is connected between P $\hat{=}$ +15 V and pin 14 (M3 $\hat{=}$ 0 V).

The current reference value I_{Soll} is conveyed to the current regulator through the following path: P/$R$100/$R$101/$R$106/K105 7/4/link SX–SZ/R166 to the comparison point for I_{Ist} with I_{Soll} at the input to the current regulator (operational amplifier N160).

In normal operation the characteristic relay K105 is de-energized (contact closed at 4/1 in the rest position). Under this condition potentiometer $R$106 has no effect. $R$101 is used to adjust the onset of current limiting in normal operation to 100% I_{Nenn} (link NE–NA inserted, factory setting).

I_{Soll} is now applied to the current regulator through the path: P/$R$100/$R$101/link NA–NE/K105 1/4/link SX–SZ/$R$166 to the input of the current regulator.

If current limiting is required to take effect at less than 100% I_{Nenn}, the links should be soldered according to the following table:

Onset of current limiting at	Link
80% I_{Nenn}	NE–NB
60% I_{Nenn}	NE–NC
40% I_{Nenn}	NE–ND

⑤ *Voltage regulator (controller) and* ⑥ *matching transformer (voltage follower)*

The voltage regulator consists of the operational amplifiers N140/1 and N140/8 connected in cascade.

The amplifier N140/1 functions as an inverting proportional amplifier. Potentiometer $R$134 is used to adjust the gain (proportionality constant). The actual voltage value applied through $R$123/$R$124 is compared with the voltage reference on $R$125/$R$126 at the inverting input 2.

The following operational amplifier N140/8 functions as a non-inverting integrator; its 'I' characteristic is determined by C141, $R$141 and $R$143. The integration time can be varied with $R$141.

The variable control voltage U_{St} passes via the following voltage follower N140/7 and N140/14 to the trigger set.

With the equipment output voltage at its rated value the difference between the reference voltage value and the actual voltage value is almost zero.

The currents flowing to the comparison point of the reference and actual values– i.e input 2 of N140/1 – are in opposite directions and are substantially equal. The output 1 of N140 is at the potential of M, since no current has to flow through

*R*120, *R*134 and R135 to the comparison point to balance the difference between the reference and actual value currents.

The following operational amplifier N140/8 is not affected by the M potential at its input 10. The output voltage at pin 8 therefore has a particular value (control voltage U_{St}), which is fed to the trigger set via the two voltage followers N140/7 and N140/14. The magnitude of the control voltage, as previously explained, determines the triggering point of the thyristors in the power section, and thus the d.c. output voltage of the rectifier.

If a disturbance occurs – a change in supply voltage or load current, for example – the difference between the reference and actual values is no longer zero, and N140/1 produces a positive or a negative voltage, according to whether the d.c. output voltage needs to be lowered or raised.

An increase in the output voltage of the equipment (arising, for example, from a reduction in load) results in an increase in the proportional negative actual voltage value. The potential of the inverting input 2 of the operational amplifier N140 becomes more negative with respect to the reference point M (BS, 0 V). An inverted, positive-going signal appears at the output 1 of the operational amplifier N140. The signal is not subject to any further inversion. The control deviation referred to therefore leads to an increased positive control voltage U_{St},

This is effected through the following path: N140, output 1, positive-going signal/*R*120/*R*144/N140, non-inverting input 10, positive-going signal/N140, output 8, positive-going signal/N140, non-inverting input 5, positive-going signal/N140, output 7, positive-going signal/*R*145/link AU–AV/*R*193/N140, non-inverting input 12, positive-going signal/N140, output 14, positive-going signal/*R*190/link AZ–AX.

The control voltage U_{St} could have been, as an example, +5.1 V and now becomes, for example, +5.2 V.

This increased control voltage is applied to the integrated triggering-pulse generators V200, V300 and V400 in the trigger set at pins 11. The trigger set now increases the delay of the triggering pulses relative to the a.c. synchronizing voltage, moving them towards the inverter stability limit α_W. This leads in turn to a reduction in the equipment output voltage, and the required actual voltage value is regained (see Fig. 1.44).

A reduction in the equipment output voltage implies a reduction in the proportional (negative) actual voltage value. The potential of the regulator input 2 of N140 becomes more positive with respect to the reference potential M. N140/1 thereupon produces a negative voltage at input 10 of N140 and thereby causes a reduction in the control voltage to the trigger set (U_{St} less positive than before). In the trigger set a lower control voltage shifts the triggering pulses towards α_G. The equipment output voltage increases and the correct level is restored (see Fig. 1.44).

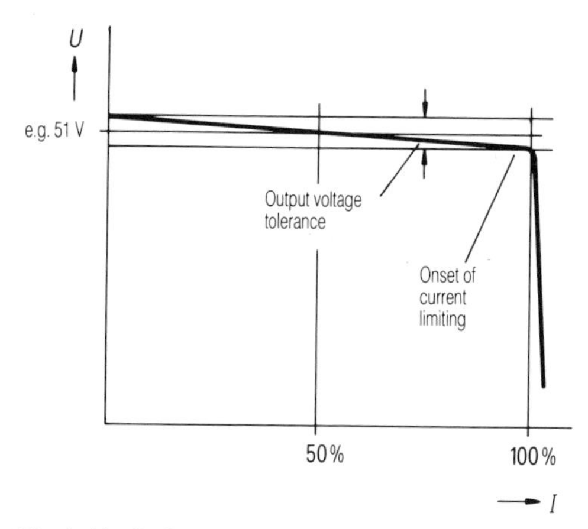

Fig. 1.44 D.C. output voltage characteristic

⑧ Current regulator (controller) and ⑦ actual current value amplifier

The current regulator is similar in design to the voltage regulator; it consists of the operational amplifiers N160/7 and N160/14; N160/7 functions as an inverting and N160/14 as a non-inverting amplifier. The current reference value is fed to N160/7 from the set-point device through $R166$.

The actual current value signal I_{Ist}, with matching preset in the assembly A10, is fed via pins 24 and 23 in A3 to the differential amplifier N150, pins 2 and 3. N150 matches (amplifies) the small actual current value to the level required for the current regulator. The voltage drop between pins 24 and 23 at 100% of the rated equipment current I_{Nenn} amounts to 30 mV, and this is amplified to about 3.7 V.

The voltage is smoothed by $R155$, C152 and N160/1. The voltage on the link R–T with respect to pin 28 is factory-set by means of $R168$ to –2.8 V at 100% I_{Nenn} from the rectifier.

The amplified and matched I_{Ist} signal is fed through $R167$ to the current reference – actual comparison point at the inverting input 6 of the operational amplifier N160, and is there compared with the I_{Soll} signal.

The 'P' feedback of the operational amplifier N160/7 (the proportionality constant) can be adjusted with potentiometer $R164$. The amplifier N160/14 has 'I' feedback, which is adjustable with potentiometer $R171$. Both potentiometers are factory-adjusted and should not be altered.

So long as the set current level at the output of the rectifier is not exceeded, the positive value I_{Soll} at the input 6 of the amplifier N160 predominates. The signal at the output is therefore negative. When the input 12 of N160 is negative, there is similarly a negative potential at its output 14; this is blocked by the diode V171.

Under normal conditions, therefore, the current regulator cannot impinge upon the control circuit and there is no current-limiting effect.

If the output current of the equipment increases, the negative signal I_{Ist} at input 6 of N160 increases in proportion to it. If the positive preset current reference is exceeded (at 100% I_{Nenn}, for example, if the current limit is set to 100% I_{Nenn}) input 6 of N160 becomes negative. The output 7 of N160 now produces a positive signal. Since input 12 of N160 is now positive, the output 14 also becomes positive and the diode V171 conducts. Since the value of the resistor $R140$ at the output of the voltage regulator (operational amplifier N140/8) is much higher than that of the resistor $R170$ at the output of the current regulator (operational amplifier N160/14), the current regulator now takes control in the control circuit and turns off the voltage regulator. Through the two voltage followers N140/7 and N140/14 the current regulator increases the positive control voltage U_{St} to the trigger set. The control voltage U_{St} is thus increased by the current regulator in dependence upon the equipment output current, and through the trigger set brings about a reduction in the equipment output voltage (onset of current limiting, see Fig. 1.44). The current remains approximately constant. An unacceptably high current cannot therefore be drawn from the equipment.

⑨ α_W *and* α_G *adjustment*

Input 5 of the amplifier N140 has a low-resistance connection, through diodes V148 and V149, to two voltages, which are set by the voltage divider $R148/R149$. These two voltages are the means of determining the limits of the control voltage U_{St} and hence the control range for the thyristors.

The voltage set at $R149$ determines the lowest possible control voltage, and thus the minimum triggering delay – the rectifier stability limit α_G (leading-edge triggering threshold) – that is, the highest possible d.c. output voltage.

The highest possible control voltage is limited by the voltage set on the potentiometer $R148$. Triggering delay is thereby limited to a maximum – the inverter stability limit α_W (trailing-edge triggering threshold) – so that the lowest possible d.c. output voltage is determined.

The potentiometers $R148$ and $R149$ are factory-set to the maximum control range and should not be altered.

⑩ *Trigger set*

The trigger set converts the control command from the regulator into trigger pulses for the thyristors in the thyristor set. The trigger pulses are shifted to a phase position – relative to the a.c. synchronizing voltage – in dependence upon the magnitude of the control voltage U_{St} (Fig. 1.45).

The trigger set consists principally of the *integrated trigger-pulse generators* V200, V300 and V400 ⑬, the following *trigger-pulse amplifier* ⑭ and a *7 kHz clock generator* ⑪.

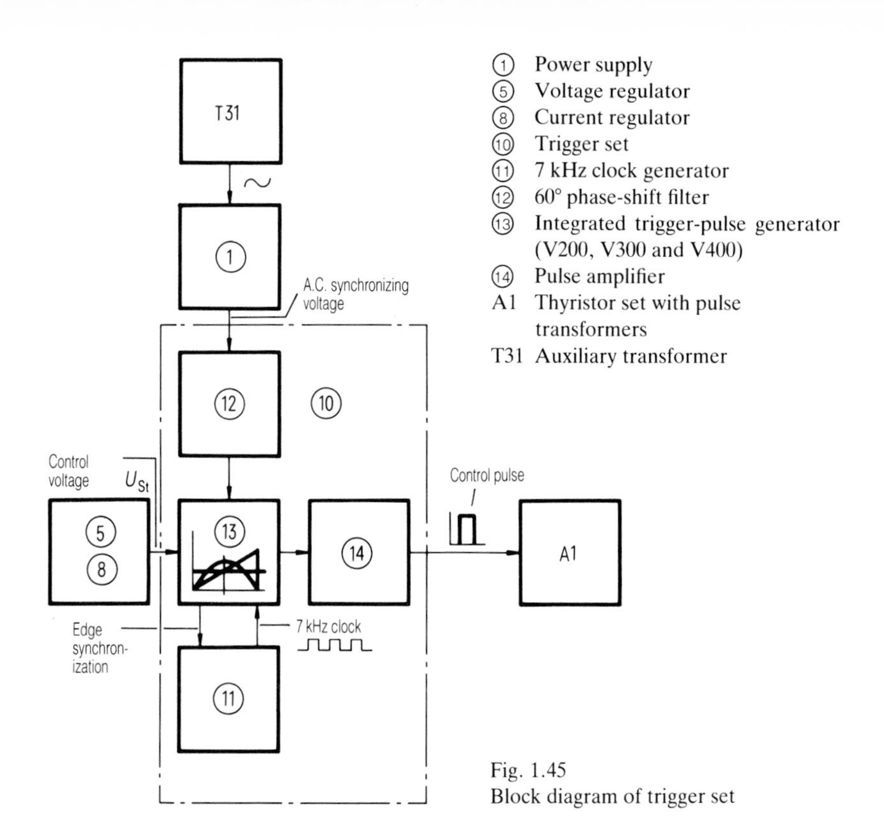

Fig. 1.45
Block diagram of trigger set

To ensure that the trigger pulses are produced in the correct time range for each thyristor, when positive voltage is applied to it, the trigger set must be synchronized to this voltage. The a.c. synchronizing voltage is supplied by transformer T31 and is taken through the internal power supply ① and the *60° phase-shift filter* ⑫ to pins 5 of V200, V300 and V400 (see Fig. 1.43). V200, V300 and V400 produce trigger pulses at the instant when the variable control voltage is equal to the constant sawtooth voltage.

The same integrated trigger-pulse generator is used in closed-loop and open-loop control A3 as in closed-loop and open-loop control A1.

The gating entailed in the formation of the trigger pulses (V200 as an example) is illustrated in Fig. 1.46.

Each integrated trigger-pulse generator is synchronized to one of the phase voltages, and produces, independently of the other two, two trigger pulses at the low trigger-pulse outputs 2 and 4 (Fig. 1.43). There is a 180° phase displacement between pins 2 and 4. The six trigger pulses occur at intervals of 60° (about

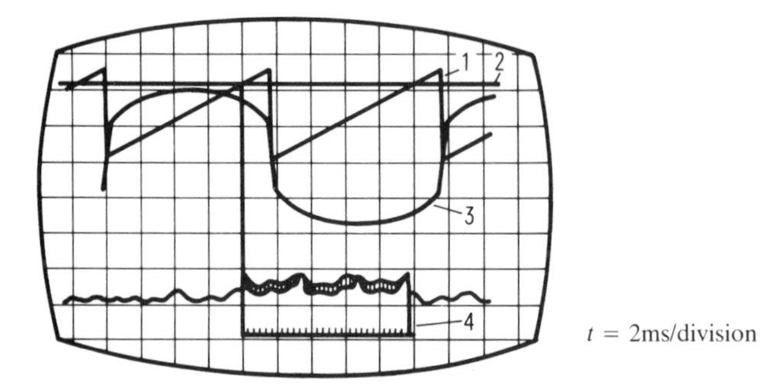

1 Sawtooth voltage on pin 10 (X9.9) with respect to X7 – 5 V/division
2 Control voltage U_{St} on pin 11 with respect to X7 – 5 V/division
3 A.C. synchronizing voltage on pin 5 with respect to X7 – 0.5 V/division
4 Trigger pulses 180°-α on socket X1 with respect to X7 – 20 V/division

Fig. 1.46 Formation of trigger pulses

3.33 ms). The phase displacement between similarly designated pins (e.g. V200, pin 2 and V300, pin 2) is 120°.

At each zero-crossing of the a.c. synchronizing voltage, a 200 μs pulse is generated in the trigger-pulse generator, which terminates the charging of the 'sawtooth capacitor' C205, C305 or C405, and brings about a rapid discharge within the 200 μs period. The charging of the capacitor then begins afresh. Thus arises a sawtooth voltage with a frequency of 100 Hz.

As in the trigger set on closed-loop and open-loop control A1, the duration of the control pulses is controlled at 180°-α by the switching of M (JMP $\hat{=}$ 0 V) onto pin 12 of V200, V300 and V400.

For a 50 Hz mains supply the link A–B is omitted (*R*520 in-circuit); it has to be soldered in for 60 Hz operation (*R*520 short-circuited).

The slope of the sawtooth waveform (charging current) on the sawtooth capacitor can be adjusted by means of the potentiometers *R*204, *R*304 and *R*404. These are factory-adjusted and should not be altered. Inequality between the settings of the charging slopes in the three trigger-pulse generators leads to unbalance, and hence to an unacceptable increase in the interference voltage at the output of the rectifier.

Pins 8 of V200, V300 and V400 are connected together. The regulated voltages of about +3.1 V in the trigger-pulse generators are thereby equalized. This ensures similar conditions for the control of all three phases.

All further details of connections, together with the internal operation of the trigger-pulse generators and the function of the 7 kHz clock generator, have previously been explained in connection with closed-loop and open-loop control A1.

Figure 1.47 shows the distribution and phase separation of the trigger pulses from the trigger-pulse generators V200, V300 and V400, and their assignment to the individual thyristors. Each main pulse for one thyristor is at the same time the auxiliary pulse for the thyristor that has received its main pulse 60° earlier.

The thyristor to which the main pulse is directed is represented within a circle (e.g. (V2)). The same trigger pulse is applied to thyristor V1 (shown without a circle) as an auxiliary pulse. Thyristors (V2) and V1 thus receive the same trigger pulse *simultaneously*.

This pulse over-coupling is acheived by means of a diode network (see Fig. 1.43) shown in the trigger set ⑩ between ⑬ and ⑭.

When the rectifier is switched on, since two thyristors immediately receive a trigger pulse at the same time, it is possible for a flow of current through the load to be initiated and for the converter to begin to deliver power. After the automatic start-up, referred to as a 'soft start', the pulse intercoupling is of no significance, since the 180°-α trigger pulses overlap.

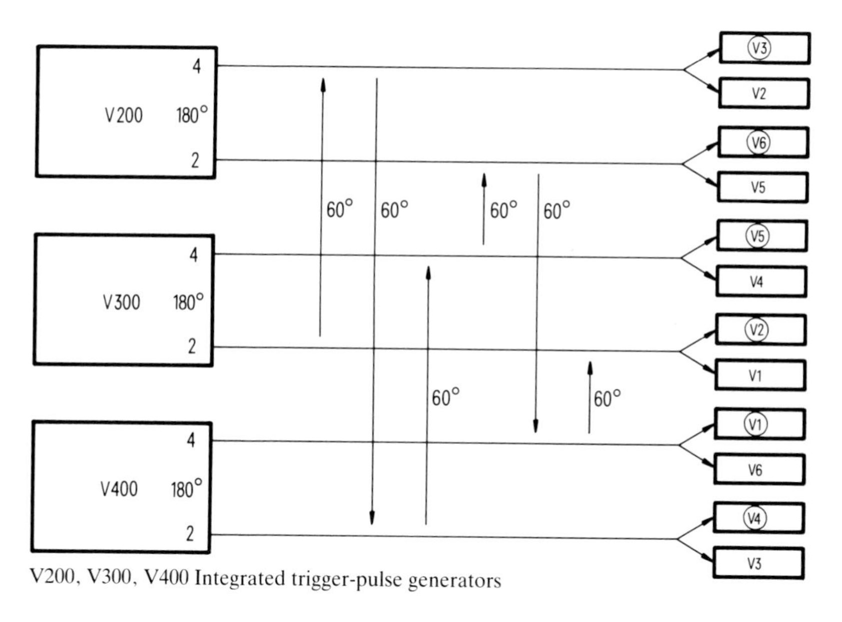

V200, V300, V400 Integrated trigger-pulse generators

Fig. 1.47 Distribution and phase relationship of trigger pulses

Now suppose, for the sake of explanation, that a trigger pulse is delivered from pin 2 of the trigger-pulse generator V300. This pulse is amplified and distributed. Thyristor Ⓥ2 receives this pulse as a main trigger pulse. Thyristor V1 receives the same pulse simultaneously as an auxiliary trigger pulse. Then 60° later the trigger-pulse generator V200 produces a pulse at output 4. This trigger pulse is applied as a main pulse to thyristor Ⓥ3 and as an auxiliary pulse to thyristor V2. Again, 60° later, the next trigger pulse is produced at the output 2 of the trigger-pulse generator V400 and applied to thyristors Ⓥ4 and V3. This represents a phase displacement of $2 \times 60° = 120°$ relative to the pulse from pin 2 of V300. The next trigger pulse is produced again in the trigger-pulse generator V300, but at pin 4. This pulse is applied to thyristors Ⓥ5 and V4. There is now a total phase displacement of 180° ($=3 \times 60°$) relative to the pulse from pin 2 of V300.

The further sequence of the trigger-pulse phase displacement is essentially exactly as described above, and as indicated in Fig. 1.47.The low signals from V200, V300 and V400 (output pins 2 and 4) are fed to inverting impedance changers, each consisting of a p–n–p transistor and two resistors (V230 to V431, *R*213 to *R*414 and *R*211 to *R*415). At the outputs of the impedance changers the pulses appear changed from low to high (see Fig. 1.43).

The impedance changers decouple the switching circuits from the following diode networks and the pulse-amplifier circuits.

The purpose of the diodes V210 to V413 is to distribute the trigger pulses and also, in conjunction with the following zener diodes V214 to V416, to ensure that the Darlington power transistors V218 to V419 are not caused to conduct spuriously by externally generated pulses. Otherwise it would be possible in certain circumstances for thyristors to be triggered at the wrong time.

Only when a trigger pulse is produced at one of the output pins 2 or 4 of V200, V300 or V400, and the associated transistor in the inverting impedance-changing stage conducts, are the thresholds of the diodes and zener diodes overcome and a trigger pulse generated.

The bases of transistors V230 to V431 are biassed positively; for example, the emitter voltage of V230 = 15 V – 0.6 V while the base voltage of V230 = 15 V via *R*211. In the same way the bases of the Darlington power transistors are biassed negatively through resistors, and thereby securely held off. In the absence of a trigger pulse, there is a current, for example, from M24 through V215, *R*210, V214 and *R*218 to N (–15 V). Consequently a voltage of about –5.8 V is applied to the base of V218.

The trigger pulses are amplified by the Darlington amplifier transistors V218 to V419 and again inverted (from high to low).

The diodes V220 to V421 connected after the Darlington transistors are for the purpose of suppressing inductive voltages. These voltages are limited by the zener diode V540 to about 10 V.

The low trigger pulses (M24 signal $\hat{=}$ 0 V) are taken from the output pins of the closed-loop and open-loop control assembly A3 and fed to the pulse transformers in the thyristor set through a multi-way plug and a control cable.

An example will illustrate the passage of the trigger pulse from the output of the trigger-pulse generator to the thyristor.

It is assumed that a trigger pulse is produced at pin 2 of V300. The transistors V331, V319 and V418 must now be turned on and triggering pulses delivered simultaneously to the two thyristors V2 and V1.

A low trigger pulse (M $\hat{=}$ JMP $\hat{=}$ 0 V) at pin 2 of V300 causes a flow of current from P through *R*325 and *R*314 to V300/2.

The base of the p–n–p transistor V331 is now driven negative with respect to the emitter, which is held at P (+15 V) through diode V531. V331 therefore turns on. The collector is at P (less the diode threshold voltage of V531 and the emitter-collector voltage of V331). Current flows from P(+15 V)/V531/V331 emitter and collector/V312/*R*312/V316/base of V319 and V331 collector/V411/*R*410/V414/ base of V418.

Since the emitters of the n–p–n transistors V319 and V418 are at M24 (0 V) and the bases have become positive, both transistors now conduct.

Through the low-resistance emitter-collector path of V319, M24 $\hat{=}$ 0 V $\hat{=}$ low signal is switched to output pin 12 of A3 via the test point X2 (X9.2). From there the low trigger pulse passes via the pulse transformer T2 to the gate of thyristor V2.

If its anode is positive with respect to its cathode, the thyristor V2 is triggered.

At the same instant that transistor V319 turns on, so also does transistor V418. M24 is thereby applied through the emitter-collector path of V418, via the test point X1 (X9.1), to the output pin 4 of A3. From there the low trigger pulse passes via the pulse transformer T1 to the gate of thyristor V1.

If its anode is positive with respect to its cathode, the thyristor V1 is triggered.

These processes are illustrated in Fig. 1.48.

If the pulse enable and the soft start have occurred at the switching-on of the rectifier, the transistor V180 in the ‘pulse enabling, pulse cancelling and automatic start-up’ functional unit is conducting. Therefore +24 V is maintained continuously on one side of the pulse transformers. V180 is accordingly designated as ‘power supply enable for the pulse transformers’.

The trigger pulses are thus generated when the Darlington power transistors

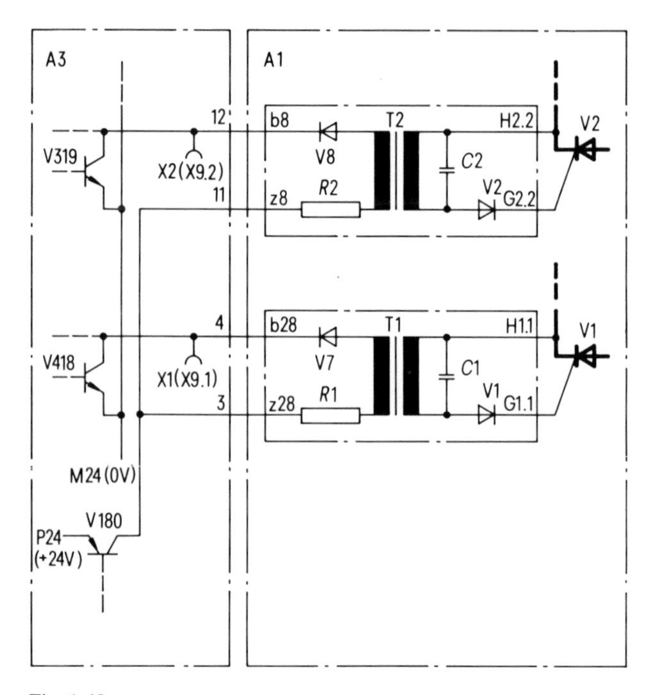

Fig. 1.48
Trigger-pulse connections from closed-loop and open-loop control assembly A3 to pulse transformers and thyristors (thyristor set A1)

are turned on and the reference potential M24 is applied to the other side of the pulse transformers.

Example: trigger pulse for thyristor V2

P24/V180 emitter and collector/pin 11 (A3)/pin z8 (A1)/*R*2/primary winding of pulse transformer T2/diode V8/pin b8 (A1)/pin 12 (A3)/test point X2 (X9.2)/ V319 collector and emitter/M24.

The trigger pulse is now delivered from the secondary side of the pulse transformer T2: pulse transformer T2 plus/diode V2/G2.2/thyristor V2 gate and cathode/H2.2/T2 to minus.

When control current flows in the primary winding of T1, a pulse is produced on the secondary side with a voltage of about 2 V.

⑮ *Pulse enabling, pulse cancelling (inhibit) and automatic start-up (soft start)*

It has to be ensured, when the rectifier is switched on, that the output voltage does not overshoot. To achieve this, the trigger pulses and the power supply for the pulse transformers and the voltage regulator are enabled with a delay.

The relays K184 'pulse enable' and K183 'automatic start-up' are initially in their rest position.

When the rectifier is switched on, the internal power supplies ±15 V and +24 V first come into operation.

Through the contact 4/1 of K184, M1 (M $\hat{=}$ 0 V $\hat{=}$ low) is applied at the output of the 7 kHz clock generator.

The following circuit is completed: P (+15 V)/*R*501/V516/K184 1/4/M1.

This effects the pulse inhibition, since all the pins 6 of the triggering-pulse generators V200, V300 and V400 are low. Consequently no pulse output is possible from pins 2 and 4 of V200, V300 and V400. Meanwhile the 'start-up capacitor' C127 is charged to –15 V (N) through the still closed contact (position 4/1) of the relay K183 (contact and capacitor are located in the voltage regulator ⑤) (see Fig. 1.43).

When the capacitor is charged there is a negative voltage at the input 2 of the inverting amplifier N140. At the output 1 there appears, therefore, a high positive voltage, which, through N140/8, N140/7 and N140/14, is received by the trigger set as a high positive control voltage U_{St} (about +15 V).

The inverting input 2 of the comparator N510 is positive with respect to the non-inverting input 3 (M1 is applied to input 3 of N510 via contact K184 4/1/V514/*R*514). The output 6 is therefore at a negative potential.

The following circuit is completed: +15 V (P)/*R*186/V512/*R*510/N510, pin 6/–15 V (N).

The zener diode V182 and transistors V185 and V180 are still in a blocking state. Relay K183 is similarly in its rest position.

After the operation of a time-delay relay – in the application example (rectifier type 48 V/100 A GR2) it is the time-delay relay K69 (switch-on delay set to 1 or 3 min) – the contactors K63 and K1 in the relay and contactor control operate (see Figs 1.9, 1.10 and 1.42). Now the N potential (–15 V) is connected to pin 25 and the pulse enable is thereby effected. Relay K184 operates, and its contact switches in position 4/7. M1 is now disconnected from the 7 kHz clock generator. The output 6 of the comparator N510 remains at N. The positive potential from the output of the 7 kHz clock generator via *R*501/V513/*R*510 to pin 6 of N510 can therefore fall towards N (as before pulse inhibit). When contact K184 opens, the capacitor C513 can be charged through the resistor *R*513.

After about 1.2 s the voltage on C513, and therefore at the non-inverting input 3 of the comparator V510, is more positive than that of the inverting input 2. The output potential switches from N to P. The 7 kHz clock generator can now supply its signal to pins 6 of V200, V300 and V400. The trigger pulses can consequently be produced at the outputs 2 and 4 of V200, V300 and V400. The zener diode V182 is rendered conducting by the signal change at the output 6 of the comparator N510; the breakdown voltage being exceeded, the two transistors V185 and V180 are turned on. Through the transistor V180 the +24 V (P24) power supply to the pulse transformers is switched on (see Fig. 1.48).

Only now can trigger pulses from the outputs of the trigger-pulse generators be thyristors.

The control voltage U_{St} is still about +15 V. This level implies a delay equal to the inverter stability limit α_W. Initially, therefore, the rectifier delivers the lowest possible d.c. output voltage.

The now-conducting transistor V180 additionally energizes the relay K183, which removes the N potential from the capacitor C127.

The capacitor C127 can now discharge at the input 2 to the amplifier N140, thereby controlling the starting process. Due to the discharge, the negative potential at the inverting input 2 of the amplifier N140, and therefore the positive control voltage U_{St} at the output of the voltage regulator, decreases. The trigger pulses are slowly advanced from α_W towards α_G, and the equipment output voltage accordingly rises slowly. The starting processes are illustrated in Fig. 1.49.

In the event of an overvoltage on the d.c. side, the trigger pulses are inhibited by a low signal (M signal, U_{IL} tolerance –0.7 to +2 V) at the pulse inhibit input 21 (see Figs 1.42 and 1.43). Pin 21 is connected, for example, to the voltage monitor with test circuit assembly A7 or in power supply installations with GR10 rectifiers, the overvoltage limiter A31 or A41, mounted in the battery switching panel with control unit or in the control panel.

If the mains monitor A4 responds to a supply failure, the pulse inhibit is effected through pin 22 (see Figs 1.42 and 1.43). The same conditions apply to pin 22 as to pin 21.

When no trigger pulses are delivered to the thyristor set A1 by assembly A3 the equipment output voltage falls to zero.

The operational amplifier N510 is reset to N at its output 6, and transistors V185 and V180 are thereby turned off. This interrupts the +24 V power supply to the pulse transformers. The automatic start-up relay K183 releases and is then ready for a subsequent re-start.

When the pulse-inhibit signal M is removed from pin 21 or 22 – assuming that the equipment is not 'locked out' by the relay and contactor control (which happens,

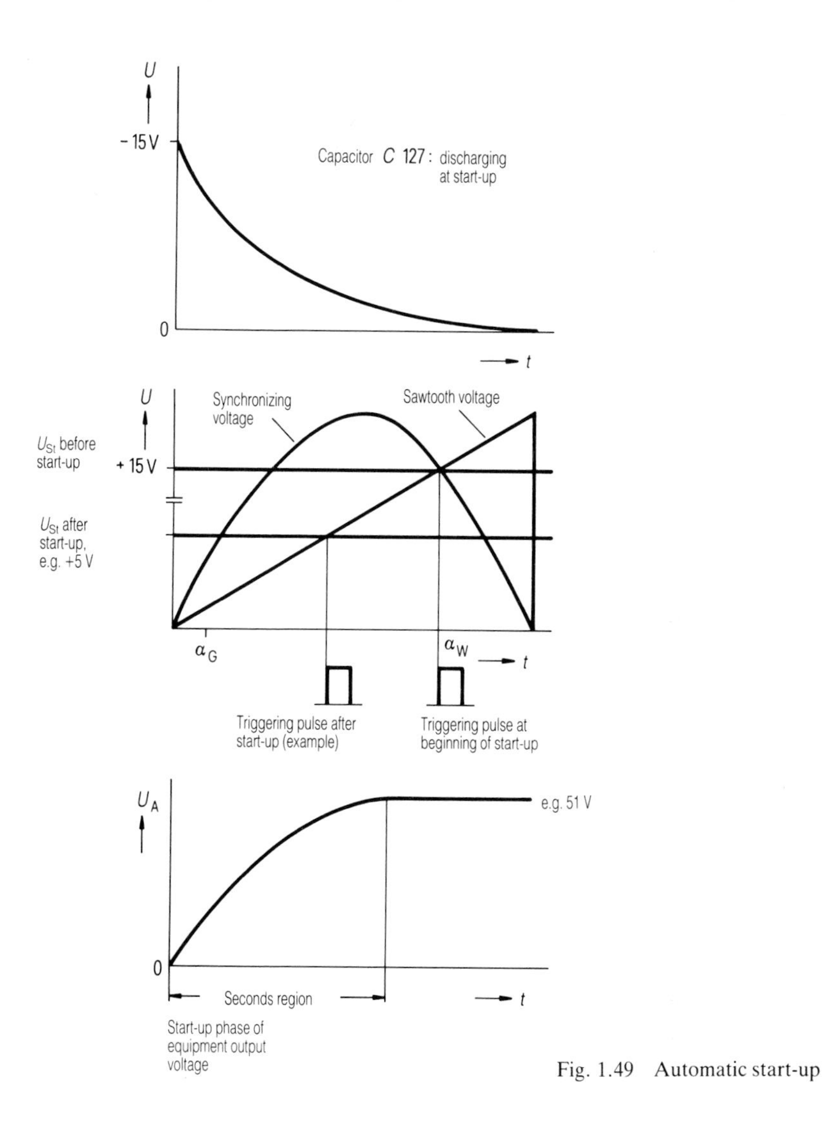

Fig. 1.49 Automatic start-up

for example, in the case of certain overvoltages on the d.c. side) – the soft-start process is initiated afresh and the equipment output voltage U_A builds up again.

1.1.3.2 Mains monitor A4

The mains monitor A4 (Fig. 1.50) protects three-phase rectifiers against phase failure, mains unbalance, mains undervoltage and mains overvoltage. 5 ms

Fig. 1.50 Mains monitor A4

after the occurrence of a supply failure, at the latest, the pulse inhibit command is fed by assembly A4 to assembly A3, so that the trigger pulses to the thyristors are interrupted and the rectifier is switched off by the fault relay.

Figure 1.51 shows the *external connections* to the assembly.

Inserting the link C–B and omitting the link B–A introduces a delay in response (about 35 ms). This variation is applied to equipments with fully controlled three-phase bridges – i.e.rectifiers up to 100 A: GR2, GR3, GR10, GR11 and GR31.

When a supply disturbance occurs, the fault is indicated and the pulse inhibit actuated only after the expiry of the delay period. If the link is moved to B–A the delay is not effective. In this form the A4 assembly can be used in an equipment with a fully controlled three-phase a.c. controller circuit, as in the rectifiers GR10 200, 500 and 1000 A.

The internal power supply of the A3 assembly provides +15 V (P15) stabilized at pin z2, M (0 V) at pin b2 and –15 V (N15) stabilized at pin z4 of assembly A4.

In the A4 assembly a contact of relay K1 is connected internally between pins 20 and 22. In normal operation K1 is continuously energized and contact K1 is therefore closed (pins 20 and 22 connected). Thus the contactor (e.g. K63 in the 48 V/100 A GR2 rectifier) in the relay and contactor control of the equipment can similarly be continuously operated.

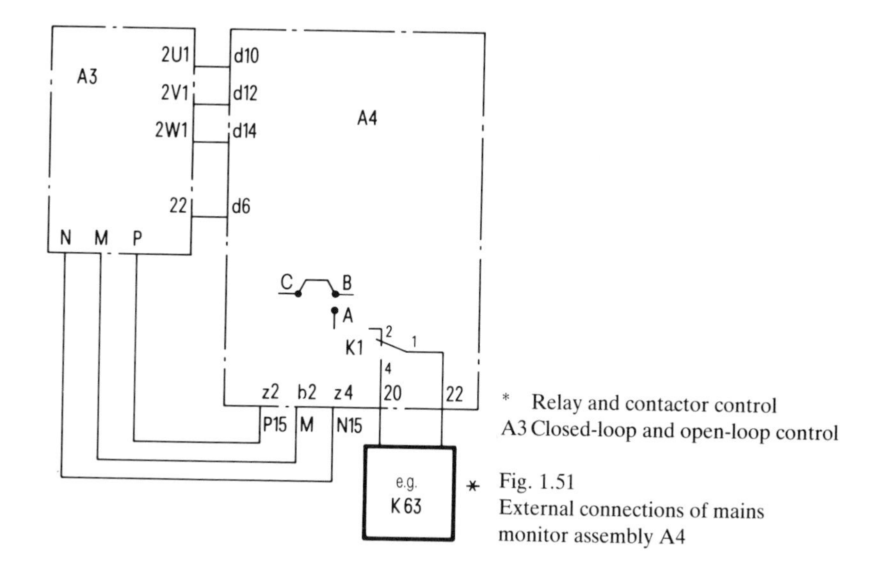

* Relay and contactor control
A3 Closed-loop and open-loop control

Fig. 1.51
External connections of mains monitor assembly A4

On failure of the supply, relay K1 releases and the changeover contact K1 changes its position from 1/4 to 1/2 (pins 20 and 22 are no longer connected). The contactor K63 releases. The mains contactor K1 therefore releases also, and the main transformer T1 is disconnected from the supply (see Fig. 1.9).

The mains monitor A4 contains the following principal functions:

▷ mains-voltage monitor,
  – undervoltage monitor,
  – overvoltage monitor,
▷ phase-failure and mains unbalance monitor,
▷ combined fault processing and
▷ power supply.

These functions are illustrated in Fig. 1.52, which also shows all the important additional functions of the mains monitor A4.

The test point X1.15 is a grouping point for all fault signals. In normal operation this point is at P (high signal). If the mains undervoltage or overvoltage monitor or the phase-failure and mains unbalance monitor responds to a supply fault, the point is switched from high to low (M) and the pulse-cancelling (inhibit) and fault trip are actuated by the following 'combined fault processing' via the relay and contactor control.

The mains monitor obtains its power supply of ±15 V with respect to M from the closed-loop and open-loop control assembly A3.

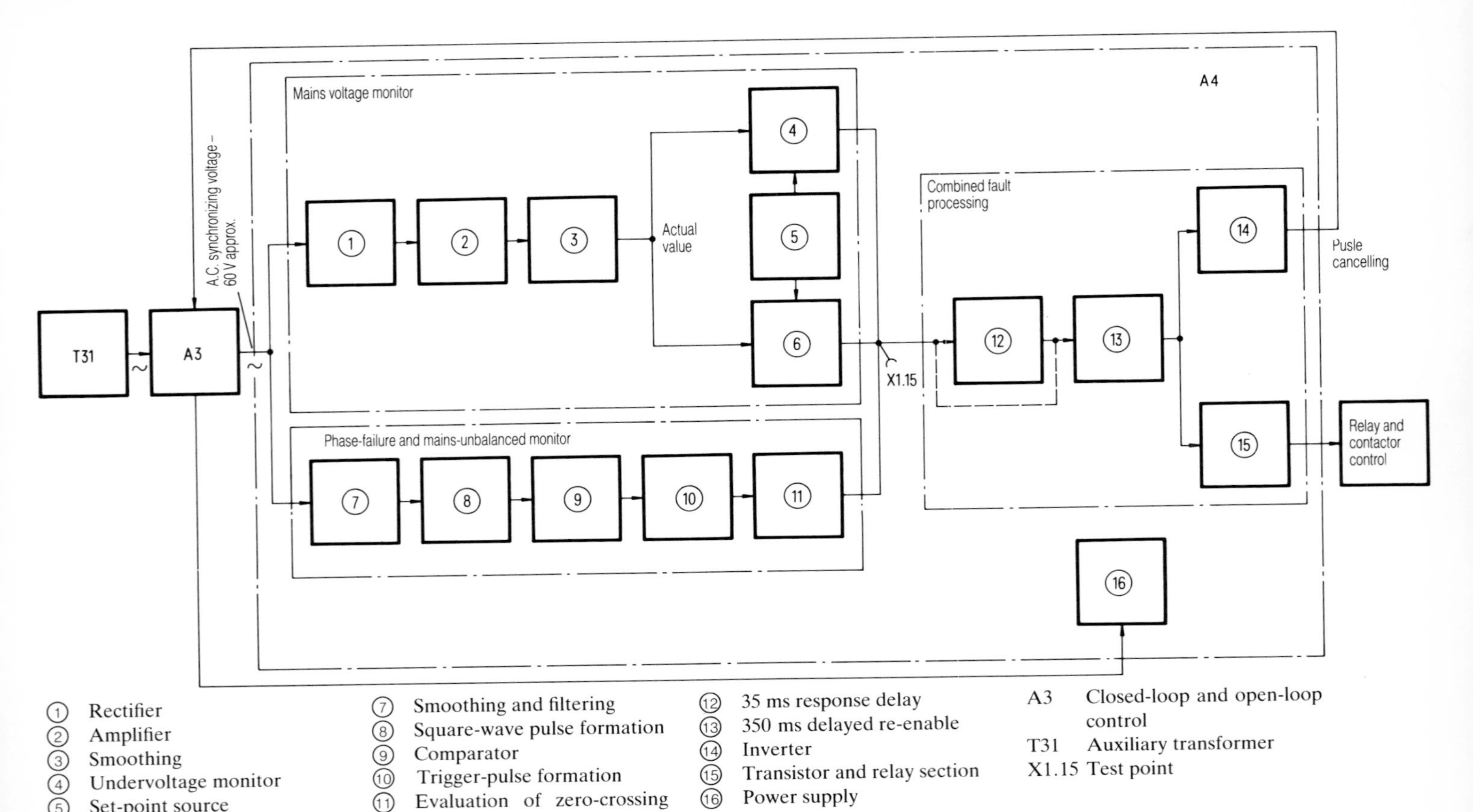

(1) Rectifier
(2) Amplifier
(3) Smoothing
(4) Undervoltage monitor
(5) Set-point source
(6) Overvoltage monitor
(7) Smoothing and filtering
(8) Square-wave pulse formation
(9) Comparator
(10) Trigger-pulse formation
(11) Evaluation of zero-crossing separation
(12) 35 ms response delay
(13) 350 ms delayed re-enable
(14) Inverter
(15) Transistor and relay section
(16) Power supply
A3 Closed-loop and open-loop control
T31 Auxiliary transformer
X1.15 Test point

Fig. 1.52 Block diagram of mains monitor A4

The circuit diagram of the mains monitor A4 (Fig. 1.53) is divided into the same functional units as the block diagram.

The a.c. synchronizing voltages is supplied to pins d10 (2U1), d12 (2V1) and d14 (2W1) via the assembly A3 from the transformer T31.

Mains voltage monitor

① *Rectifier*

The a.c. synchronizing voltage is rectified by the rectifier V1 to V3. Since the diodes pass the negative half-cycles of the three-phase a.c. voltages, a negative (undulating) direct voltage appears at the test point X1.9 (about –40 V).

② *Amplifier*

The operational amplifier N15/7 is connected as an inverting amplifier with negative feedback ($v = 0.1$). At the test point X1.7, therefore, appears the positive, rectified and attenuated synchronizing voltage (ratio 10:1).

③ *Smoothing*

The following active low-pass filter – consisting of the voltage follower N15/8, *R*47 and C5 – delivers the mean value of this direct voltage. Since N15/8 is non-inverting, a positive voltage appears likewise at the test point X1.12, which is fed as an actual value to the undervoltage and overvoltage monitor.

④ *Undervoltage monitor*

The operational amplifier N15/1 is connected as a non-inverting comparator; it compares the signal at the non-inverting input 3 with the fixed potential M connected to the inverting input 2.

With 'normal' supply voltage (not less than the lower limit of voltage tolerance), a positive voltage is applied to pin 3 of N15/1 (actual value > reference value). The output 1 is therefore similarly P (+15 V). This positive voltage can be measured with respect to M at the test point X1.8.

Since diode V10 is non-conducting, P (+15 V $\hat{=}$ high signal) appears at the test point X1.15 through resistor *R*35. The high signal does not represent a fault signal to the following 'combined fault processing'.

An undervoltage (below the lower limit of supply voltage tolerance) causes the acutal voltage value at test point X1.12 to decrease.

When the polarity at pin 3 of the operational amplifier N15/1 changes from positive to negative (reference value > actual value), the output 1 (X1.8) is

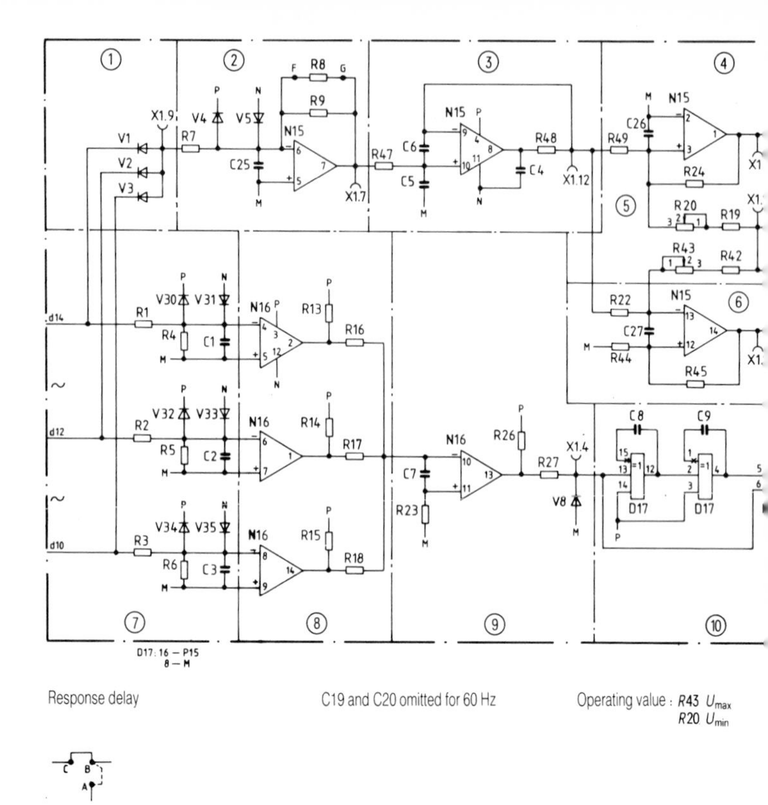

Fig. 1.53 Mains monitor A4

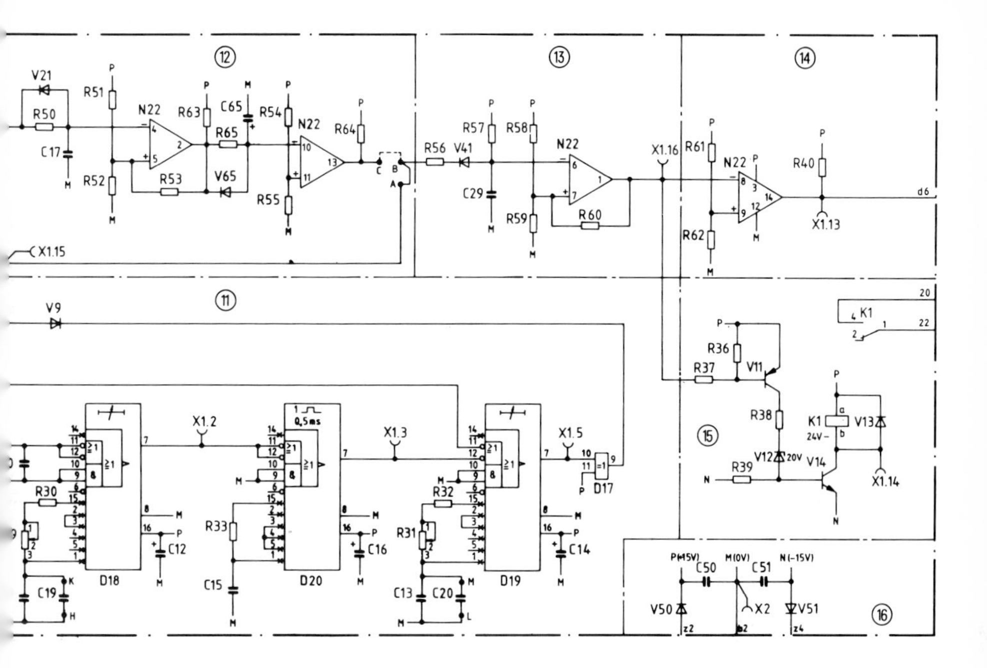

Mains-unbalance monitor: $R\,29\ f>$
$R\,31\ f<$

Mains-voltage monitor	Phase failure and mains unbalance monitor	Combined fault processing	Power supply
(1) Rectifier circuit (2) Amplifier (3) Smoothing (4) Undervoltage monitor (5) Set-point source (6) Overvoltage monitor	(7) Smoothing and filtering (8) Square-wave pulse formation (9) Comparator (10) Trigger-pulse formation (11) Evaluation of zero-crossing separation	(12) 35 ms response delay (13) 350 ms delayed re-enable (14) Inverter (15) Transistor and relay section	(6) Power supply

switched to N (–15 V); since the cathodes of V7 and V10 are more negative than M by about 0.7 to 0.8 V (the diode threshold voltage), current flows from M through V7 and *R*25 to N (N15/1). Test point X1.15 is thus pulled down to M (low). M (low signal) represents a fault signal to the following 'combined fault processing'.

To make the resetting level of the monitoring system different from the operating level, hysteresis is introduced by means of *R*24.

⑤ *Set-point source*

The negative stabilized reference voltage for the two voltage monitoring systems is obtained by means of the zener diode V40. The set points are thus independent of the power supply voltage tolerance. The voltage –6.2 V is on the test point X1.10.

The two reference potentiometers, *R*20 for the undervoltage monitor (normal setting: nominal supply voltage –15%) and *R*43 for the overvoltage monitor (normal setting: nominal supply voltage +15%), are factory-set and should not be altered.

⑥ *Overvoltage monitor*

The overvoltage monitor is arranged similarly to the undervoltage monitor. Here the operational amplifier N15/14 is connected as an inverting comparator. If the supply voltage is within the positive tolerance range (normal voltage), the inverting input N15/13 is negative (reference value > actual value) and the output 14 (X1.11) is positive. The test point X1.15 carries a high signal via *R*35 (as explained in reference to undervoltage monitor).

In the case of a mains overvoltage the positive tolerance limit is exceeded and the positive actual voltage value at the test point X1.12 increases. When the polarity at pin 13 of amplifier N15 changes from negative to positive (actual value > reference value) the output 14 (X1.11) switches to N (–15 V). The test point X1.15 assumes the M potential (low signal), which is interpreted by the 'combined fault processing' as a fault signal.

Phase failure and mains unbalance monitor

The phase failure and mains unbalance monitor checks the separation of the zero-crossings of the a.c. synchronizing voltage. This enables not only interruptions of phase voltages but also asymmetry of the three-phase system to be detected.

⑦ *Smoothing and filtering*

As previously mentioned, the a.c. synchronizing voltage of about 60 V is taken to pins d14, d12 and d10. It is then smoothed and phase-shifted by 30° by RC networks. This lag renders the circuit less sensitive to commutation notches in the supply. The diodes protect the operational amplifiers against overvoltages.

⑧ *Square-wave pulse formation*

Square-wave voltages with an amplitude of about 30 V are produced at the outputs of the operational amplifiers N16/2, N16/1 and N16/14, mutually phase-displaced by 120° (Fig. 1.54a). If the frequency of the synchronizing voltage is 50 Hz, for example, the frequency of the square-wave voltage is similarly 50 Hz.

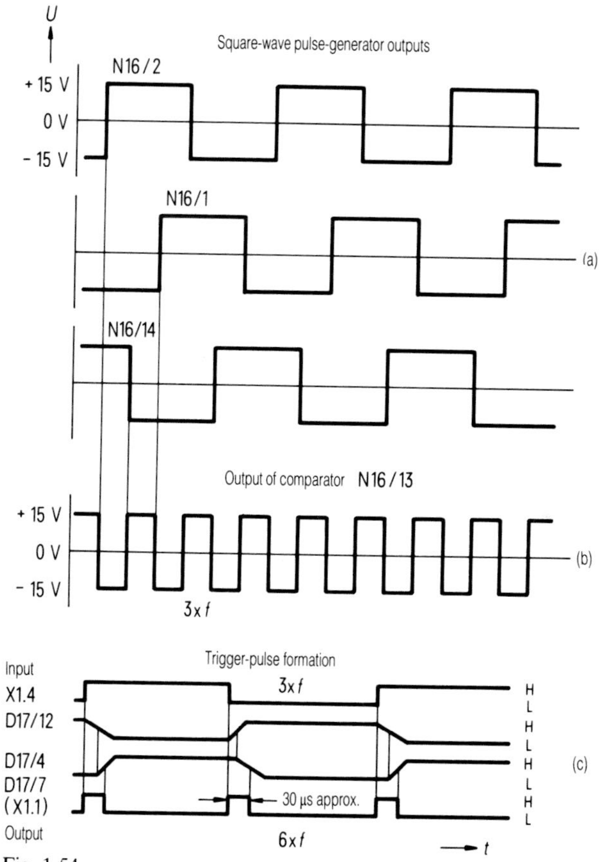

Fig. 1.54
Pulse diagram – square-wave pulse formation, comparator and trigger-pulse formation

⑨ *Comparator*

The incoming square-wave pulses are summed and compared by the operational amplifier N16/13. Each zero-crossing of one of the three square-wave voltages at input 10 of N16 produces a corresponding zero-crossing at the output 13. The output at pin 13 is thus three times the input frequency (Fig. 1.54b). A square wave then appears at test point X1.4 with reduced amplitude, at the frequency of the output 13 of N16 ($3 \times f$).

⑩ *Trigger-pulse formation*

The trigger-pulse formation consists of three EXCLUSIVE–OR gates D17/12, D17/4 and D17/7 and two capacitors C8 and C9; the purpose of these is to generate, at each zero-crossing of the square-wave voltage at the input (test point X1.4), a short trigger pulse of about 30 μs duration at the output D17/7 (test point X1.1) (Fig. 1.54c).

The frequency at the output is twice that at the input. While the beginning of each trigger pulse at test point X1.1 of D17/7 occurs at the zero-crossing of the square-voltage at the test point X1.4, the square-wave voltage is inverted twice by the EXCLUSIVE–OR gates D17/12 and D17/4 and its transitions retarded by the effect of C8 and C9, so that from the square waveform a voltage of trapezoidal waveform is produced.

The inversion and edge retardation have the effect of terminating the trigger pulse after about 30 μs. This takes place when the corresponding low and high thresholds are attained in the waveforms.

⑪ *Evaluation of zero-crossing separation*

The functional unit for the evaluation of the zero-crossing separation consists essentially of the two storage elements D18 and D19 connected as retriggerable monostable toggle stages (pulse delay), the non-retriggerable monostable toggle stage D20 (output pulse about 0.5 ms), the EXCLUSIVE–OR gate D17/9 connected as an inverter and associated RC networks.

The purpose of the functional unit is to check the zero-crossings (Fig. 1.55).

If after one pulse at time t_1 the next pulse does not arrive between T_{min} and T_{max} (at t_3, for example), a fault signal is produced, because D19 is not retriggered.

The permissible minimum and maximum separation intervals between two consecutive zero-crossings is determined by the two storage elements D18 and D19. Adjustment is possible by means of the potentiometer *R*29 (shorter interval T_{min} between zero-crossings) and R31 (longer interval T_{max} between zero-crossings). This represents a tolerance band for the nominal mains frequency of about ±7 Hz. Both potentiometers are factory-set and should not be altered.

The monostable toggle stage D20 resets the timing stage D19 in normal oper-

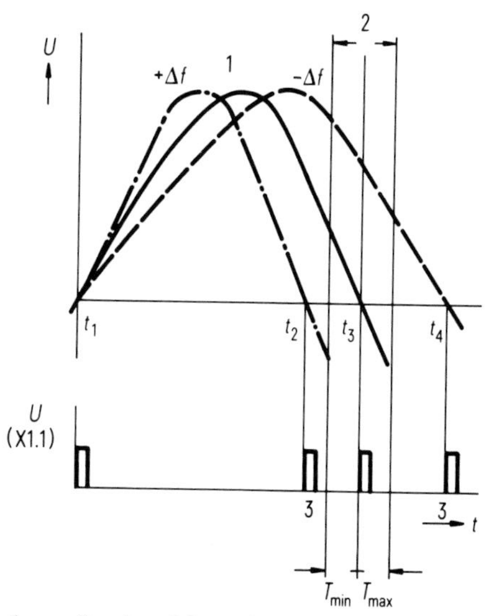

1 Synchronizing voltage, normal
2 Tolerance range of supply symmetry
3 Possible pulses under fault conditions (e.g. frequency deviation)
T_{min} Frequency tolerance range extended upwards by *R*29
T_{max} Frequency tolerance range extended downwards by *R*31

Fig. 1.55 Evaluation of zero-crossing separation

ation. As a result, in disturbance-free operation, a low signal (M) is produced continuously at the output 7 of D19 (test point X1.5).

If a trigger pulse appears too early or too late for D18 (due to a displaced zero-crossing), the output condition of the timing stage D19 changes from low to high (about +15 V). If, due to a phase failure, the relative trigger pulse disappears completely, the effect is the same as when it appears too late.

If the trigger pulse (Fig. 1.55, instant t_1) is followed by the next in sequence at instant t_3, or at least within the permitted tolerance range, no fault signal is produced.

If the next trigger pulse occurs at t_2 (due, for example, to an excessively high frequency) or not until t_4 (due, for example, to an excessively low frequency), a fault signal is generated.

This will be explained with the aid of three examples with reference to the pulse diagram of Fig. 1.56. The diagram similarly shows all the significant signal conditions at the test points X1.2, X1.3 and X1.5.

The first example illustrates *normal operation* (Fig. 1.56a). The timing stages D18 and D20 are reset by the signal change from low to high of the first trigger pulse (left in the diagram). After about 30 μs the trigger pulse switches back from high

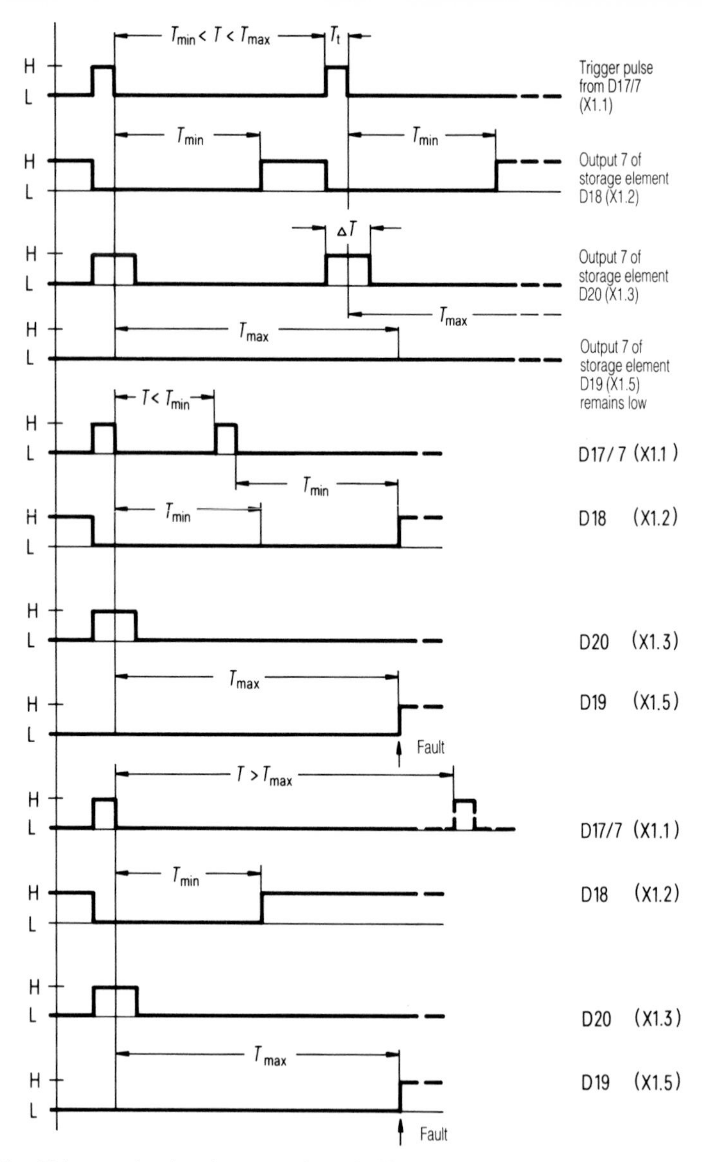

T_t Trigger-pulse duration approximately 30 μs
$\triangle T$ Approximately 0.5 ms
(a) Normal operation
(b) The separation interval between two consecutive trigger pulses is too short ($f > f_N$)
(c) The separation interval between two consecutive trigger pulses is too long ($f < f_N$)

Fig. 1.56 Pulse diagram – evaluation of zero-crossing separation

to low. At this point the timing periods T_{min} in D18 and T_{max} in D19 start simultaneously. When the next trigger pulse appears, after the end of T_{min} but before the end of T_{max}, D18 is again reset by the trigger-pulse transition from low to high. A pulse of about 0.5 ms appears at the output of D20.This retriggers D19 (the timing period T_{max} starts afresh) and the output 7 of D19 remains low (disturbance-free operation). Thus the governing conditions for disturbance-free operation are:

▷ The next trigger pulse must have occurred within the required time range $T_{min} < T < T_{max}$ and

▷ a signal change from low to high and back must have taken place at the output of D20.

In the second example (Fig. 1.56b) *the separation interval between two consecutive trigger pulses is too short.*

In this case it will be observed that the next trigger pulse appears before the end of the timing period T_{min}. At the transition from high to low of this trigger pulse D18 is retriggered, the timing period T_{min} of D18 starts again and the output of D18 remains low. Consequently the 0.5 ms pulse does not appear at the output of D20, and since D19 is now retriggered by D20 the signal at the output of D19 changes from low to high at the expiry of the timing period T_{max}. The fault signal is thereupon fed to the following 'combined fault processing'.

Finally, in the third example (Fig. 1.56c), *the separation interval between two consecutive trigger pulses is too long.*

Here the situation is the reverse of that in the second example.

In this case the next trigger pulse would occur after the end of the time interval T_{max} if it were not inhibited.

The signal at the output of D18 changes from low to high at the end of T_{min}. Since, however, the next trigger pulse does not appear within the required range of time, no signal change is produced at the output of D20. This means that D19 is again not retriggered and the fault signal is produced at the expiry of the time interval T_{max}.

Pin 11 of the EXCLUSIVE–OR gate D17/9 is held permanently high (P $\hat{=}$ +15 V). D17/9 consequently functions as an inverter.

In normal operation (disturbance-free) the output of D19/7 (test point X1.5) is continuously low (M). The output 9 of D17 is therefore high (+15 V). Diode V9 is non-conducting and the required P (high signal) appears at test point X1.15 via *R*35.

On the occurrence of a fault the signal at the output 7 of D19 (test point X1.5) changes from low to high. Since the input 10 of D17 now carries a high signal, the signal at output 9 changes from high to low (M). The diode V9 conducts and test point X1.15 is at M (low signal). The fault signal is now generated.

Combined fault processing

The functional unit 'combined fault processing' has to deal with three possible fault signals from the three monitoring systems described above.

The signal through V9 from the phase failure and mains unbalance monitor has the states low (M) or high (P). The signal level of the voltage monitor (N or P) is matched to these signal conditions by means of *R*25, V7, V10 and *R*41, V6, V23.

A response in one of the three monitoring systems,

▷ mains undervoltage monitor,
▷ mains overvoltage monitor or
▷ phase failure and mains unbalance monitor,

is registered in the 'combined fault processing'.

This is followed by pulse inhibit and also the switching off of the rectifier by the relay and contactor control. In the case of a supply fault the two other monitoring systems are decoupled by blocking diodes.

If the *35 ms response delay* ⑫ is not required, the link between the soldering lugs A–B is retained and no link is inserted between B and C. The signal then passes without a delay to the following functional units (in equipments with fully-controlled three-phase a.c. controller circuits).

In order that the mains monitor A4 should not respond immediately to brief deviations of voltage or frequency, such as can arise from switching processes (the switching-on surge from a rectifier, for example), there is a facility, in equipments with fully-controlled three-phase bridges, to delay the response. For this purpose the link is transferred from A–B to B–C.

Under disturbance-free conditions, as explained above, the signal at test point X1.15 is high (P). The capacitor C17 is charged and pin 4 of the comparator N22 is at P. An M signal is present at the output 2, capacitor C65 is discharged and pin 10 of N22 is similarly at M. The output 13 of N22 therefore carries a high signal (P).

In the following *350 ms delayed re-enable* ⑬, diode V41 is non-conducting. Capacitor C29 is charged from P towards M through *R*57. When the capacitor is charged, the input pin 6 of the operational amplifier N22 is at P and the output 1 (test point X1.16) is at M.

The comparator N22/14 is connected as an *inverter* ⑭.

The input 8 is at M; the output 14 (test point X1.13) is therefore high (P). The output pin d6, in normal operation, provides a high signal (no pulse inhibit) to assembly A3.

From the test point X1.16 the M signal is also fed to the *transistor and relay section* ⑮. This causes transistor V11, zener diode V12 and transistor V14 to conduct (N signal on test point X1.14). Relay K1 is thereby energized and contact K1, in normal operation, links the external pins 20 and 22 connected to the relay and contactor control.

If a fault signal (low $\hat{=}$ M) appears at test point X1.15, capacitor C17 is rapidly discharged through V21, and N22, pin 4, similarly receives an M signal. Output 2 of N22 switches to P and C65 is now charged through *R*63 and *R*65. When the capacitor voltage reaches a particular threshold (P), pin 10 of N22 becomes positive with respect to pin 11. The output 13 then switches to M. Diode V41 now conducts and C29 is discharged through V41 and *R*56. The input 6 of N22 similarly assumes the M potential, and the output 1 changes to P. This P signal is applied to input 8 of N22 via test point X1.16 and the output 14 changes to M (low signal) (test point X1.13). The pulse-inhibit command is now delivered to the closed-loop and open-loop control A3 through pin d6.

At the same time the positive signal at X1.16 passes to the transistor and relay section, where transistor V11, zener diode V12 and transistor V14 become non-conducting.

Relay K1 releases and the contact K1 switches to the rest position 1/2. The connection between output pins 20 and 22 of the mains monitor is thus broken and the rectifier is switched off by the relay and contactor control which is connected to them.

If no further failure occurs, the circuits are reset to the conditions described for normal operation. The enabling of the triggering pulses and the fault relay follows about 350 ms after the end of the fault condition. This switch-on delay ensures the correct subsequent switching sequence of the contactor control and the delayed pulse enable.

V41 blocks and C29 is charged through *R*57. N22/6 is then at P. After inversion the pulse-enable signal is produced at N22/14. V11, V12 and V14 conduct again and relay K1 switches the equipment on again through the operation of the relay and contactor control.

Power supply

The input to the *power supply* ⑯ for the mains monitor A4, ±15 V with respect to M, is obtained from the closed-loop and open-loop control assembly A3.

The diodes V50 and V51 provide reverse-polarity protection and also prevent the capacitors C50 and C51 from discharging back into the power supply of the closed-loop and open-loop control A3. The test point X2 (M $\hat{=}$ 0 V) serves as a reference point for the great majority of measurements on the A4 assembly.

1.1.3.3 Voltage limiter A2

The voltage limiter A2 (Fig. 1.57) is used in single-phase rectifiers (see Section 1.1.1, Fig. 1.4).

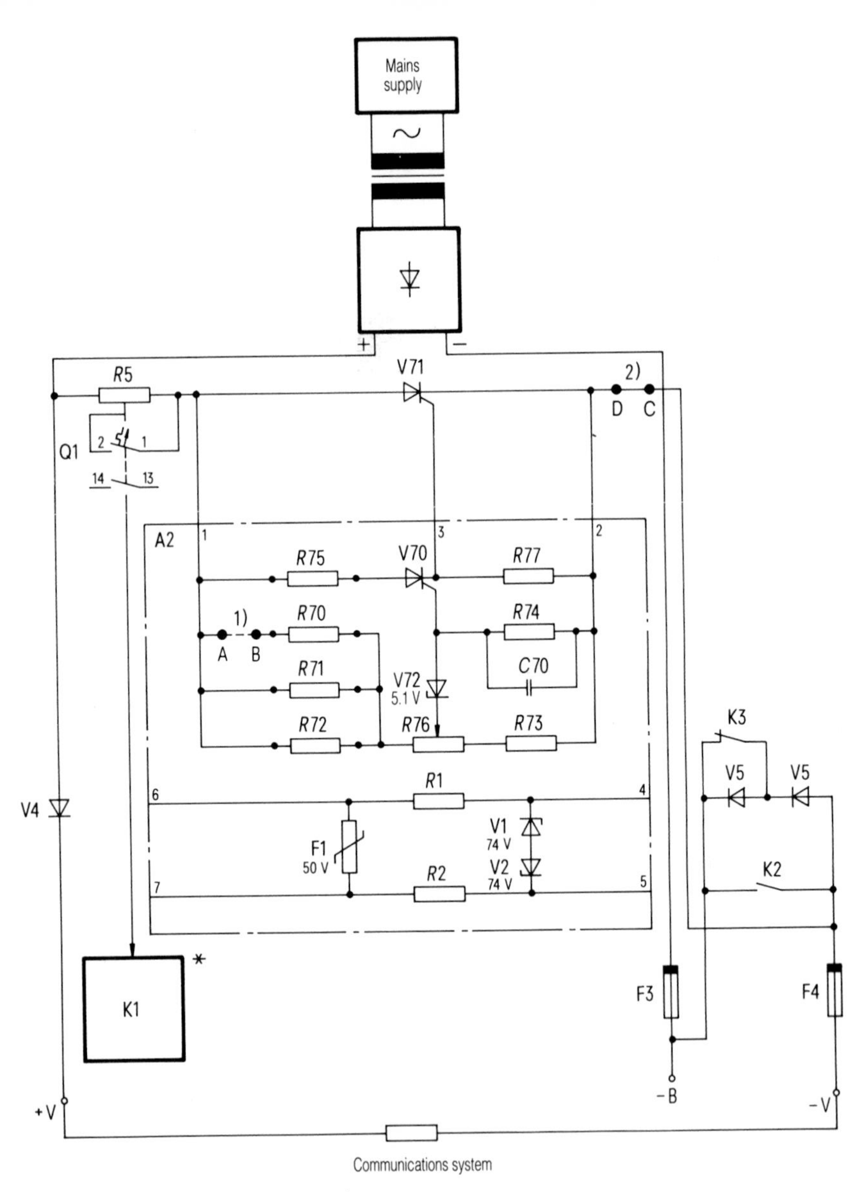

1) Link soldered in for 48 V equipments
2) Link not required for initial charging (fuse F4 removed)
* Relay and contactor control

Fig. 1.57 Voltage limiter A2

Faults in a regulating system lead to deviations of the d.c. output voltage. These may be so large as to be unacceptable for the load. Supply overvoltages – due to atmospheric discharges, for example – can also result in an unacceptable rise in the d.c. output voltage of the rectifier.

If the preset operating level is exceeded, the zener diode V72 conducts and the auxiliary thyristor V70 is triggered; this in turn triggers the voltage-limiting thyristor V71. The current that flows through V71 is limited by the series resistor *R*5 and the equipment output voltage is held down to an acceptable level. At the same time the automatic circuit breaker Q1 connected in parallel with *R*5 operates and switches off the equipment via K1 (relay and contactor control).

The operating levels of the circuit can be adjusted by means of the potentiometer *R*76 as follows:

▷ 73 V for 60 V systems,
▷ 60 V for 48 V systems.

For 48 V systems coarse adjustment is effected by soldering in the link A–B to bring *R*70 into circuit.

In later A2 assemblies, additional components – varistor F1, resistors *R*1, *R*2 and suppressor diodes V1, V2 – have been introduced; they are connected between the transformer T1 and the closed-loop and open-loop control A1 (see Fig. 1.35). These prevent mains overvoltages from reaching assembly A1.

1.1.3.4 Reducing-diode control A5 and voltage monitor A6, A8

This assembly is used in single-phase and three-phase equipments (GR2, GR3, GR11 and GR31).

In using the assembly as a reducing diode control A5, the links E–F and G–H are soldered in. When the assembly is used as voltage monitor A6 or A8, these links are omitted.

The basic operation has been explained with reference to Figs 1.4, 1.9, 1.16 and 1.18.

The assembly A5, A6, A8 (Figs 1.58 and 1.59) consists of the functional units:

① selection of voltage limits,
② internal power supply –24 V and set-point source,
③ lower switching-limit monitor,
④ upper switching-limit monitor.

The varistor F100 protects the assembly against overvoltage spikes.

Fig. 1.58
Reducing-diode control A5 and voltage monitor A6, A8

① *Selection of voltage limits*

The setting of the lower voltage-monitoring limit is carried out by inserting the plug X3 into the appropriate socket. Fine adjustment is possible with the potentiometer *R*280.

To set the upper voltage-monitoring limit, plug X4 is inserted into the appropriate socket. Fine adjustment is available with potentiometer *R*270.

② *Internal –24 V power supply and set-point source*

The operational amplifiers in the voltage monitor require a voltage of –24 V. From the equipment output voltage at pins 6 (+) and 7 (–) a constant base voltage is applied to the transistor V104 through *R*103 and the zener diode V103.

Zener diodes V101 and V102 are consequently supplied with a constant current determined by *R*104. A stabilized voltage of –24 V (with respect to M) can thus be

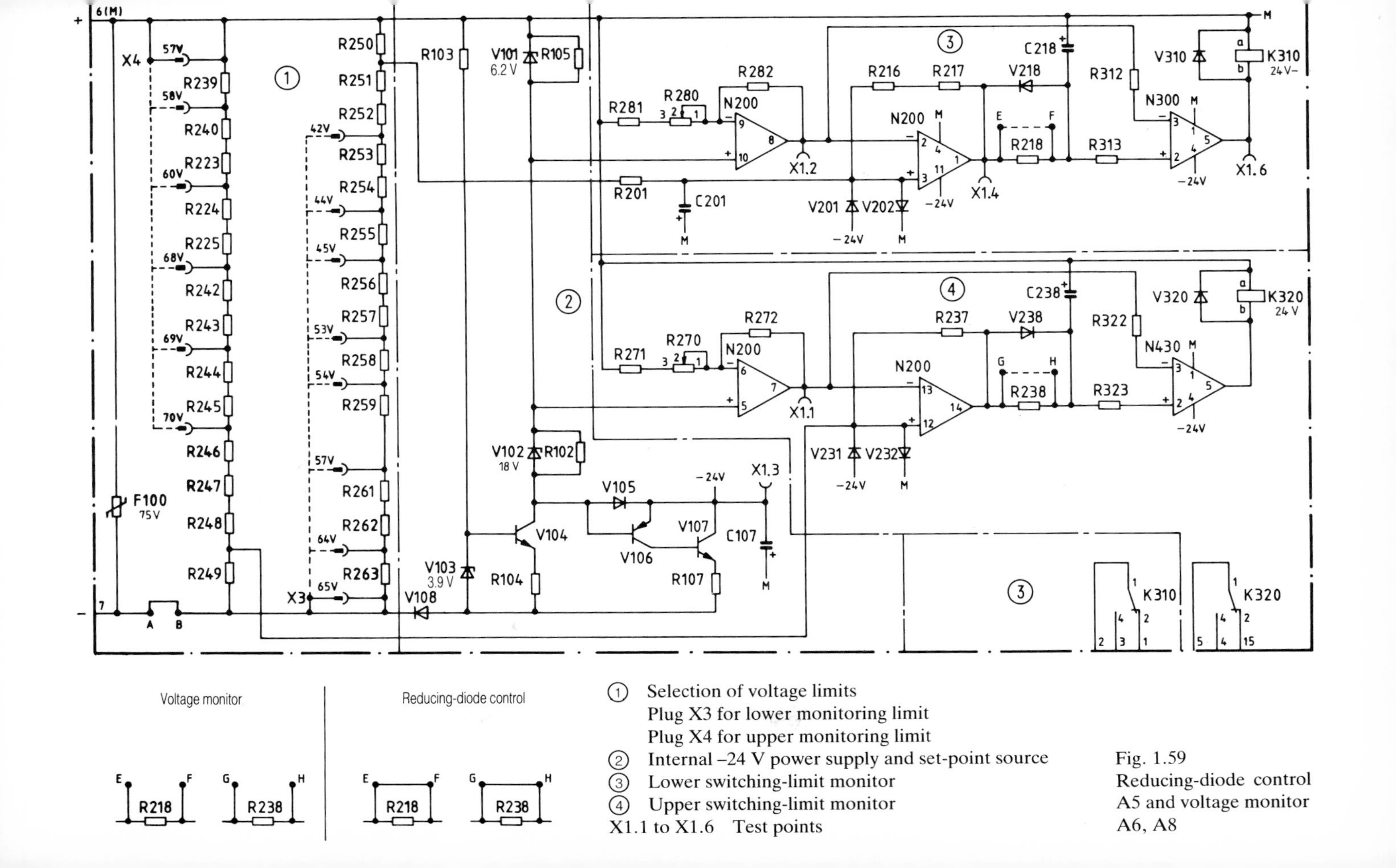

① Selection of voltage limits
Plug X3 for lower monitoring limit
Plug X4 for upper monitoring limit
② Internal –24 V power supply and set-point source
③ Lower switching-limit monitor
④ Upper switching-limit monitor
X1.1 to X1.6 Test points

Fig. 1.59
Reducing-diode control
A5 and voltage monitor
A6, A8

obtained at the collector of V104. The influencing of the stabilizing circuit by load current is prevented by the blocking effect of V105. Voltage variations in the 24 V circuit cause the transistor V107 to conduct to a greater or lesser extent, under the control of the transistor V106, so that the variations are limited. The voltage of the –24 V supply can be measured with respect to M at the test point X1.3.

The zener diodes V101 and V102 provide references for the two voltage-monitoring systems.

The monitoring of the lower and upper switching limits will be explained by reference to the application of the assembly as the reducing-diode control A5.

③ *Lower switching-limit monitor*

The voltage-monitoring reference is tapped off between the zener diodes V101 and V102 and amplified to approximately 12 V by the operational amplifier N200/8. From the output 8 of the operational amplifier N200 (test point X1.2) the reference voltage value is applied on the one hand to input 2 of N200 and on the other hand via *R*312 to input 3 of the operational amplifier N300.

The actual voltage value, matched by the voltage divider, is applied to the input 3 of N200 via *R*201. The reference is compared with the actual value at inputs 2 and 3 of the operational amplifier N200. So long as the actual value does not exceed the prescribed limit, it is more positive than the reference, and the output 1 of the operational amplifier N200 is at M (test point X1.4).

The input 3 of N300 is consequently negative with respect to input 2. The output 5 (test point X1.6) is similarly at M; relay K310 is released.

When the prescribed operating level is exceeded, the actual value becomes negative relative to the reference. The output 1 of the operational amplifier N200 switches to the –24 V supply voltage. The input 2 of operational amplifier N300 becomes negative with respect to input 3, and the output 5 similarly switches to the negative potential of the power supply. The relay K310 now operates and the contact switches to position 1/4. If the input voltage drops below the set operating limit (about 1 V lower), the operational amplifiers N200 and N300 switch back to M at their outputs 1 and 5. The relay K310 releases again.

④ *Upper switching-limit monitor*

The circuit for the monitoring of the upper switching limit is designed in the same way as that for the lower limit. The relay K320 operates when the voltage limit selected by X4 is exceeded and releases at a level about 1 V lower.

Table 1.10 lists the settings of the switching levels for the reducing diode groups with different systems and numbers of battery cells.

In the use of the assembly as a voltage monitor A6 the description given above

Table 1.10 Settings of switching levels for reducing-diode groups

System/number of cells	Plug X3 First reducing-diode group inserted at	Plug X4 Second reducing-diode group inserted at
48 V/25 cells	53 V	58 V
60 V/30 cells	64 V	69 V
60 V/31 cells	65 V	70 V

under ③ 'lower switching-limit monitor' and ④ 'upper switching-limit monitor' is essentially applicable; the only features to be expected are the energization of K310 and the de-energization of K320 at 'normal' voltage. In this case K310 releases on undervoltage, while K320 is operated on overvoltage (see Section 1.1.3.5).

1.1.3.5 Voltage monitor with test circuit A7

The voltage monitor with test circuit A7 is similarly designed to the A5, A6, A8 assembly (see Section 1.1.3.4). The overvoltage monitor of assembly A7 is provided with an additional test circuit.

The rectifier type 48 V/100 A GR2 (Section 1.1.2.1, see Fig. 1.9) will be considered as the basis for an example. The assembly is also used in GR3 (GR31) equipments and, if required , in GR11 equipments (see Figs 1.16 and 1.18).

The A7 assembly (Figs 1.60 to 1.65) has two functions. If the load voltage falls below the prescribed limit of, for example, 45 V, the *undervoltage monitoring* responds and a visual 'voltage deviation' alarm is given. The rectifier remains in operation (Fig. 1.61).

If the upper limit is exceeded – e.g. a load voltage >57 V – the *overvoltage monitoring* responds and the test circuit is brought into operation.

In the case of an overvoltage, in addition, the output voltage of the rectifier is reduced by the triggering of the voltage-limiting thyristor V200 and the triggering pulses are inhibited.

The *test circuit* provides that in the event of supply voltage transients the rectifier is not 'locked out': i.e. switched off so that it can be switched on again only by a manual operation.

When on overvoltage has occurred, the rectifier is automatically restored to operation by the test circuit, by means of a trigger pulse enable, after a disconnect time (switch-off period) of 3 min (Fig. 1.62). If thereafter no further overvoltage occurs within a testing time (monitoring period) of 6 min (Fig. 1.62a), the test circuit reverts to its original condition and the information 'first overvoltage' in the test circuit memory is erased.

Fig. 1.60 Voltage monitor with test circuit A7

If a further overvoltage occurs within the testing time (Fig. 1.62b), the voltage-limiting thyristor is again triggered and the triggering pulses blocked. Now, in addition the rectifier is switched off by the relay and contactor control and the visual 'voltage-deviation alarm' given. The situation described in reference to Fig. 1.62(b) could arise, for example, as a result of a fault in the control system of the rectifier.

An overvoltage that occurs after the expiry of the testing time is treated again as a 'first overvoltage' (Fig. 1.62c).

It should be noted that the 'first overvoltage' is not signalled; the information is stored in the test circuit and the relay and contactor control is not affected by it.

Figure 1.63 shows the external connections and Fig. 1.64 the functional diagram of the A7 assembly.

To avoid damage to components by overvoltages, which might be generated by, for example, switching processes, the assembly is protected on the input side by a varistor F100.

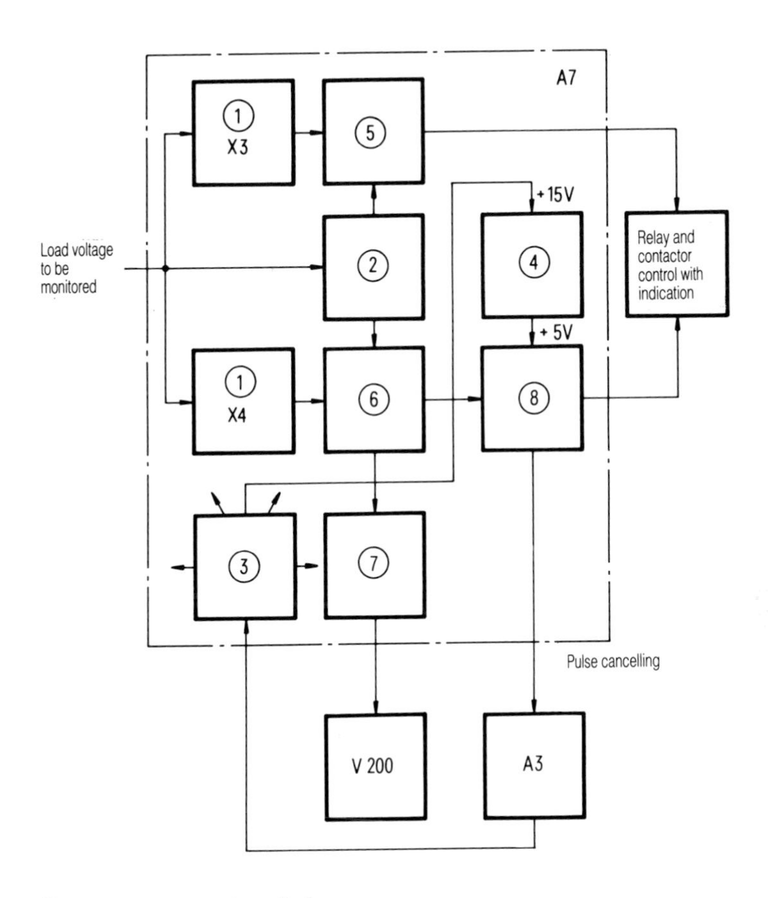

①	Selection of voltage limits Plug X3 for undervoltage monitoring Plug X4 for overvoltage monitoring
②	Internal –24 V power supply and set-point source
③	Internal ±15 V power supply
④	Internal +5 V power supply
⑤	Undervoltage monitor
⑥	Overvoltage monitor
⑦	Limiting-thyristor drive
⑧	Test circuit
V200	Voltage-limiting thyristor (in power section)
A3	Closed-loop and open-loop control

Fig. 1.61
Block diagram of voltage monitor with test circuit A7

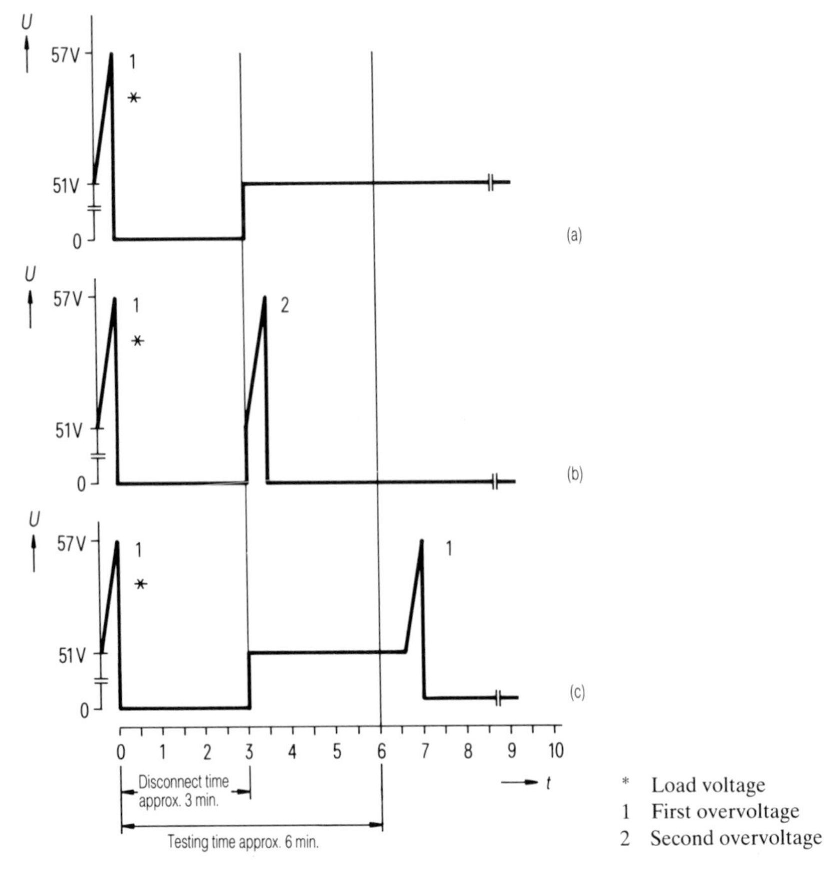

Fig. 1.62
Voltage monitor with test circuit, first and second overvoltages

The functional units ① *selection of voltage limits* and ② *internal –24 V power supply and set-point source* have been described with reference to Fig. 1.59.

The ±*15 V internal power supply* ③ is provided by assembly A3 and the *+5 V internal power supply* ④ is obtained by means of a zener diode from the +15 V supply (P15).

The *undervoltage monitor* ⑤ has similarly been explained in connection with Fig. 1.59. C218 and its associated circuit provide a switching delay of about 500 ms. The switching point (<45 V) is selected by means of the plug X3. Fine adjustment is possible with the potentiometer *R*280. The relay K310 is energized at normal voltage and releases on undervoltage.

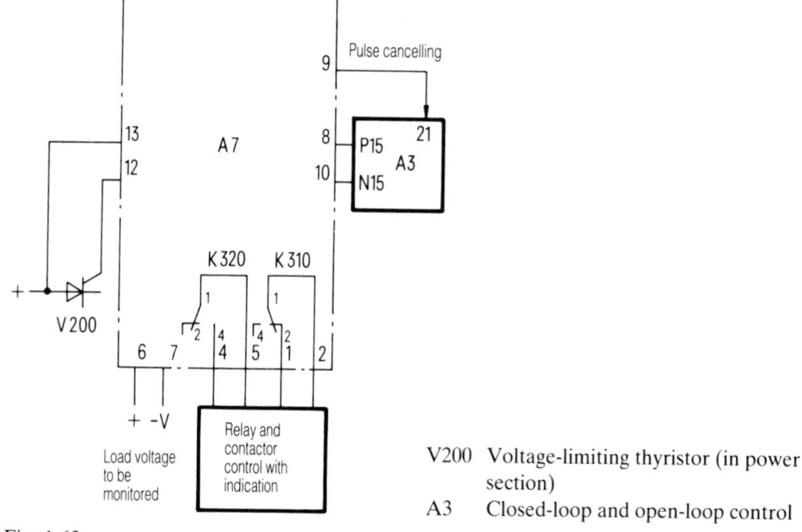

V200 Voltage-limiting thyristor (in power section)
A3 Closed-loop and open-loop control

Fig. 1.63
External connections of voltage monitor with test circuit A7

The *overvoltage monitor* ⑥ is similar in design to the undervoltage monitor. The reference voltage is supplied from N200/7 (test point X1.1) through *R*410 to the input 6 of N400. The actual value passes from the actual value voltage divider 'selection of voltage limit' ① (for overvoltage) through *R*411 to the input 5 of N400. The switching point (>57 V) is selected by the plug X4. Fine adjustment is carried out in this case on the potentiometer *R*270.

In normal operation pin 5 of N400 is positive with respect to pin 6, and the output 7 (test point X1.5) is at a potential of about +15 V. Transistor V500 in the *limiting thyristor drive* ⑦ is consequently non-conducting and the voltage-limiting thyristor V200 (in the power section) is not triggered. The operational amplifier N400/8 functions as an inverter, the non-inverting input 10 being connected directly to M (0 V). Since pin 9 of N400 carries a high signal (+15 V) via X1.5, the output 8 is at about –15 V. This condition represents the information 'no overvoltage – operational normal' for the following *test circuit* ⑧. The test point X1.9 and the output pin 9 are high at about +15 V. This means that no pulse inhibition takes place.

If an overvoltage occurs (>57 V), input 5 of N400 becomes negative relative to input 6 and the output 7 changes to about –15 V (test point X1.5). Transistor V500 conducts and the voltage-limiting thyristor V200 (in the power section) is thereby triggered.

Input 9 of N400 is now at –15 V. This signal appears inverted at output 8 (+15 V) and actuates the test circuit. The diode V405 couples the P15 signal from N400/8

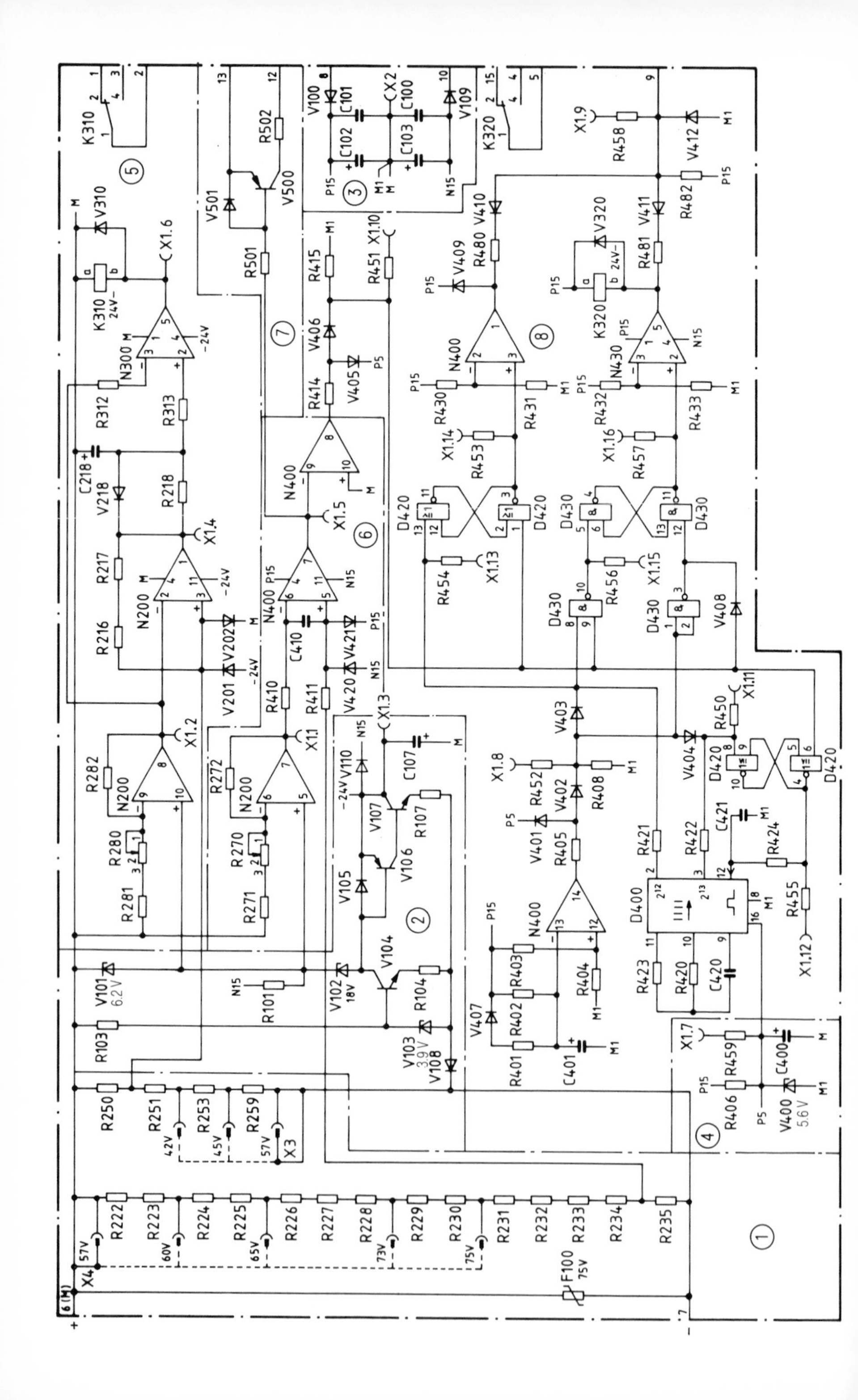

① Selection of voltage limits
Plug X3 for undervoltage monitoring
Plug X4 for overvoltage monitoring
② Internal –24 V power supply and set-point source
③ Internal ±15 V power supply
④ Internal +5 V power supply
⑤ Undervoltage monitor
⑥ Overvoltage monitor
⑦ Limiting thyristor drive
⑧ Test circuit

X1.1 to X1.16 Test points
X2 (M, M1, terminal 6, $\hat{=}$ 0 V) Reference point for all measurements
D420, D430: Pin 14 P5 (+5 V); Pin 7 M1

Fig. 1.64 Voltage monitor with test circuit A7

to the +5 V (P5) supply voltage. In the subsequent processing in the test circuit the high signal assumes this P5 potential. Among other things this triggers the RS flip-flop D420/3 low. The following amplifier N400/1 similarly switches to low.

The output pin 9 (test point X1.9) is now switched from high to low (about 0.8 V) with virtually no delay. The pulse-inhibit command is thereby passed to assembly A3. The timing periods for disconnecting and testing in the test circuit now start. As a result of the triggering of V200 and the pulse inhibition, the overvoltage disappears and the output N400/7 switched to +15 V.

The function ⑧ *test circuit* (automatic re-energization) will be explained below.

So that the counter and the memories shall be in the 'correct condition' (i.e. in defined states) at switch-on, when the auxiliary voltage builds up, the test circuit includes a so-called *switch-on setting*.

The capicitor C401 is charged from P15 through *R*402 (100 k). Since the voltage at the junction of *R*403 and *R*404 increases without delay with the build-up of the power supply voltage, in this condition pin 12 of N400 is positive with respect to pin 13. The output 14 is therefore at P15 and current flows to P5 through *R*405 and V401.

The anode of V401 is at a voltage higher than P5 by the diode threshold voltage, and P5 (high) is therefore applied to test point X1.8 through V402. This high signal persists only until the capacitor C401 is charged ('steering pulse', about 0.01 s).

The states of the gates (digital switching elements) with P5 (+5 V), high, applied and with M (0 V), low, applied are described below.

The high signal at test point X1.8 passes to the gate inputs in part directly and in part through diodes V403 and V404. This results in the following switching states.

The test points X1.11 and X1.13 are high. Inputs 1 and 2 of the NAND gate D430 are similarly high. D430/3 functions as an inverter and its output 3 is therefore low. This low signal appears via V408 and *R*451 at X1.10.

The RS flip-flop consisting of the NOR gates D420/10 and D420/4 is set by the high signal at the 'set' input 8 (S) and the low signal at the 'reset' input 6 (R). This produces a high signal at the output 4 and at the test point X1.12, which is applied to the enable input of the oscillator and counter D400. The counter is reset and blocked. The outputs 2 and 3 are low.

The RS flip-flop consisting of the NOR gates D420/11 and D420/3 is set by the high signal at the 'set' input 13 and the low signal at the 'reset' input 1. A high signal appears at the output 3 and thence via R453 at the test point X1.14.

The input 3 of the operational amplifier N400 is more positive than input 2. The output 1 is at +15 V (P15) and the diode V410 is non-conducting. The test point X1.9 and the output pin 9 receive and carry a high signal (P15, +15 V) through R842. This means no pulse inhibition.

The input 8 of the NAND gate D430/10 is high and input 9 is low. The output 10, and therefore the test point X1.15, via *R*456, are high.

The RS flip-flop consisting of the NAND gates D430/11 and D430/4 is set by the low signal at the 'set' input ($\bar{S}$) 12 and the high signal at the 'reset' input ($\bar{R}$) 5. The output 11 and the test point X1.16, via *R*457, are high.

Input 2 of the operational amplifier N430 is more positive than input 3 and the output 5 is therefore high (P15, +15 V). Relay K320 remains in the rest position (released). Diode V411 is non-conducting and X1.9 and the output pin 9 therefore receive a high signal (P15, +15 V) through R482. This means no pulse inhibition.

When the capacitor C401 has charged to a voltage higher than about +10 V, pin 13 of N400 becomes more positive than pin 12. The output 14 of the operational amplifier switches to N15 (–15 V).

The diodes V401 and V402 block, and M1 ($M \hat{=} 0$ V) is applied to test point X1.8 through *R*408 and *R*452. The switch-on setting is therefore disconnected. The RS flip-flops remain in the states described (Table 1.11).

These switching states apply in normal operation until an overvoltage occurs.

On the occurrence of an overvoltage, N400/8 switches to the high state. The two RS flip-flops D420/3 and D420/4 accordingly switch their outputs from high to low.

To provide the switch-off timing period of 3 min and the testing period of 6 min, on overvoltage an *oscillator* is started, consisting of two inverters which are contained in D400 and the RC network *R*423, *R*420 and C420.

The oscillator is enabled on overvoltage by a low signal (from D420/4) on pin 12 of D400.

The frequency of the oscillator is so determined, by the RC network, that a switch-off time of about 3 min is obtained when the logic signal at the output 2 of the counter D400 switches from low to high after 2^{12} cycles. At the end of this period the RS flip-flop D420/3 is reset to high. The pulse inhibition is removed and the rectifier can supply power again.

A similar description applies to output 3 of D400 (2^{13}) with the time (testing time) doubled. The testing period is about 6 min. The equipment is not switched on again until the switch-off period (3 min) has expired, so that the actual monitoring takes place only during the second three minutes of the testing period.

Table 1.11 Signal states after switch-on setting

X1.8	L (M)	X1.11	L (M)	X1.14	H (P5)
X1.9	H (P15)	X1.12	H (P5)	X1.15	H (P5)
X1.10	L (M)	X1.13	L (M)	X1.16	H (P5)

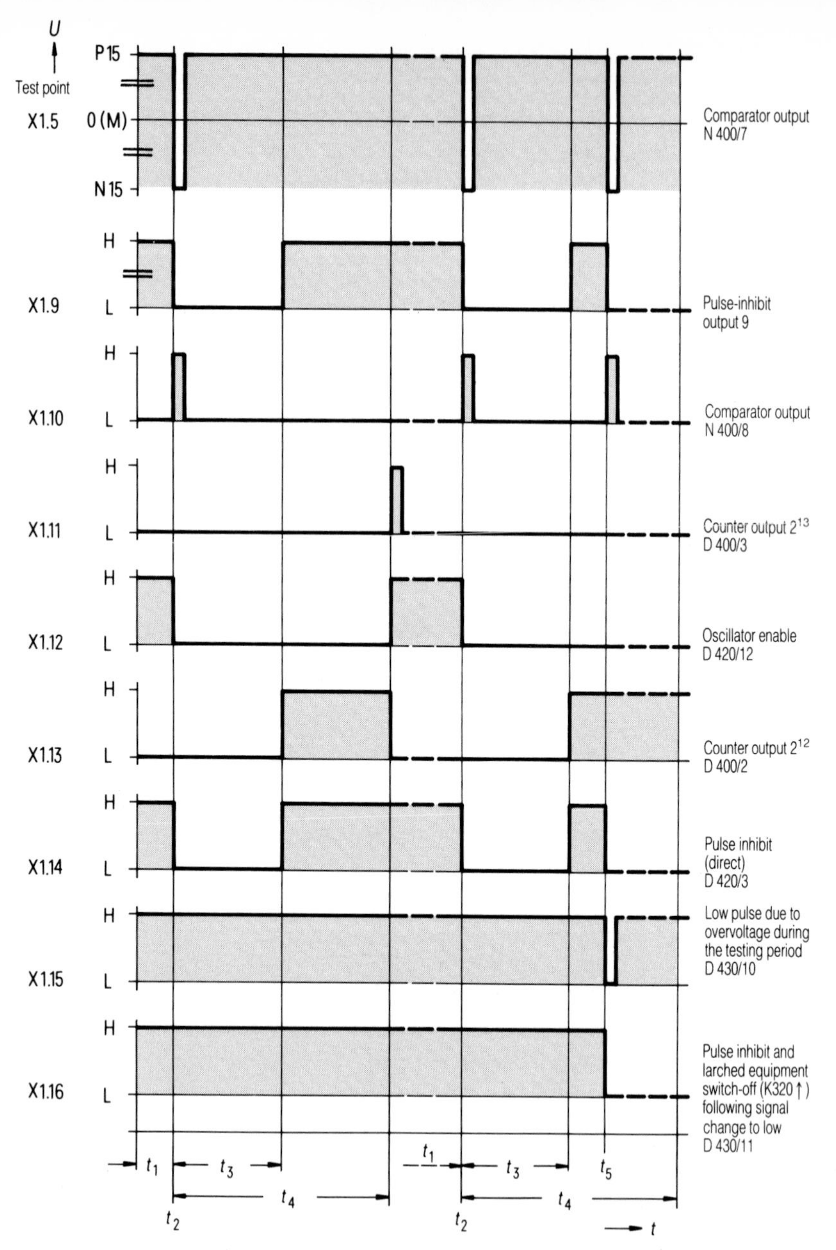

t_1 After switch-on setting
t_2 Overvoltage
t_3 Disconnect time approximately 3 min
t_4 Testing time approximately 6 min
t_5 Overvoltage before the expiry of the testing time

Fig. 1.65
Pulse diagram – test circuit

At the end or the testing period the counter resets itself with its high signal at output 3 fed back to its input 12 via the RS flip-flop D420/4. If a further overvoltage occurs during the testing period, on the other hand, the output 10 of the NAND gate D430 switches to low, because its input 8 has been driven high by D400/2 since the expiry of the switch-off period. The low signal at the reset input 5 of the RS flip-flop D430 switches its output 11 similarly to the low state. The amplifier N430 now produces N15 at its output 5. This causes relay K320 to operate and pulse inhibition is instituted via output pin 9.

The rectifier is switched off by the relay K320, and the relay and contactor control locks out. Switching on the equipment again is possible only by turning the switch S32 to 'off' and then to 'on' again.

The pulse diagram, Fig. 1.65 (left-hand part), applies to the case where an overvoltage occurs but is not repeated within the testing period (see Fig. 1.62a). In the right-hand part of the pulse diagram the effect is illustrated when a further overvoltage occurs within the testing period (see Fig. 1.62b).

1.1.3.6 Current balancing and current monitoring A31

The purposes and the basic functions of the current-balancing and current-monitoring assembly A31 (Figs 1.66 and 1.67) have been explained with the aid of the basic circuit diagram (Fig. 1.24).

The A31 assembly is plugged into the closed-loop and open-loop control assembly A3.

The functional diagram (Fig. 1.67) of the A31 assembly shows the following

Figure 1.66 Current balancing and current monitoring A31

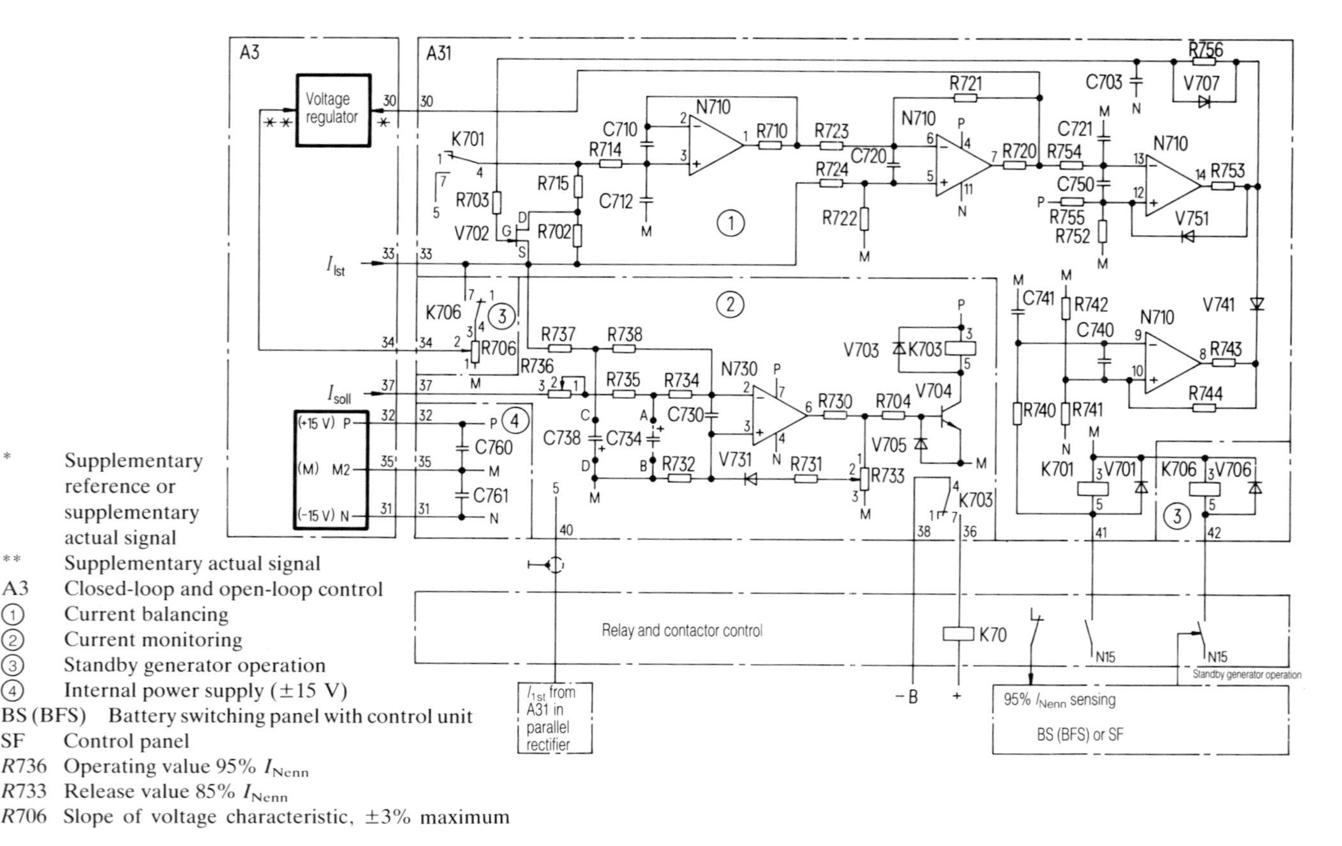

*	Supplementary reference or supplementary actual signal
**	Supplementary actual signal
A3	Closed-loop and open-loop control
①	Current balancing
②	Current monitoring
③	Standby generator operation
④	Internal power supply (±15 V)
BS (BFS)	Battery switching panel with control unit
SF	Control panel
*R*736	Operating value 95% I_{Nenn}
*R*733	Release value 85% I_{Nenn}
*R*706	Slope of voltage characteristic, ±3% maximum

Fig. 1.67 Current balancing and current monitoring A31 (application example: rectifier type GR10)

functions:

① current balancing,
② current monitoring,
③ standby generator operation,
④ internal power supply (±15 V).

The diagram also shows the pin connections of the A31 and A3 assemblies, and the way in which the A31 assembly is connected to the battery switching panel with control unit, or the control panel, via the relay and contactor control, to the parallel-connected rectifiers and to the relay and contactor control itself.

① *Current balancing*

The current-balancing (load-sharing) functional unit is used, for example, in the GR10 and GR11 rectifiers.

In this function the actual current value in the respective rectifier is compared with the average of all the actual current values.

The deviation from the average is fed to the voltage regulator, and through the regulator so adjusts the equipment output voltage that all the equipments that are operating in the same mode supply the same current (tolerance $\leqq 20\% \ I_{\mathrm{Nenn}}$).

Terminal 40 carries a voltage value which represents the sum of the currents of all the parallel-connected rectifiers divided by the number of rectifiers (0 to –2.8 V). Terminal 33 carries the actual current value of the equipment as a voltage (0 to –2.8 V).

In the following discussion it will be assumed that a number of parallel-connected rectifiers are operating and that an additional equipment is switched onto the system.

Through the relay and contactor control the application of N15 (–15 V) causes the relay K701 to operate, and the contact K701 connects terminal 40 to R714 and R715. In addition, if the potential of terminal 41 changes from M towards N, the initially discharged capacitor C741 is charged through *R*740.

So long as the voltage at N710/9 is more positive than that at N710/10, N710/8 is at N and C703 is discharged through V707 and V741. The field-effect transistor V702 is consequently turned off. Terminal 40 has a high-resistance connection through *R*715 (10 k) and *R*702 (470 k) to the 'own' actual current signal.

The high-resistance input of the actual current value of the equipment prevents a rapid fall in the output voltage of the equipments that are switched on. The fewer the equipments already operating at the moment of switching on, the more important this is.

The connection through *R*715 and *R*702 is for the purpose of formulating the sum of all the equipment currents divided by the number of equipments. Because of the high-resistance coupling when an equipment is switched in, this value is only slightly affected.

Via the voltage follower N710/1, with smoothing *R*714 and C712 and resistors *R*710, *R*723, the voltage on terminal 40 is applied to the input 6 of the differential amplifier N710. The other input 5 of N710 receives the I_{Ist} of the equipment, which is initially zero at the instant of switching on.

The voltage difference is amplified 1:1 and supplied as a supplementary reference value to the voltage regulator of the equipment (on A3) through terminal 30. This causes the equipment to increase its output voltage and supply power. When the equipment is supplying about 20% I_{Nenn}, N710/14 switches from N to P. C703 is charged slowly through *R*753 and *R*756. When the voltage on C703 exceeds about +2 V, V702 conducts and *R*702 is by-passed. The actual current value of the equipment is now coupled by a low resistance to terminal 40 (average value) through *R*715 (10 k).

Through V751 the comparison point N710/12 is shifted from about +0.5 V to about +4.4 V, so that the state of N710/14 is fixed. Complete current balancing between all the rectifiers involved now takes place.

Example

A rectifier equipment is supplying too much power. I_{Ist} at input 5 of N710 is therefore negative with respect to the average signal at input 6 of N710. The output 7 becomes negative. Through terminal 30 a negative supplementary actual value signal is impressed upon the voltage regulator (on A3). The voltage regulator adjusts the voltage somewhat downwards and the equipment in question becomes more lightly loaded.

② *Current monitoring*

The current-monitoring function is used in GR10 rectifiers.

The relay K703 is operated or released according to the comparison between I_{Soll} and I_{Ist}.

Relay K703 operates when I_{Ist} rises above 95% of I_{Nenn} and releases again when I_{Ist} drops below 85% of I_{Nenn}. The operating and release levels are adjustable by means of potentiometers.

If relay K703 operates due to $I_{Ist} > 95\%\ I_{Nenn}$, its contact changes to position 4/7, and relay K70 in the relay and contactor control is energized. Through the opening of the K70 contact the information 'I > 95%' is passed to BS (BFS) or SF and stored there. When all the 62 V equipments in, for example, a 60 V power supply system arranged for changeover mode are loaded to more than 95% I_{Nenn}, the installation switches to parallel operation. The charging rectifiers, and if necessary the battery as well, can then be applied to supplying the loads.

The (positive) reference current value is applied to terminal 37 and the (negative) actual current value to terminal 33. The actual current value is smoothed by C738 and *R*737, *R*738.

The non-inverting input 3 of the comparator N730 is at M via *R*732. To the inverting input 2 of this comparator is applied the positive voltage resulting from the $I_{Soll} - I_{Ist}$ comparison (about +2.2 V with *R*736 in the mid-position). The output of the comparator N730/6 is switched to N. Transistor V704 is non-conducting and relay K703 is released.

If I_{Ist} increases beyond the 95% I_{Nenn} value (adjustable with the potentiometer *R*736), the output 6 of the comparator N730 switches to P. Transistor V704 now conducts and the relay K703 is energized.

Through the positive feedback loop (*R*733, *R*731 and V731) a current flows through *R*732 to M and increases the comparison voltage at N730/3 by about +0.25 V, thereby introducing hysteresis into the switching characteristic of the comparator. The degree of hysteresis is adjusted by the potentiometer *R*733; it is generally set so that the resetting (release) level of relay K703 is about 85% I_{Nenn}.

③ *Standby generator operation*

In operation with a standby power supply system (standby generating set), there is a facility for introducing a 'current-dependent' slope in the voltage characteristic. This enables oscillation effects between the standby generator and the rectifier to be reduced. For this purpose the relay K706 is energized from BS (BFS) or SF through the relay and contactor control of the rectifier. Through output 34 the slope of the rectifier is varied up to a maximum of ±3%, depending on the equipment output current and the setting of potentiometer *R*706, by an increment to the actual voltage value in the voltage regulator (on A3).

④ *Internal power supply (+15 V)*

For the power supply, the stabilized ±15 V supply is fed to the A31 assembly from assembly A3.

1.2 Panels

1.2.1 Mains distribution switchboards and mains switch panels

Rectifiers up to 50 A and also 100 and 200 A can be connected to the supply mains through *mains distribution switchboards* NV; these contain the appropriate supply fuses and the mains supply switch.

The following mains distribution switchboards are available:

for 60 V systems	Types 63, 100 and 250 A NV10,
for 48 V systems	Types 35, 63, 150 and 250 A NV11,
for shelters	Types 40 and 63 A NV11C.

Mains switch panels NF are used to connect larger rectifiers (e.g. GR10, 500 and 1000 A) to the supply (cubicle construction, Figs 1.68 to 1.71).

Type NF10 mains switch panels are available for 60 V systems and type NF11 for 48 V systems. It is possible to disconnect the power supply installation from the mains with the load-breaking (disconnector) switch Q1 (mains supply switch) (among other purposes, for emergency disconnection from the mains). The rectifiers are connected through NH (HRC low-voltage high-power) fuse load

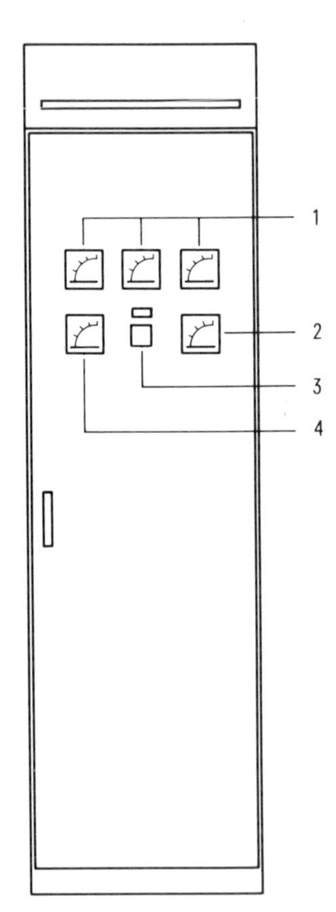

1 Ammeter
2 Frequency meter
3 Voltmeter measurement switch
4 Voltmeter

Fig. 1.68
Mains switch panel 380 V/600 A NF10, exterior

1 NH fuse/load disconnector

Fig. 1.69
Mains switch panel 380 V/600 A NF10 with door open

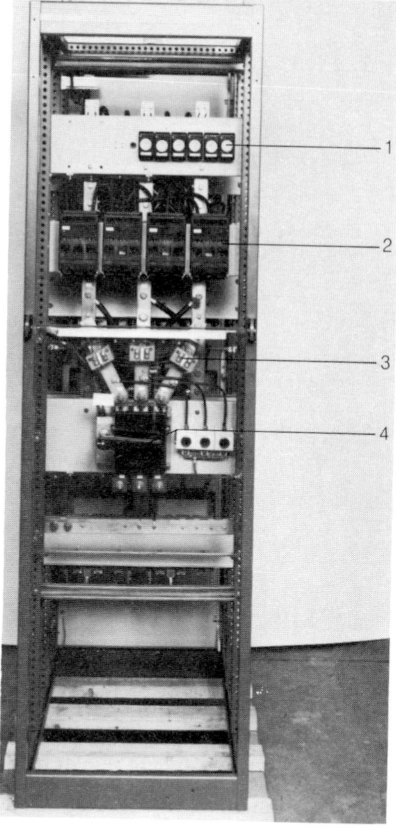

1 Fuses
2 NH fuse load disconnector tor small loads
3 Current transformer
4 Load-breaking switch

Fig. 1.70
Mains switch panel 380 V/600 A NF10, back view

disconnectors, which permit the safe disconnection of individual equipments from the mains supply. A maximum of eight rectifiers of any size can be connected. It is important that the current rating of the mains switch panel should not be exceeded (see Table 1.12).

In a 48 V system with a supply voltage other than 380 V or a supply frequency of

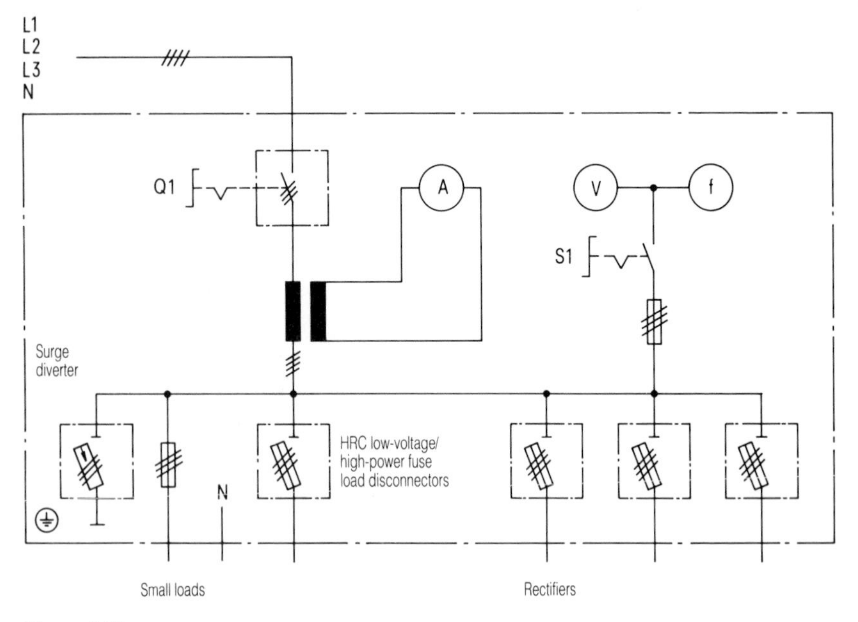

Figure 1.71
Basic circuit of mains switch panel type NF10

60 Hz, the control circuits in the battery switching panel and in the rectifier are supplied from a matching transformer built into the mains switch panel.

Since it cannot always be predicted whether protection against overvoltages, which may be generated by the interruption of short-circuits in neighbouring systems or by atmospheric disturbances, will be installed in the mains supply, surge diverters are included in the mains distribution switchboards and mains switch panels.

Technical data

The principal technical data relating to mains switch panels are listed in Table 1.12.

1.2.2 Battery switching panels

1.2.2.1 Battery switching panel with control unit

Power supply installations up to 2000 A – with rectifiers in the GR10 series – are provided with battery switching panels with control units BS (BFS) (cubicle

Table 1.12 Technical data for mains switch panels NF10 and NF11

Type		For 60 V systems		For 48 V systems		
		380 V/600 A NF10	380 V/1000 A NF10	400 A NF11	600 A NF11	1000 A NF11
Supply voltage	(V)	3 × 380		3 × 220/ 3 × 346/ 3 × 380/ 3 × 400/3 × 415		
Frequency	(Hz)	50		50 (60)		
Rated current	(A)	600	1000	400	600	1000
Ammeter	(A)	3 × 600/1	3 × 1000/1	400	600	1000
Voltmeter	(V)	250		250		
Frequency meter	(Hz)	47–53		47–53 (57–63)		
Load-breaking switch, 3-pole, for mains input (bottom entry)	(A)	630	1000	630		1000
NH fuse/load disconnector, 3-pole/ fuses for load outputs (bottom exit)	(A)	8 × 3N P5, 250 3 × 3N P4, 125		6 × 3N P5, 250 2 × 3N P4, 125 3 fuses 25 or 10		
NH fuse/load disconnector, 3-pole, for surge diverters	(A)	1 × 3N P4, 125 provided with surge diverter NH VA 280		1 × 3N P4, 125 provided with surge diverter NH VA 280		
Dimensions ($H \times W \times D$)	(mm)	Cubicle construction: front with door, back open: 2000 × 600 × 1200. Height with headboard: 2150				

construction, Figs 1.72 to 1.74). In addition to the power circuits, these contain the control unit, the indication unit and the monitoring systems.

All the necessary battery and load fuses are incorporated, and also measuring instruments and shunts.

The BS contains the following modules:

▷ control unit A9 with monitoring assemblies,
▷ auxiliary rectifier A80 (only in 60 V installations).

The control unit embraces, among other things:

▷ control section A10 with voltage monitoring and overvoltage limiting,
▷ charging control for the batteries,
▷ characteristic control for the rectifiers,
▷ indication unit for aural and visual alarms and for remote indications.

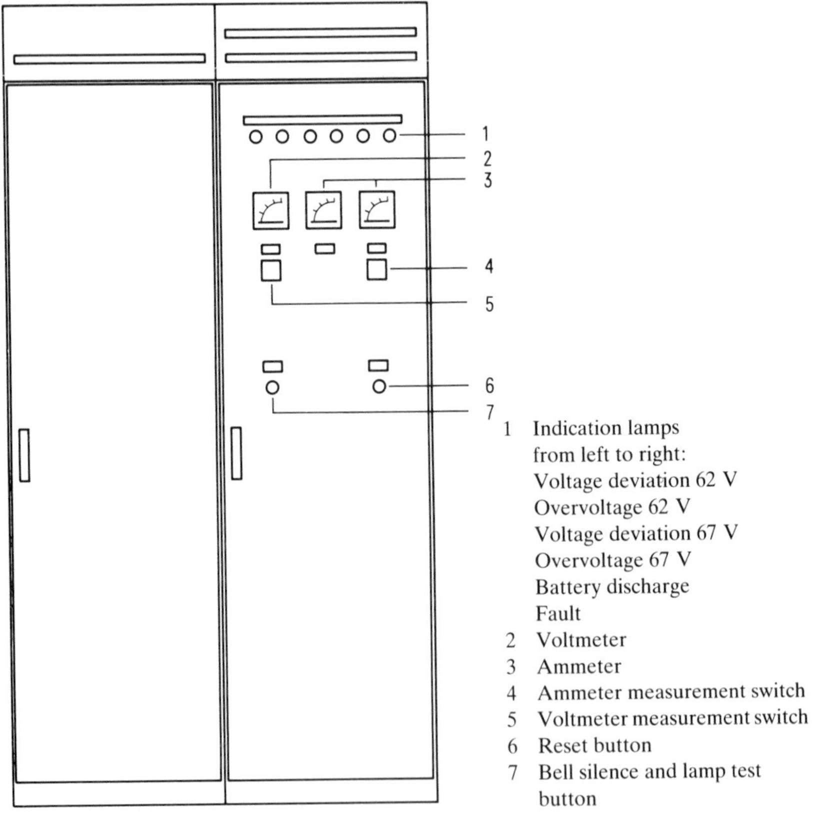

Fig. 1.72
Battery switching panel with control unit 60 V/1000 A BS10, exterior

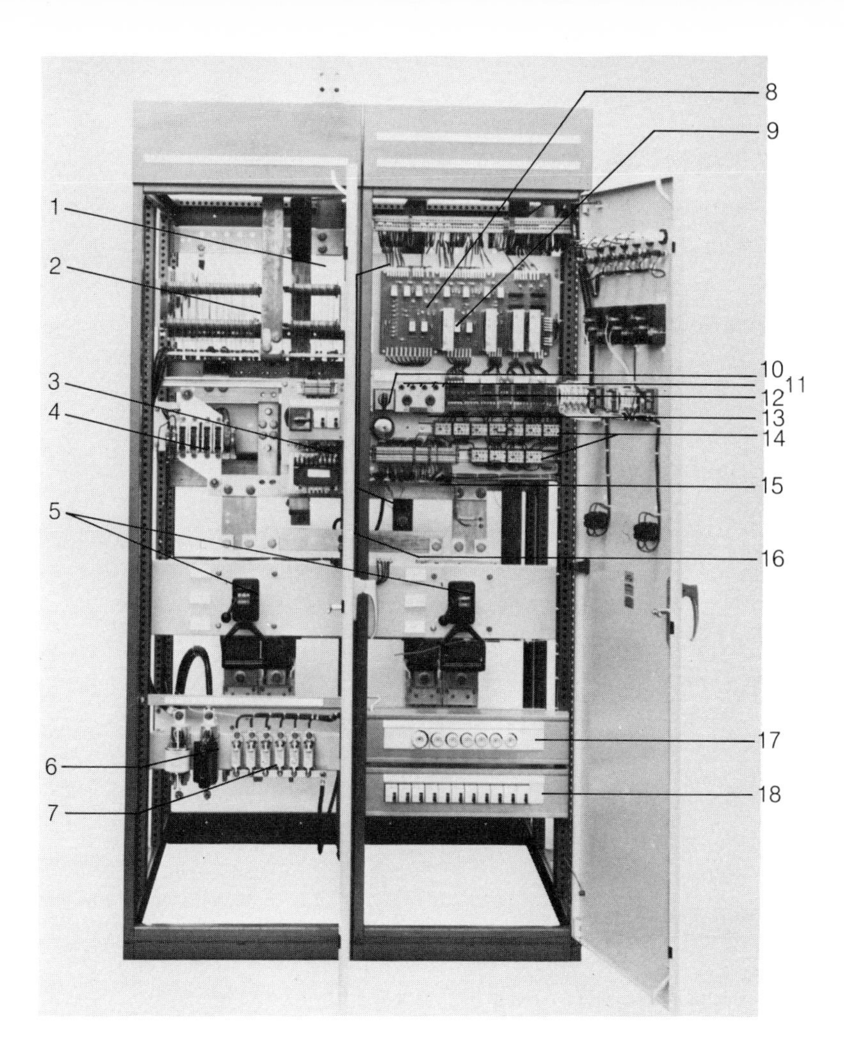

1 Tapping diode
2 Decoupling diode
3 Auxiliary rectifier A80
4 Battery-discharge contactor
5 Battery switch
6 Connection for mobile power plant
7 Fuses for small loads
8 Control section A10
9 Monitoring assemblies A11, A21, A31, A41
10 Characteristic switch
11 Test switch
12 Time-delay relay
13 Test plug
14 Auxiliary contactor
15 Bell
16 Control unit A9
17 Fuses
18 Automatic circuit breakers

Fig. 1.73
Battery switching panel with control unit 60 V/1000 A BS10 with doors open

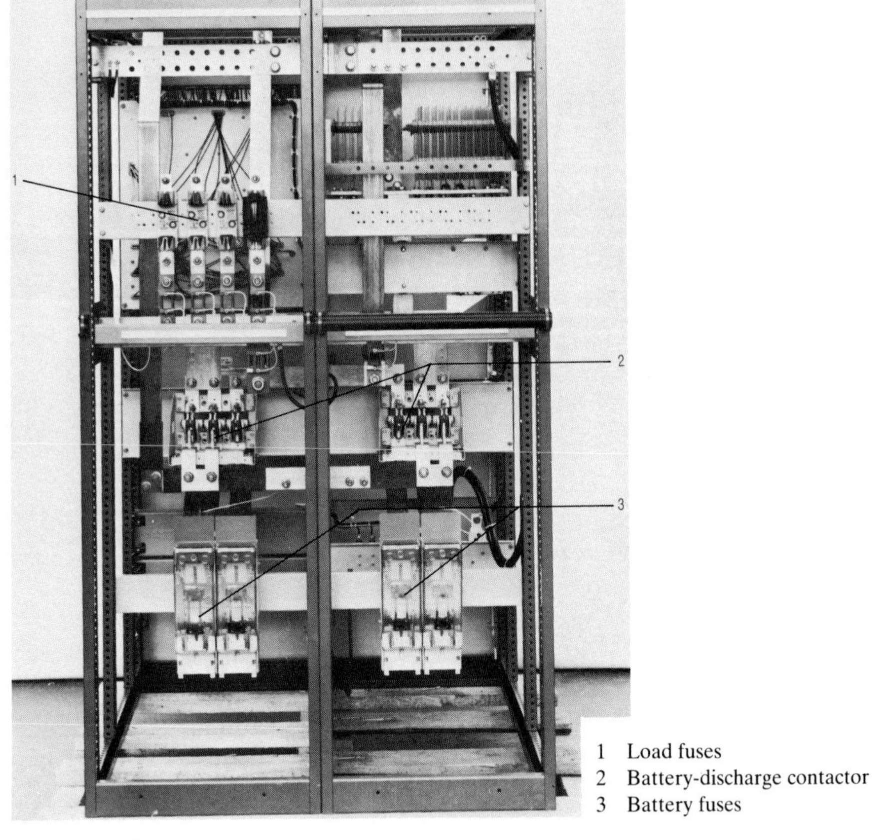

1 Load fuses
2 Battery-discharge contactor
3 Battery fuses

Fig. 1.74
Battery switching panel with control unit 60 V/1000 A BS10, back view

The battery switching panels with control unit are designed for connecting two batteries to a power supply installation with a maximum of eight rectifiers.

Figure 1.75 shows the block diagram of a 60 V battery switching panel with control unit. The control section is variously equipped, depending upon the type of installation (Table 1.13).

Figure 1.76 shows the block diagram of a control unit.

Figure 1.77 shows the basic circuit diagram of a battery switching panel with control unit.

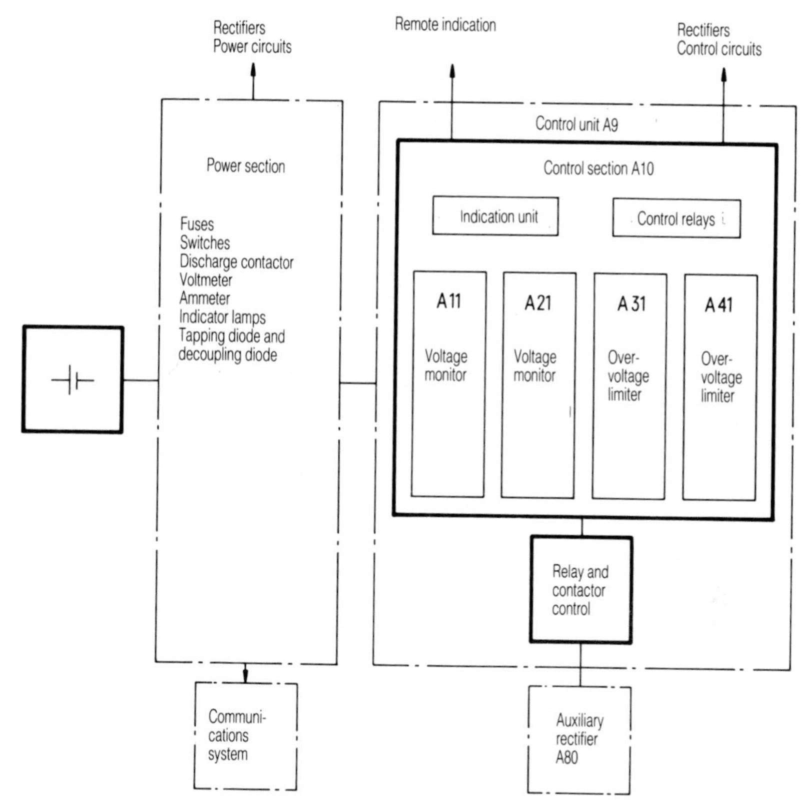

Fig. 1.75
Block diagram of a battery switching panel with control unit; 60 V system for combined standby parallel and changeover mode

The voltage measurement switch S1 (see Fig. 1.72) enables the voltmeter P1 to be switched to the following circuit points:

1 62 V busbar,
2 67 V busbar,
3 battery 1 and
4 battery 2.

The total current drawn by the 67 V load is measured by means of a shunt *R*2 with a directly connected ammeter P4.

The total current and the currents in the individual 62 V load outlets are measured by the ammeter P2 in conjunction with shunts and the switch 62 V, S2 (see Fig. 1.72).

Table 1.13 Monitoring assemblies and systems

Monitoring assemblies	60 V systems				48 V systems
	Standby parallel mode with reducing diodes with 62 V loads (and 67 V if required)	Standby parallel mode without reducing diodes with 67 V loads	Changeover mode with 62 V loads	Combined standby parallel and changeover mode with 62 V and 67 V loads	Change-over mode with 51 V loads
A11 $U\uparrow\downarrow$	X[2])		X[2])	X[2])	X[1])
A21 $U\uparrow\downarrow$	X[4])	X[5])	X[3])	X[5])	X[3])
A31 $U\uparrow\uparrow$			X[2])	X[2])	
A35 $U\downarrow$					X
$U\downarrow$					
A36 $U\uparrow$	X[6])				
$U\uparrow$					
A41 $U\uparrow\uparrow$	X[4])	X[5])	X[3])	X[5])	X[1])

A11, A21 } Voltage monitors (contain undervoltage and overvoltage monitors)

A31, A41 } Overvoltage limiters

A35 Control of battery-discharge contactor K11

A36 Voltage-reducing diode control

[1]) For 51 V loads.
[2]) For 62 V loads.
[3]) For battery.
[4]) For loads and battery.
[5]) For 67 V loads and battery.
[6]) Not contained in control section A10, but included in control unit A9.

The battery 1 and battery 2 switches Q1 and Q2 (rotary or lever switches) enable the batteries to be switched on or off, to trickle charging at 2.23 V/cell, to charging at 2.33 V/cell or to charging up to 84 V (e.g. for a lead-acid battery with 30 cells).

With the characteristic switch S61 (see Fig. 1.73) the power supply installation can be switched to the following operating conditions and characteristics:

1 continuous parallel operation,
2 automatic operation,
3 continuous trickle charging (2.23 V/cell),
4 continuous charging (2.33 V/cell).

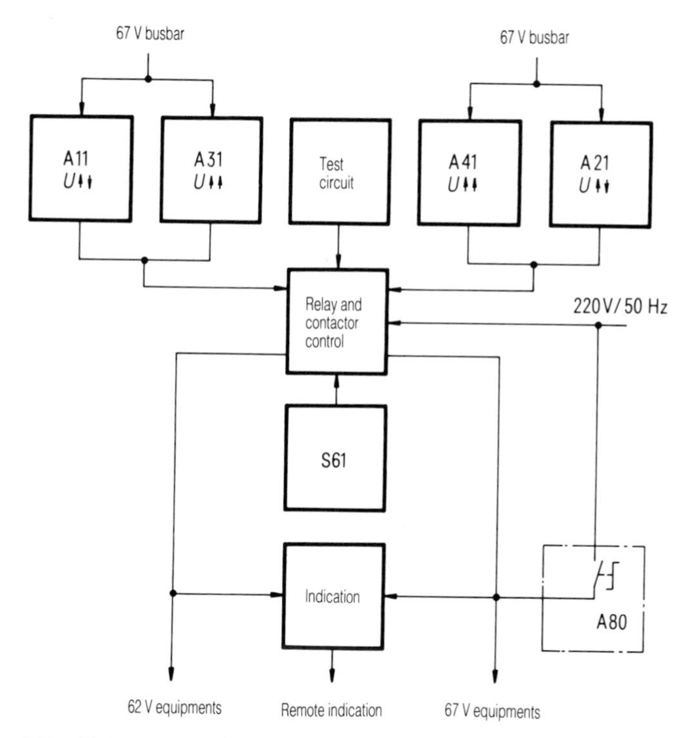

A11	Voltage monitor 62 V	A41	Overvoltage limiter 67 V
A21	Voltage monitor 67 V	A80	Auxiliary rectifier
A31	Overvoltage limiter 62 V	S61	Characteristic switch

Fig. 1.76
Block diagram of a control unit for a battery switching panel with control unit BS10; 60 V system for combined standby parallel and changeover mode

This switch is normally set at position 2; all the operating conditions are then governed automatically. It is possible to switch to position 1 from any operating condition; all the rectifiers are thus switched in parallel with the loads and the battery. The timing processes of the charge-instruction and charge-time relays are then cancelled. In switch position 3 the operating conditions are governed automatically, as in position 2; the charging of the battery (2.33 V/cell) is simply omitted. Similarly, in switch position 4 the operating conditions proceed automatically as in position 2, but with the restriction that automatic switching to trickle charging (2.23 V/cell) is inhibited. It is possible to switch to this operating mode from any operating condition for the purpose of boost-charging the battery.

The release pushbutton (unlocking facility) S81 (see Fig. 1.72) is used to unlock the voltage-monitoring and overvoltage-limiting systems.

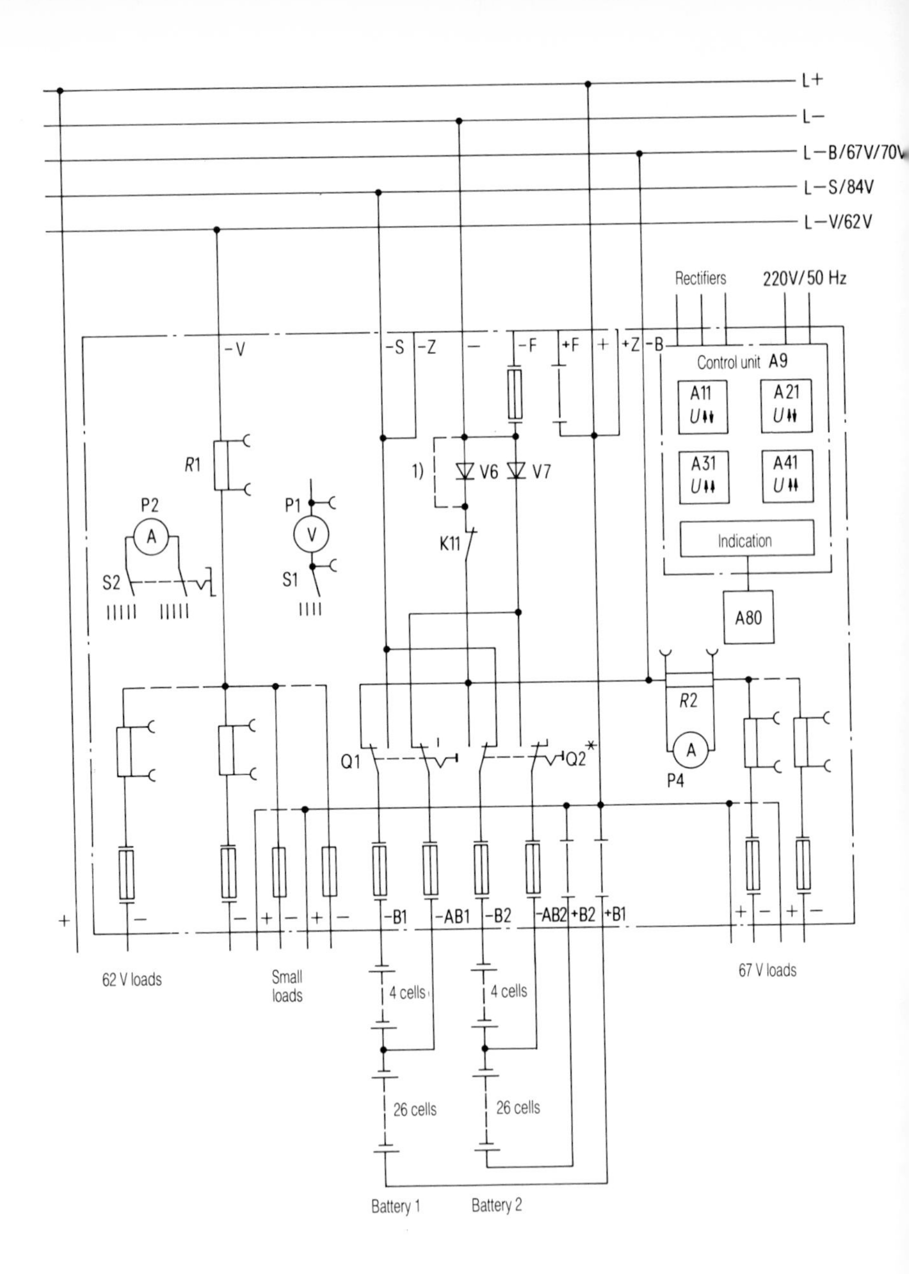

Fig. 1.77
Basic circuit of a battery switching panel with control unit BS10; 60 V system for combined standby parallel and changeover mode

With pushbutton S82, bell disconnection and lamp testing (see Fig. 1.72), the acoustic alarm is inhibited; at the same time all the indication lamps can be tested.

The test switches S15, S25, S35 and S45 (see Fig. 1.73) facilitate the testing and adjustment of the voltage-monitoring and overvoltage-limiting systems with a variable-test voltage (voltage stabilizer) applied through the test plug X25 or X45.

The test switches have the following positions:

1 operation,
2 test.

The various faults are indicated by signal lamps (see Fig. 1.72). Table 1.14 shows

Table 1.14 Indication lamps

Indication lamps	60 V systems				48 V systems
	Standby parallel mode with reducing diodes	Standby parallel mode without reducing diodes	Changeover mode	Combined standby parallel and changeover mode	Changeover mode
H81	Voltage deviation – 62 V		Voltage deviation – load	Voltage deviation – 62 V	Voltage deviation – load
H82			Overvoltage – load	Overvoltage – 62 V	Overvoltage load
H83	Voltage deviation – 67 V	Voltage deviation – 67 V	Voltage deviation – battery	Voltage deviation – 67 V	Voltage deviation – battery
H84	Overvoltage – 67 V	Overvoltage – 67 V	Overvoltage – battery	Overvoltage – 67 V	
H85	Fault	Fault	Fault	Fault	Fault
H86	Battery discharge	Battery discharge	Battery discharge	Battery discharge	Battery discharge

◁ 1) In power supply systems without a compensator it may be necessary to by-pass V6
+F, –F Connections for mobile power plant
+Z, –Z Connections for load resistor
–AB Battery tap
* Set for initial charging
A11 Voltage monitor 62 V
A21 Voltage monitor 67 V
A31 Overvoltage limiter 62 V
A41 Overvoltage limiter 67 V
A80 Auxiliary rectifier

the significance of the indication lamps in the various power supply systems. In addition to the visual indication the alarm bell H501 is actuated (see Fig. 1.78).

Alarms and indications

Figure 1.78 gives a general impression of the alarm and indication system (60 V systems). An urgent (dr) alarm signal coming from the rectifiers or the control (via V502) is passed on directly through the diode V509 from the main alarm

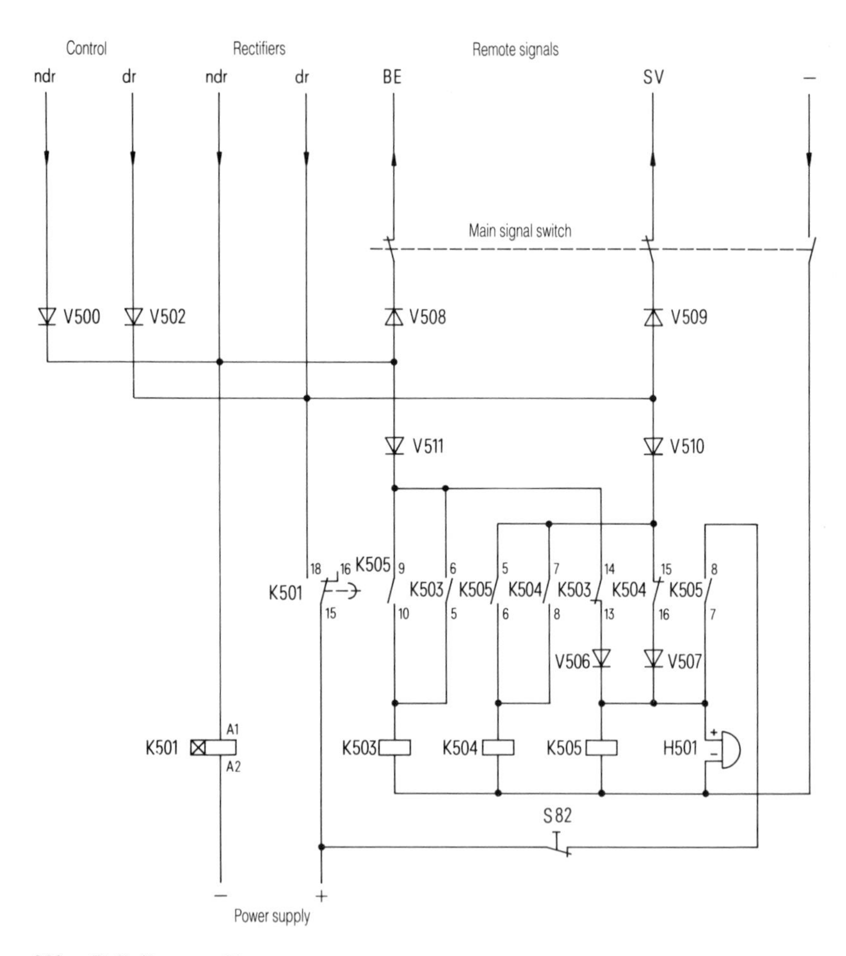

S82 Bell silence and lamp test
BE Remote signal: battery discharging
SV Remote signal: power supply

Fig. 1.78 Indication in 60 V systems

switch (in position 2: remote indication on) as a remote signal SV. If the main alarm switch is in one of the positions 1, 3 or 4, the alarm bell H501 and relay K505 are energized via diodes V510 and V507. The relay K505 is held, with the alarm bell, over the contact K505 7/8. The audible alarm is silenced by the pushbutton S82, which breaks the holding circuit of relay K505 and thereby switches off the alarm bell H501.

A non-urgent signal (ndr) received from the rectifiers (through the operation of the mains monitor A4) or the control (through V500) is passed on directly through the diode V508 and the main alarm switch (in position 2: remote alarm on) as a remote signal BE. If the main alarm switch is in position 1, 3 or 4, the alarm bell H501 and relay K505 are energized through diodes V511 and V506. The further effects are as described for the urgent signal.

At the same time, the non-urgent signal starts the operation of the time-delay relay K501, which, after the set timing period (about 3 h), provides an additional urgent signal (dr) through diode V509 and the main alarm switch (in position 2: remote alarm on) as a remote signal SV. Appropriate steps can then be taken to maintain the security of the power supply – e.g. the starting of a standby generating set.

The control, monitoring and alarm circuits are protected by automatic circuit breakers. If one trips, the indicator contact (except Q83) energizes the alarm bell H501 and the fault lamp H85 (not shown) and initiates an urgent alarm signal (dr).

The supplies for the whole control system of the power supply installation, the voltage-measuring circuits and the control of the ventilating fans are protected by screw-fitting fuses.

The main alarm switch (usually only in installations for the German Federal Republic Postal Administration, DBP) is located in the power supply room near the entrance. It enables the visual and aural alarms of the power supply installation to be switched on and the remote alarms to be disabled.

The main alarm switch has the following positions:

1 remote alarms off (lamps and audible alarm on),
2 remote alarms on (lamps and audible alarm off),
3 test remote signal ‘power supply’ SV (lamps and audible alarms on),
4 test remote signal ‘battery discharging’ BE (lamps and audible alarms on).

To reduce the risk of forgetting to switch over to remote alarms when leaving the power room, a yellow reminder lamp is provided, which shows when the switch is set to position 1, 3 or 4.

Alarm extension

If required, an alarm extension MZ can be connected to the control unit (Fig. 1.79).

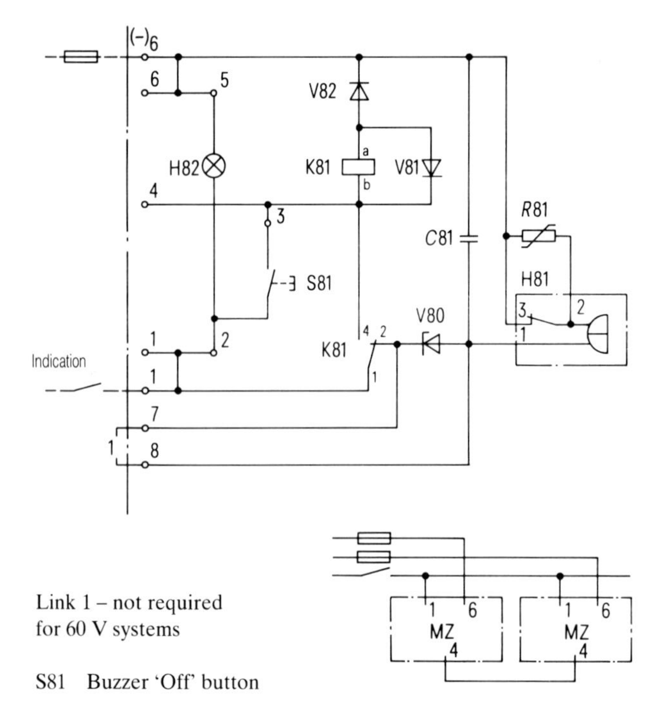

Fig. 1.79 Alarm extension 60 V (48 V) MZ11

The alarm extension is installed in a room occupied by staff. It combines the urgent SV (dr) and non-urgent BE (ndr) remote signals transmitted by the power supply installation and applies them to an audible alarm (buzzer H81) and a visual indication (indication lamp H82).

The acoustic alarm can be turned off with the pushbutton S81. Operation of S81 energizes the relay K81, whose changeover contact disconnects the signal voltage from the buzzer H81 and functions as a holding contact.

The visual alarm persists until the termination of the remote indication (the elimination of the fault); K81 releases and the alarm extension is again ready for operation. The varistor R81 protects the buzzer H81 against overvoltage spikes.

It is possible to connect several alarm extensions in parallel.

1.2.2.2 Auxiliary rectifier

An auxiliary rectifier A80 (Fig. 1.80) is built into the 60 V battery switching panel with control unit (BS10). This permits the switching on of the power supply system after a power failure with a deeply discharged battery, or switching on

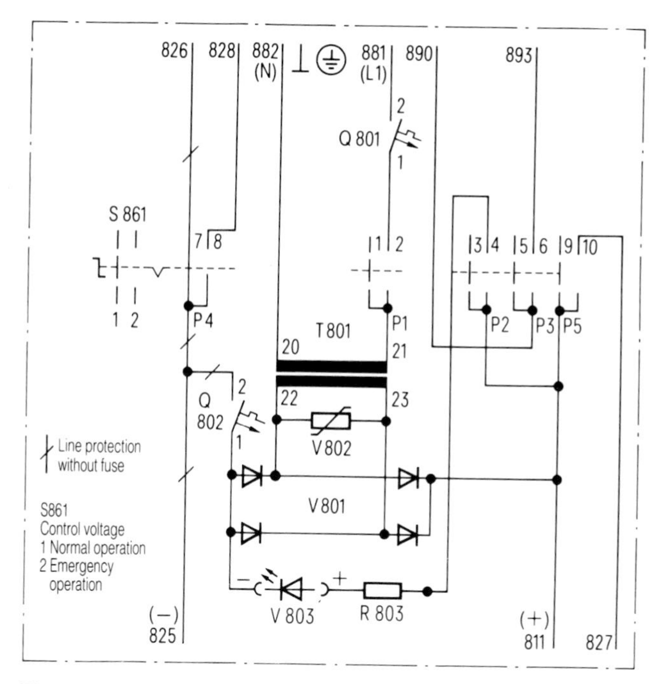

Fig. 1.80 Auxiliary rectifier A80 (type 60 V/5 A HGR10)

without a battery. The auxiliary rectifier is switched on by throwing the switch S861 'control voltage' from the normal operation position to emergency operation. In the emergency operation position the following switching functions are effected:

▷ the input of the auxiliary rectifier is connected to 220 V/50 Hz and the light-emitting diode V803 is lit,
▷ in the rectifiers the fault relays K80, K80A and K81 are energized from the auxiliary rectifier. An additional contact on the switch S861 bridges the outputs of the monitoring assemblies A21 and A41, and, in rectifiers switched to '67 V operation' only, energizes the contactors K60. The decoupling diode V6 (in the BS) is thereby protected from possible damage. Because of the bridging of the A21 and A41 assemblies, any response in them is ineffective; the pulse inhibit in the overvoltage limiter A41 remains operational, however, so that if an overvoltage occurs the trigger pulses for the thyristors in the rectifiers are switched off,
▷ the supplies to the indication curcuits are taken over by the auxiliary rectifier,
▷ the fault indication lamp H85 is switched on as a reminder.

Technical data

The principal technical data for the battery switching panel with control unit are listed in Tables 1.15 and 1.16.

Table 1.15a Technical data for battery switching panels with control unit BS10 and rectifier with control unit GR10

Type		60 V systems, standby parallel mode with reducing diodes	
		60 V/100 A BS10	60 V/25 A GR10[1])
Rated current	(A)	100	25
Ammeter (total current) for 62 V	(A)	100	
loads	(mV)	60	
Ammeter (equipment current)	(A)		40
	(mV)		60
NH fuse bases for battery fuses (minus)		2 × NH00	1 × NH00
NH fuse bases for 62 V loads		4 or 6 × NH00	1 × NH00
NH fuse bases for 62 V small loads		4 or 6 × NH00	3 × NH00
NH fuse bases for 67 V loads		2 or 0 × NH00	1 × NH00
NH fuse bases for connection of a transportable rectifier		2 × NH00	1 × NH00
Dimensions ($H \times W \times D$)	(mm)	Cubicle construction: 2000 × 600 × 600 Height with headboard: 2150	

[1]) The battery switching panel and the control unit are incorporated into the 25 A individual rectifier.

Table 1.15b Technical data for battery switching panels with control units

Type		60 V systems Standby parallel mode without voltage-reducing diodes, changeover or combined standby parallel and changeover					
		60 V/100 A BS10	60 V/200 A BS10	60 V/400 A BS10	60 V/600 A BS10	60 V/1000 A BS10	60 V/2000 A BS10
Rated current	(A)	100	200	400	600	1000	2000
Ammeter, dual-range (total current and individual measurement) for 62 V loads	(A) (mV)	60/100 60	150/250 60	150/400 60	250/600 60	600/1000 60	1000/2500 60
Ammeter (total current) for 67 V loads	(A) (mV)	60 60	100 60	250 60	400 60	600 60	1000 60
NH fuse bases for battery fuses (minus and tap)		4 × NH2	4 × NH2	4 × NH2	4 × NH3	4 × NH4	6 × NH4
NH fuse bases/fuses for 67 V loads	(A)	2 × NH2/36 to 63 or 5 × NH00/6 to 100	2 × NH2/36 to 100 or 5 × NH00/6 to 100	2 × NH2/80 to 250 or 5 × NH00/6 to 100	2 × NH2/80 to 400 or 5 × NH00/6 to 100	4 × NH2/80 to 400 or 10 × NH00/6 to 100	3 × NH2/80 to 250 3 × NH3/224 to 630 or 10 × NH00/6 to 100 3 × NH3/224 to 630
NH fuse bases/fuses for 62 V loads	(A)	3 × NH2/36 to 63	3 × NH2/80 to 200	3 × NH2/80 to 400	4 × NH2/80 to 400	2 × NH3/80 to 400 2 × NH3/224 to 630	3 × NH2/80 to 250 3 × NH3/224 to 630 1 × NH4/800 to 1000
NH fuse bases/fuses for small 62 V loads	(A)	3 × NH00/10 3 × NH00/16					
NH fuse bases/fuses for the connection of a mobile power plant	(A)	2 × NH3/425					
Dimensions ($H \times W \times D$)	(mm)	Cubicle construction: front with door, back open: 2000 × 900 × 600	2000 × 900 × 600	2000 × 900 × 600	2000 × 900 × 1200	2000 × 1200 × 1200	2000 × 1800 × 1200
		Height with headboard: 2150					

Table 1.16 Technical data for battery switching panels with control units BS11

Type		48 V systems: changeover mode		
		48 V/600 A BS11	48 V/1000 A BS11	48 V/2000 A BS11
Rated current	(A)	600	1000	2000
Ammeter	(A)	600	1000	2500
	(mV)	60	60	60
NH fuse bases for battery fuses/fuses (minus and tap)	(A)	4 × NH3/224 to 630	4 × NH4/630 to 1000	6 × NH4/630 to 1000
NH fuse bases/ fuses for loads	(A)	5 × NH2/36 to 250	5 × NH3/36 to 400	8 × NH3/36 to 630
NH fuse bases/ fuses for small loads	(A)	5 × NH00 16 1 × NH00 16 (for emergency lighting)	4 × NH00 16 2 × NH00 16 (for emergency lighting)	
Dimensions ($H \times W \times D$)	(mm)	Cubicle construction: front with door, back open: 2000 × 900 × 1200	2000 × 1200 × 1200	2000 × 1800 × 1200 Height with headboard: 2150

1.2.2.3 Battery switching panel without control unit

The battery switching panel BF (cubicle construction) is used in power supply installations of 2000 to 10 000 A with rectifiers in the GR10 series for the connection of *one* battery (Figs 1.81 to 1.83). Unlike the BS it contains no control unit. That is located, in these systems, in the control panel SF (one per installation) (see Section 1.2.3).

The basic circuit diagram of a 60 V battery switching panel is shown in Fig. 1.84. This includes, among other things:

- ▷ fuses,
- ▷ battery switch Q1,
- ▷ battery discharge contactor K11,
- ▷ decoupling diode V6,
- ▷ tapping diode V7,
- ▷ possible connection to a mobile power plant,
- ▷ disconnectors and
- ▷ shunt with measuring amplifier, supplying a measuring signal to the control panel for the formulation of the total current.

The battery associated with a particular battery switching panel can be switched to

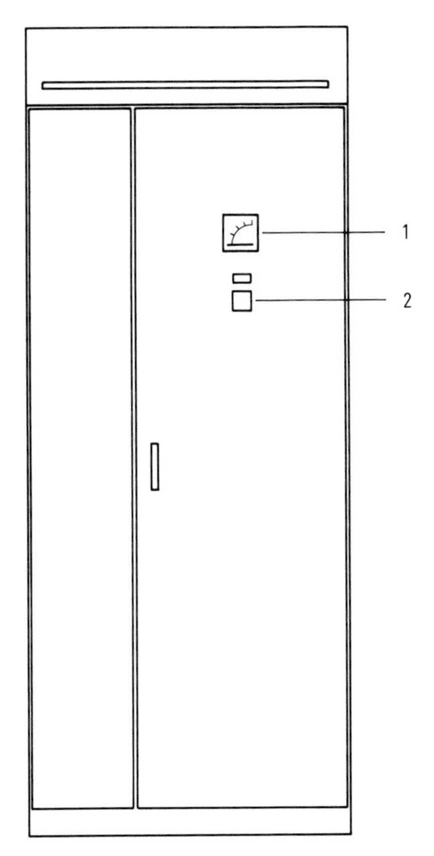

1 Ammeter
2 Ammeter measurement switch

Fig. 1.81 Battery switching panel 60 V/2000 A BF10, exterior

trickle charging or to charging at up to 84 V (or to 2.7 V/cell) by means of the battery switch Q1 (lever switch).

In the 'off' position (lever down) the negative terminal and the tap of the battery are disconnected from the power supply system. With the switch in this position it is possible with the preselector located at the side of the operating head to select switch position 1, trickle charging, or position 2, charging to 84 V (or to 2.7 V/cell).

The preselector can be moved only with the lever switch in the 'off' position, and stays only in the upper or lower position.

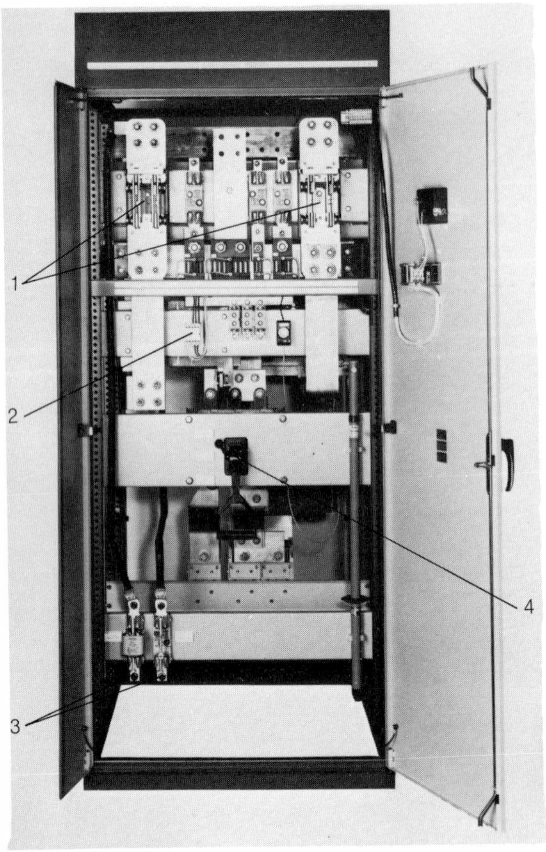

1 Disconnector
2 Measurement amplifier A1
3 Connection for mobile power plant
4 Battery switch

Fig. 1.82
Battery switching panel 60 V/ 2000 A BF10 with doors open

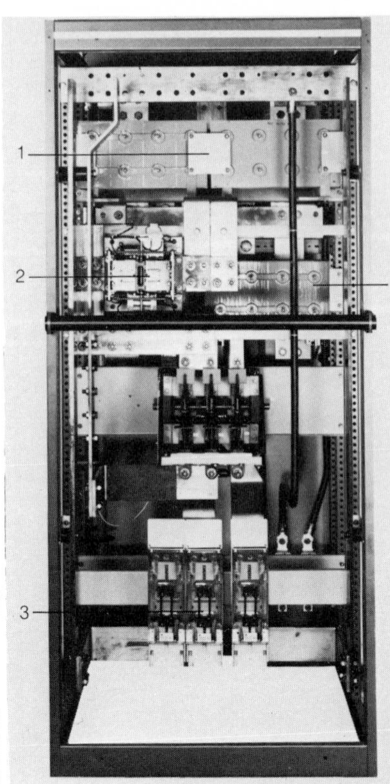

1 Decoupling diode
2 Battery-discharge contactor
3 Battery fuses
4 Tapping diode

Fig. 1.83
Battery switching panel 60 V/ 2000 A BF10, back view

With the switch position preselected, the switch can be moved to 'on' with the lever up.

In the 60 V/2000 A battery switching panel, for the purpose of measuring the total load current, the 2500 A/60 mV shunt $R1$ is connected to a measurement amplifier A1, whose output signal (20 mA impressed current) is fed to the control panel for the formulation of the total current signal.

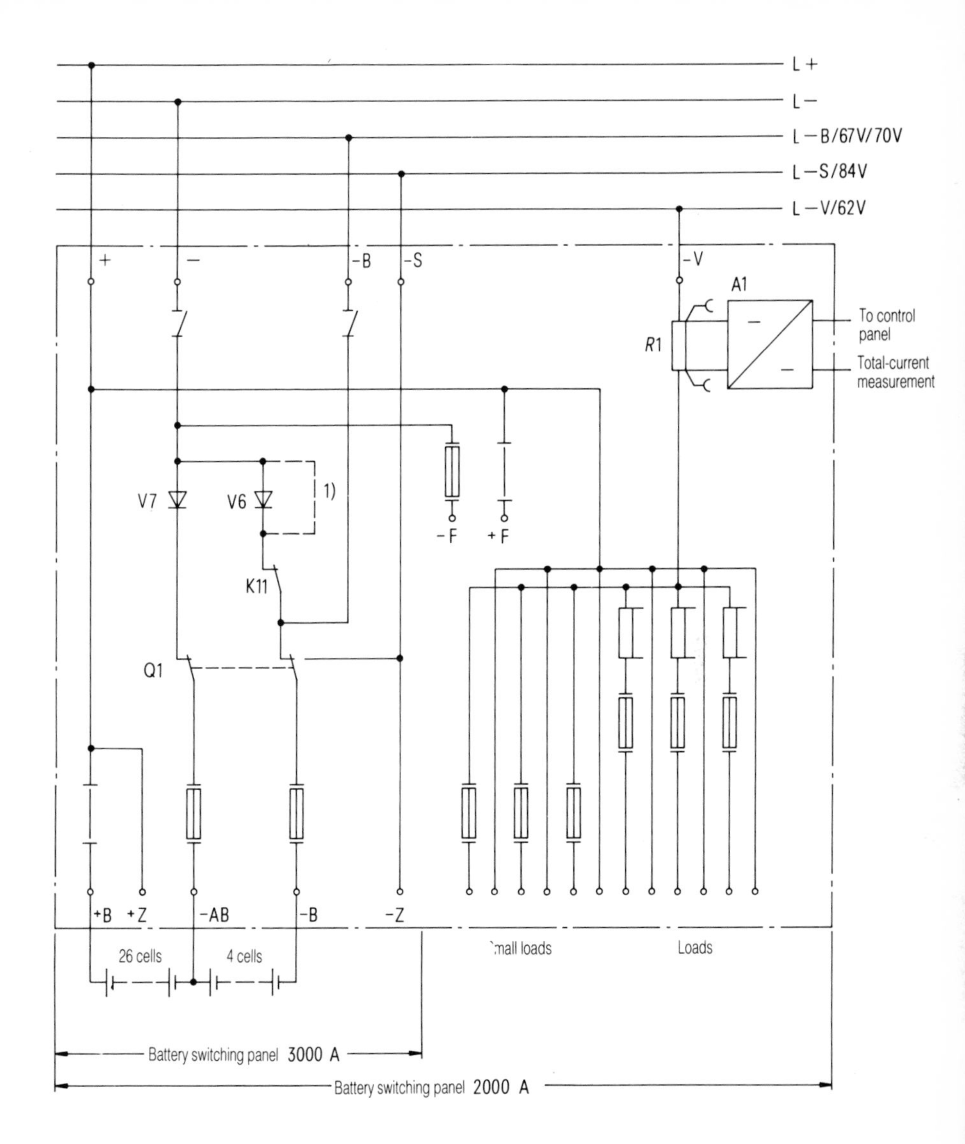

+F, −F Connection for mobile power plant
+Z, −Z Connection for load resistor
−AB Battery tap
A1 Measurement amplifier

Fig. 1.84 Basic circuit of a 60 V battery switching panel BF10

[1]) In power supply systems without a compensator it may be necessary to by-pass V6.

Table 1.17 Technical data for battery switching panels BF10

Type		60 V systems	
		60 V/2000 A BF10	60 V/3000 A BF10
Rated current	(A)	2000	3000
Ammeter, dual range	(A) (mV)	100/250 60	
Measurement amplifier	(Qty)	1	
NH fuse bases for battery fuses/fuses (minus and tap)	(A)	3 × NH7/500 to 1000	5 × NH7/1000 (500 to 1000)
NH fuse bases/fuses for loads	(A)	1 × NH7/500 to 1000 (224 to 630) 1 × NH7/224 to 630 1 × NH7/224 to 250	
NH fuse bases/fuses for small loads	(A)	3 × NH7/6 to 100	
NH fuse bases/fuses for connection of a mobile power plant	(A)	2 × NH7/500	
Dimensions ($H \times W \times D$)	(mm)	Cubicle construction: front with door, back open: 2000 × 900 × 1200 Height with headboard: 2150	

Table 1.18 Technical data for battery switching panels BF11

Type		48 V systems	
		48 V/2000 A BF11	48 V/3000 A BF11
Rated current	(A)	2000	3000
NH fuse bases for battery fuses/fuses	(A)	4 × NH4/1000 630 to 1000	5 × NH4/1000 630 to 1000
Dimensions ($H \times W \times D$)	(mm)	Cubicle construction: front with door, back open: 2000 × 900 × 1200 Height with headboard: 2150	

The currents in the individual load outputs are measured by the 250/1000 A (60 mV) ammeter. The shunts in the load outputs are provided with sockets for a portable measuring instrument.

The 60 V/3000 A and 48 V/2000 A and 3000 A battery switching panels have no measuring devices.

Technical data

The principal technical data for battery switching panels are listed in Tables 1.17 and 1.18.

1.2.3 Control panels

The control panel SF is used in power supply installations of from 2000 to 10 000 A in conjunction with rectifiers of the GR10 series (Figs 1.85 and 1.86),

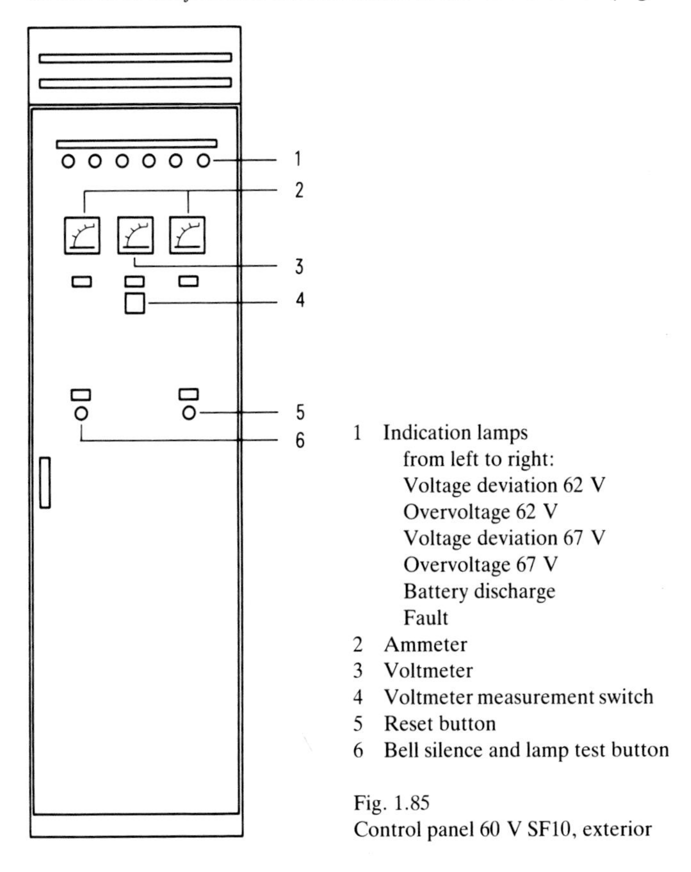

1 Indication lamps
from left to right:
Voltage deviation 62 V
Overvoltage 62 V
Voltage deviation 67 V
Overvoltage 67 V
Battery discharge
Fault
2 Ammeter
3 Voltmeter
4 Voltmeter measurement switch
5 Reset button
6 Bell silence and lamp test button

Fig. 1.85
Control panel 60 V SF10, exterior

1 Shunts
2 Chart recorder
3 Automatic circuit breakers
4 Fuses
5 Contactor
6 Time-delay relay
7 Power supply for measurement amplifier (A61)
8 Control unit A9 (see Fig. 1.76)
9 Auxiliary rectifier A80

Fig. 1.86 Control panel 60 V SF10 with door open

one being required for each installation of this type as a central facility. It provides for the control and monitoring of the power supply system, which may consist of up to a maximum of twelve rectifiers and the associated battery switching panels.

The control panel contains the control for the charging regimes, voltage-monitoring, overvoltage-limiters and voltage-measuring systems, an indication unit for visual and aural alarms, and in 60 V installations a total-current measuring system.

The control unit A9 and the auxiliary rectifier A80 (used only in 60 V systems) are incorporated in the SF, as in the BS (see Section 1.2.2.1). In regard to the functions of the switches and pushbuttons, the information given in connection with the BS is equally applicable.

Special features of the 60 V control panel are considered below.

Total-current measurement in the 60 V control panel

Measurement of the total of the load currents is included in installations with SF.

The measurement points are located outside the control panel in the distribution panel or battery switching panel. The current signals are supplied to the control panel through measurement amplifiers. The amplifiers convert the voltage drop

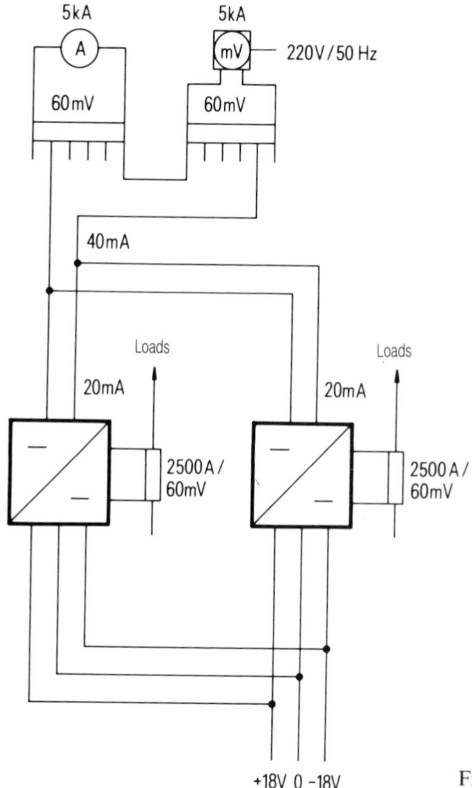

Fig. 1.87 Total-current measurement

Table 1.19 Total-current measurement

Number of 60 mV/2500 A measurement points and measurement amplifiers	Connection for R2, R4, R21, R22	Range Ammeter	Current recorder
		P2, P4	P21, P22
1	20 mA	2.5 kA	60 mV ≙ 2.5 kA
2	40 mA	5.0 kA	60 mV ≙ 5.0 kA
3	60 mA	7.5 kA	60 mV ≙ 7.5 kA
4	80 mA	10.0 kA	60 mV ≙ 10.0 kA
5	100 mA	12.5 kA	60 mV ≙ 12.5 kA

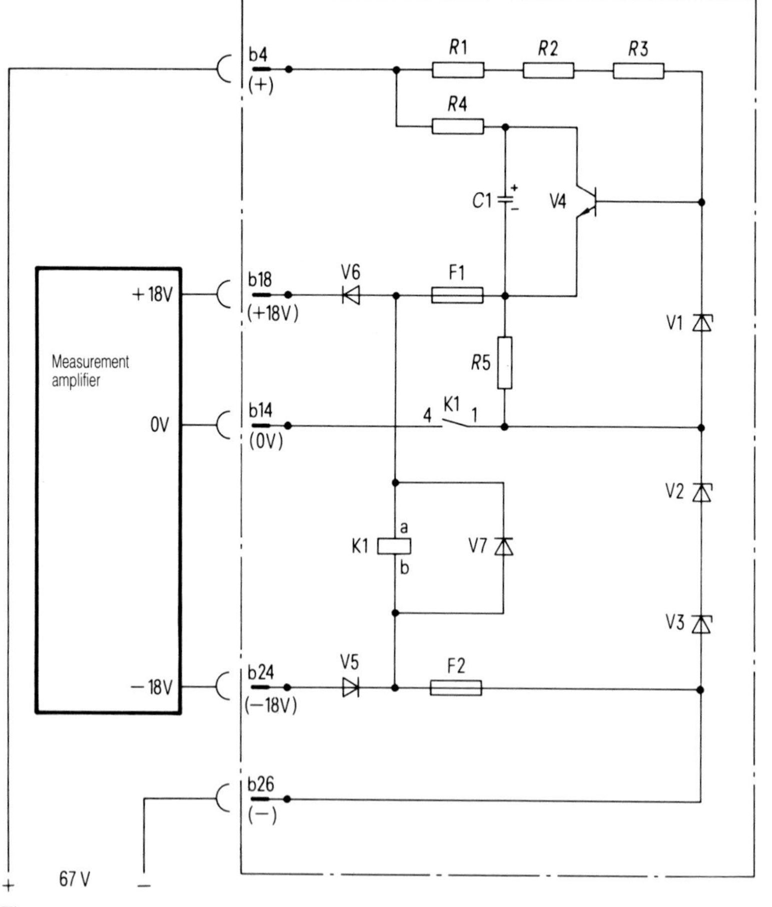

Fig. 1.88 Power supply for measurement amplifier (A61)

produced at the measurement point into an impressed current, so that considerable distances between the measurement points and the control panel can be spanned without measurement errors (Fig. 1.87).

The taps to be selected on the meter shunts and the scale ranges required for the ammeter and the current recorder can be obtained from Table 1.19. The total current to the 62 V loads is measured by the ammeter P2 and the current recorder P21. The instruments for the total current of the 67 V loads are the ammeter P4 and the current recorder P22.

Power supply for measurement amplifiers (A61)

The 18 V d.c. supply required by the measurement amplifiers is provided by the power supply A61 (Fig. 1.88). The reference for the +18 V supply is established at the base of transistor V4 by means of the zener diode V1, and the drive for V4 is provided by the actual voltage value developed across $R5$. The –18 V supply voltage is stabilized by the zener diodes V2 and V3. If one of the fuses (F1 or F2) blows, relay K1 releases and opens the common line M of the power supply, protecting it against overload.

Technical data

The principal technical data of the control panels are listed in Table 1.20.

Table 1.20 Technical data for control panels SF10 and SF11

Type		60 V systems 60 V SF10		48 V systems 48 V SF11
Supply voltages				
Alternating voltage	(V)	220 +10 to –15%		220 +10 to –15%
Frequency	(Hz)	50 ±5%		50 ±5%
Direct voltage	(V)	54 to 70		45 to 60
Voltmeter (switchable)	(V)	100		100
Total-current ammeter, dual or single-range	(kA/mV)	5.0/2.5/60 or 7.5/5.0/60 or 10.0/7.5/60 or 12.5/60	1 each for 62 and 67 V loads	
Total-current recorder, 60 mV (corresponds to 2.5 to 12.5 kA according to connections)	(mV) (V/Hz)	Signal 60 Point spacing 72 s. Paper feed 10 or 60 mm/day. Mains supply 220/50	1 each for 62 and 67 V loads	
Dimensions ($H \times W \times D$)	(mm)	Cubicle construction: front with door, back open (not equipped): 2000 × 600 × 1200 Height with headboard: 2150		

1.2.4 Distribution panels

Distribution panels VF (cubicle construction) are used in power supply installations of from 2000 to 10 000 A, with rectifiers in the GR10 series and battery switching panels BF 3000 A, to distribute power to the loads; they also carry the load fuses.

The distribution panel type VF10 (Figs 1.89 and 1.90) is used in 60 V systems for 62 or 67 V loads. One input supply can be distributed to eight circuits of from 36 to 1000 A, or to 24 circuits of from 6 to 100 A. A distribution panel may have one or two input supplies, depending on the type. The total input supply current is measured by means of a 2500 A/60 mV shunt *R*1 with a directly connected 2500 A ammeter. To the same shunt is connected a measurement amplifier A1, whose output signal (20 mA impressed current) is fed to the control panel for the purpose of formulating the total-current signal.

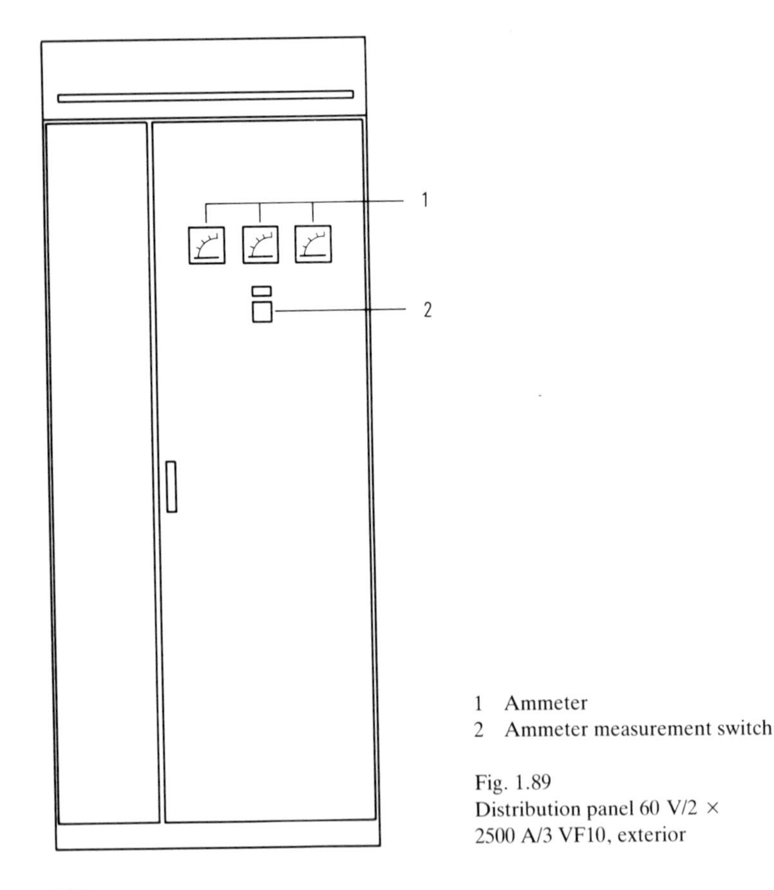

1 Ammeter
2 Ammeter measurement switch

Fig. 1.89
Distribution panel 60 V/2 × 2500 A/3 VF10, exterior

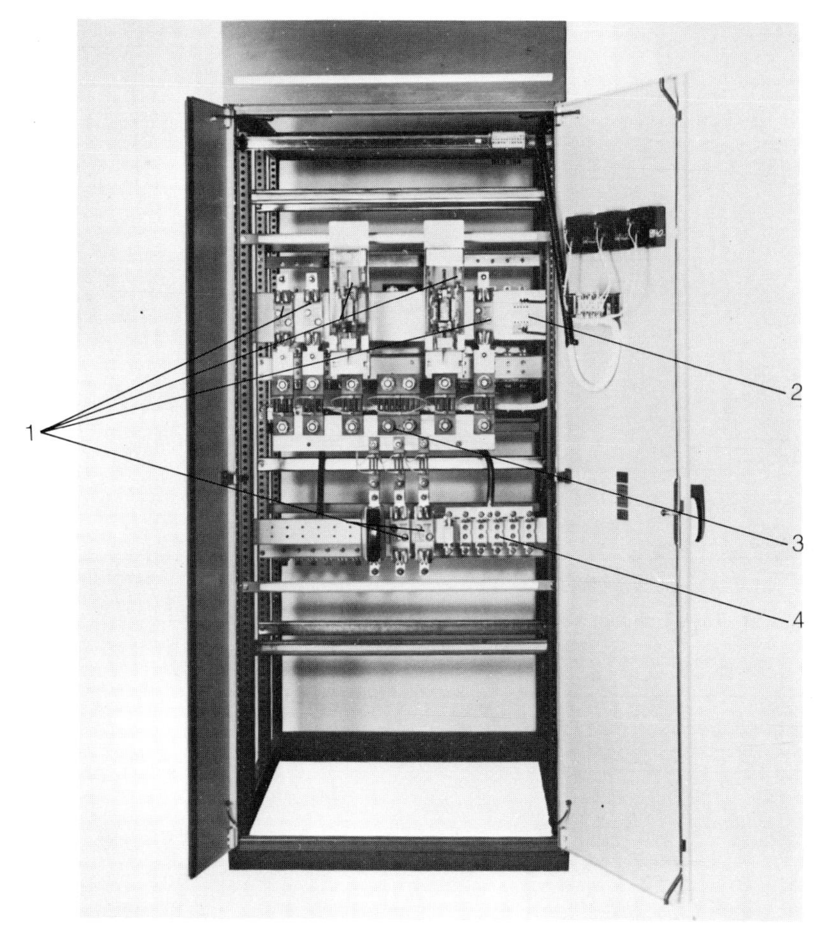

1 Load fuses
2 Measurement amplifier A1
3 Shunt in power circuit
4 Fuses for small loads

Fig. 1.90 Distribution panel 60 V/2 × 2500 A/3 VF10 with doors open

The currents in the individual load outputs are measured, depending on the equipment of the distribution panel, either by the built-in ammeters or with portable instruments. In distribution systems with eight circuits the individual load output currents can be measured with one 250/1000 A (60 mV) ammeter. In distribution systems with 24 circuits the individual load output currents are measured with a portable instrument directly on the shunts.

Figure 1.91 shows the basic circuit diagram of a 60 V distribution panel.

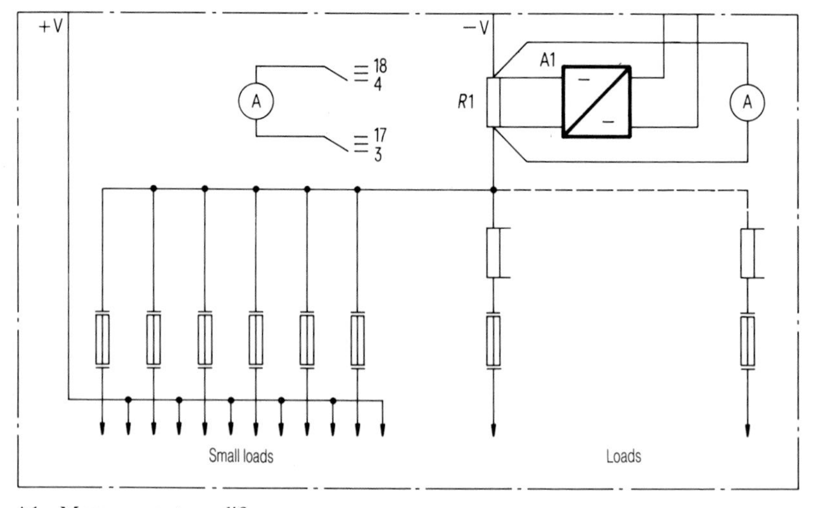

A1 Measurement amplifier

Fig. 1.91 Basic circuit of a 60 V distribution panel VF10

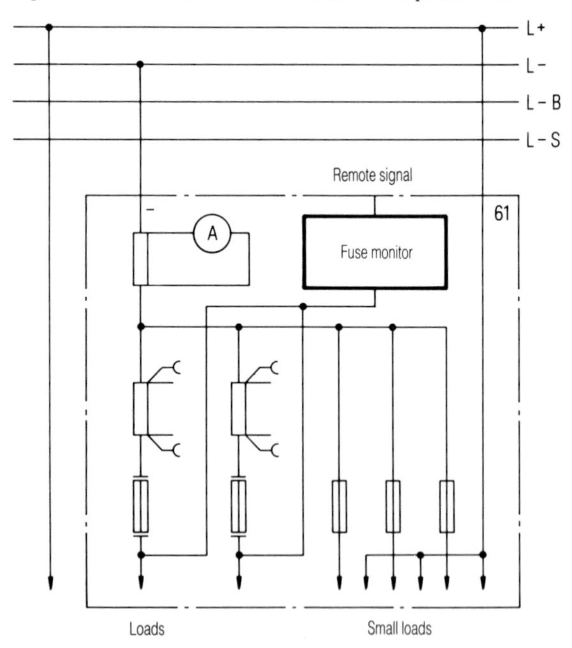

Fig. 1.92 Basic circuit of a 48 V distribution panel VF11

The distribution panel type VF11 (Fig. 1.92) is used in 48 V systems. In these panels the total current of an input supply is measured by means of a 3000 A/60 mV shunt with a directly connected 3000 A ammeter. Shunts are incorporated into the ten load outputs if required, whereby the currents can be measured with a portable instrument.

Distribution panels 2 × 3000 A are double equipped (in the front and at the back). There are two input supplies in this case, each provided with a built-in 3000 A ammeter.

Fuse monitor

Fuse monitoring is included in the 48 V distribution board type VF11. If one of the load fuses blows, an auxiliary relay (K1) is connected to the positive line (through the respective load) via the corresponding decoupling diode V (not shown in Fig. 1.92). K1 operates and switches on the visual fault alarm indication in the distribution panel, as well as the audible alarm in the control unit, and also provides a remote signal.

Technical data

The main technical data for the distribution panels are listed in Table 1.21.

Table 1.21 Technical data for distribution panels VF10 and VF11

Type	60 V systems					48 V systems	
	60 V/ 2500 A/1 VF10	60 V/ 2500 A/2 VF10	60 V/2[1]) × 2500 A/3[2]) VF10	60 V/2 × 2500 A/4 VF10	60 V/2 × 2500 A/5 VF10	60 V/48 V/ 3000 A/1 VF11	60 V/48 V/2 × 3000 A/2 VF11
Ammeter – total current (input supply) (A/mV)	1 × 2500/60	1 × 2500/60	2 × 2500/60	2 × 2500/60	2 × 2500/60	1 × 3000/60	2 × 3000/60
Measurement amplifiers (Qty)	1	1	2	2	2		
Ammeter – switchable to individual load outputs (A/mV)	1 × 1000/60		2 × 250/1000/ 60 (dual range)	1 × 250/1000/ 60 dual range)			
NH fuse bases/fuses for loads (A)	2 × NH4/800 to 1000 3 × NH2/224 to 630 3 × NH3/36 to 250	24 × NH00/6 to 100	4 × NH4/800 to 1000 6 × NH3/224 to 630 6 × NH1/36 to 250	2 × NH4/800 to 1000 3 × NH3/224 to 630 3 × NH1/36 to 250 24 × NH00/6 to 100	48 × NH00/6 to 100	10 × NH3/80 to 630	2 × 10 × NH3/80 to 630
NH fuse bases/fuses for small loads (A)	6 × NH00/6 to 100					3 × NH00/16	2 × 3 × NH00/16
Dimensions ($H \times W \times D$) (mm)	Cubicle construction: front with door, back open: 2000 × 900 × 1200 Height with fascia: 2150					2000 × 600 × 1200	

[1]) Example of number of inputs: two.
[2]) Example of design: no. 3.

1.2.4.1 Crossbar distribution panels

A special variant of the distribution panel is represented by the crossbar distribution panels KV (cubicle construction, Figs 1.93 to 1.95); they are envisaged only for 60 V systems.

The crossbar distribution panel makes it possible to transfer the 62 V load outputs between different compensators so that the individual compensators are equally loaded, or, in the event of the loss of a compensator, important loads can be switched over to the remaining units. Also, if the communications system is extended and additional compensators are introduced, the loads can be distributed between the compensators without any difficulty.

The total current in each input supply busbar is measured by means of a 2500 A/60 mV shunt on a 2500 A ammeter via a measurement-point selector switch. A measurement amplifier is also connected to each shunt and its output (20 mA impressed current) fed to the control panel for the formulation of the total-current signal.

To measure the current in the individual load outputs a dual-range 400/1000 A

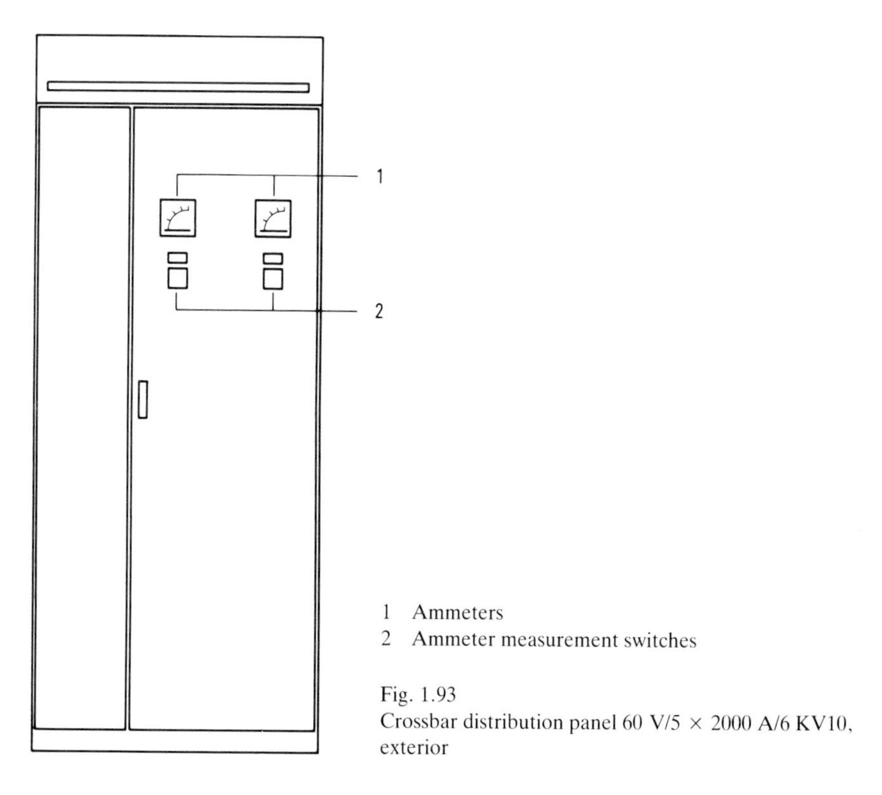

1 Ammeters
2 Ammeter measurement switches

Fig. 1.93
Crossbar distribution panel 60 V/5 × 2000 A/6 KV10, exterior

ammeter is provided. In special cases a current recorder can also be incorporated for the purpose.

The equipment of the panel depends upon the installation. A maximum of five input supply bars and six output supply bars can be incorporated. Thus the maximum provision entails 30 plug positions. If more than six load outputs are to be connected, further crossbar distribution panels can be added.

1 Mounting location for additional filtering
2 Differential voltmeter
3 Crossbar distribution panel
4 Fuses for small loads

Fig. 1.94
Crossbar distribution panel 60 V/5 × 2000 A/6 KV10 with doors open

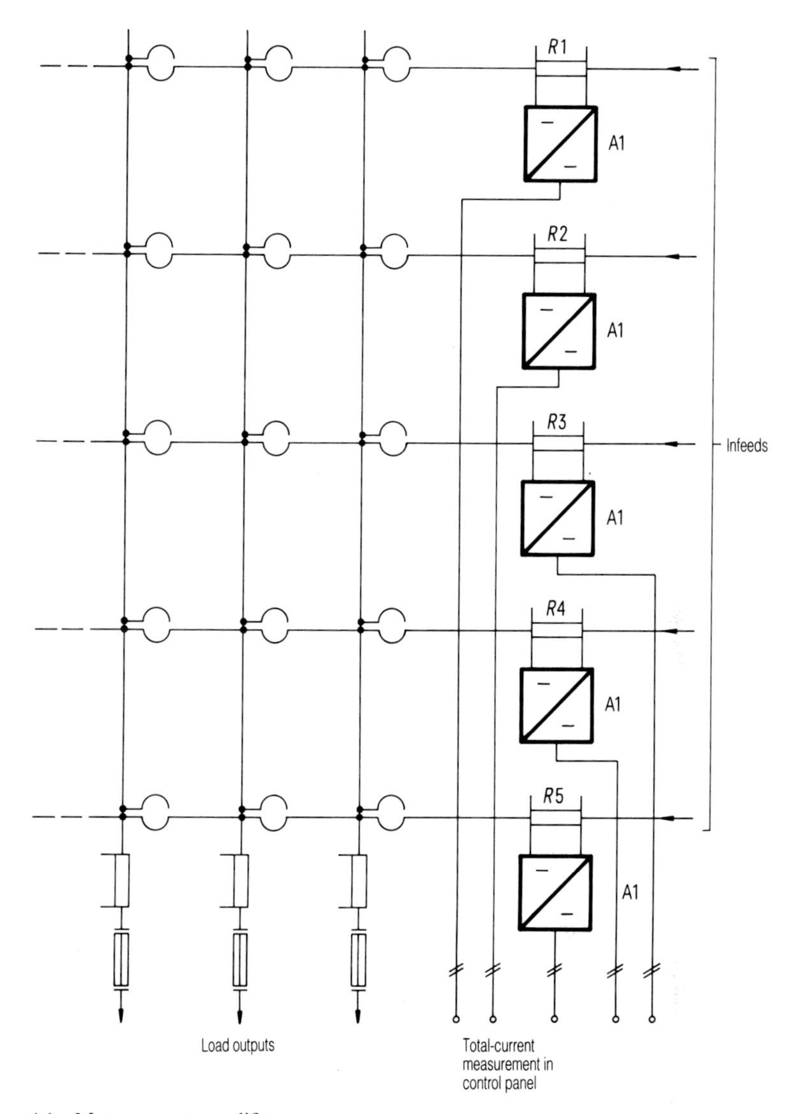

A1 Measurement amplifiers

Fig. 1.95
Basic circuit of a crossbar distribution panel KV10

Loads can only be transferred from one compensator to another if the power supply system is switched to continuous parallel operation and the operating mode selector switches of the compensators involved are set to 'off'.

Before loads are transferred from one input supply bar to another, the voltage difference between the two bars should be checked with the built-in 7.5–0–7.5 V voltmeter. The voltage difference during the plugging operation should be less than 0.5 V.

A load that is to be transferred must first be connected to the new input supply bar with the extra plug. The two input supplies are thus connected in parallel. Only then may the connection to the original input bar be broken.

Technical data

The principal technical data for the crossbar distribution panel are listed in Table 1.22.

Table 1.22 Technical data for crossbar distribution panels KV10

Type	60 V systems			
	60 V/2 × 2000 A/ 3 KV10 .../4 KV10 .../5 KV10 .../6 KV10	60 V/3[1]) × 2000 A/ 4[2]) KV10 .../5 KV10 .../6 KV10	60 V/4 × 2000 A/ 5 KV10 .../6 KV10	60 V/5 × 2000 A/ 6 KV10
Input supply (A)	2 × 2000	3 × 2000	4 × 2000	5 × 2000
Measurement amplifiers (Qty)	2	3	4	5
Ammeter – switchable to each input supply line (A/mV)	2500/60			
Ammeter with dual range – switchable to each load output (A/mV)	400/1000/60			
Voltmeter – switchable to individual input supply lines (V)	7.5–0–7.5			
NH fuse bases/ fuses for loads (A)	Maximum 6 × NH3/80 to 400 or maximum 6 × NH4/500 to 1000 (according to requirements)			
NH fuse bases/ fuses for small loads (A)	6 × NH00/6 to 100			
Dimensions (*H* × *W* × *D*) (mm)	Cubicle construction: front with door, back open: 2000 × 900 × 1200 Height with headboard: 2150			

[1]) Example of number of inputs: three.
[2]) Example of number of load outputs: four.

1.2.5 Monitoring assemblies

1.2.5.1 Voltage monitors A11, A21, battery-discharge contactor (K11) control A35, reducing-diode control A36

The assemblies A11, A21 and A35 are of similar design, and the assembly A36 is essentially similar to the A11, A21 and A35; they are used in 48 and 60 V power supply installations (see Section 1.2.2).

Each assembly contains two monitoring units, each with a light-emitting diode for indication.

The *voltage monitor A11* monitors the power supply voltage (51 or 62 V according to the system). An excess or shortfall relative to the operating levels results in the switching off of the associated rectifier group and an appropriate indication.

The *voltage monitor A21* monitors – according to the system – the 67 V power supply voltage and the battery voltage (60 V systems in the standby parallel or combined standby parallel and changeover mode) or only the battery voltage (48 and 60 V systems in the changeover mode). An excess or shortfall relative to the operating levels results in the switching off of the associated rectifier group and an appropriate indication.

If, for example, as a result of a mains supply failure the 48 V power supply system changes from rectifier supply to battery supply, the battery is available with 26 cells at the trickle charging voltage (58 V) or the charging voltage (61 V). These voltages are too high for the loads. The loads are therefore supplied in the first instance from 22 cells of the battery through the tapping diode V7 until the battery voltage falls to 46 V. Only then are the loads connected through the battery-discharge contactor K11 to all 26 cells of the battery.

The voltage-dependent switching from 22 to 26 cells is controlled by the *battery-discharge contactor (K11) control A35*.

For reasons of security, the voltage-responsive control is duplicated, the operating level of the first voltage-monitoring unit being set at 46 V while that of the second unit is set at 45.5 V. Also provided in this assembly are two light-emitting diodes for checking the two voltage-limiting levels.

The *reducing-diode control A36* for rectifier series GR10-25 A and BS100 A (see Section 1.1.2.4) controls the voltage-dependent insertion or by-passing of the two groups of voltage-reducing diodes. This enables the load voltage to be kept within the permissible tolerance under all operating conditions and operating sequences.

The function of these assemblies will be examined more closely with the aid of an example (assembly A11).

In the voltage monitor A11, two similar voltage-monitoring units are provided, with different settings for the upper and lower operating levels. The operating

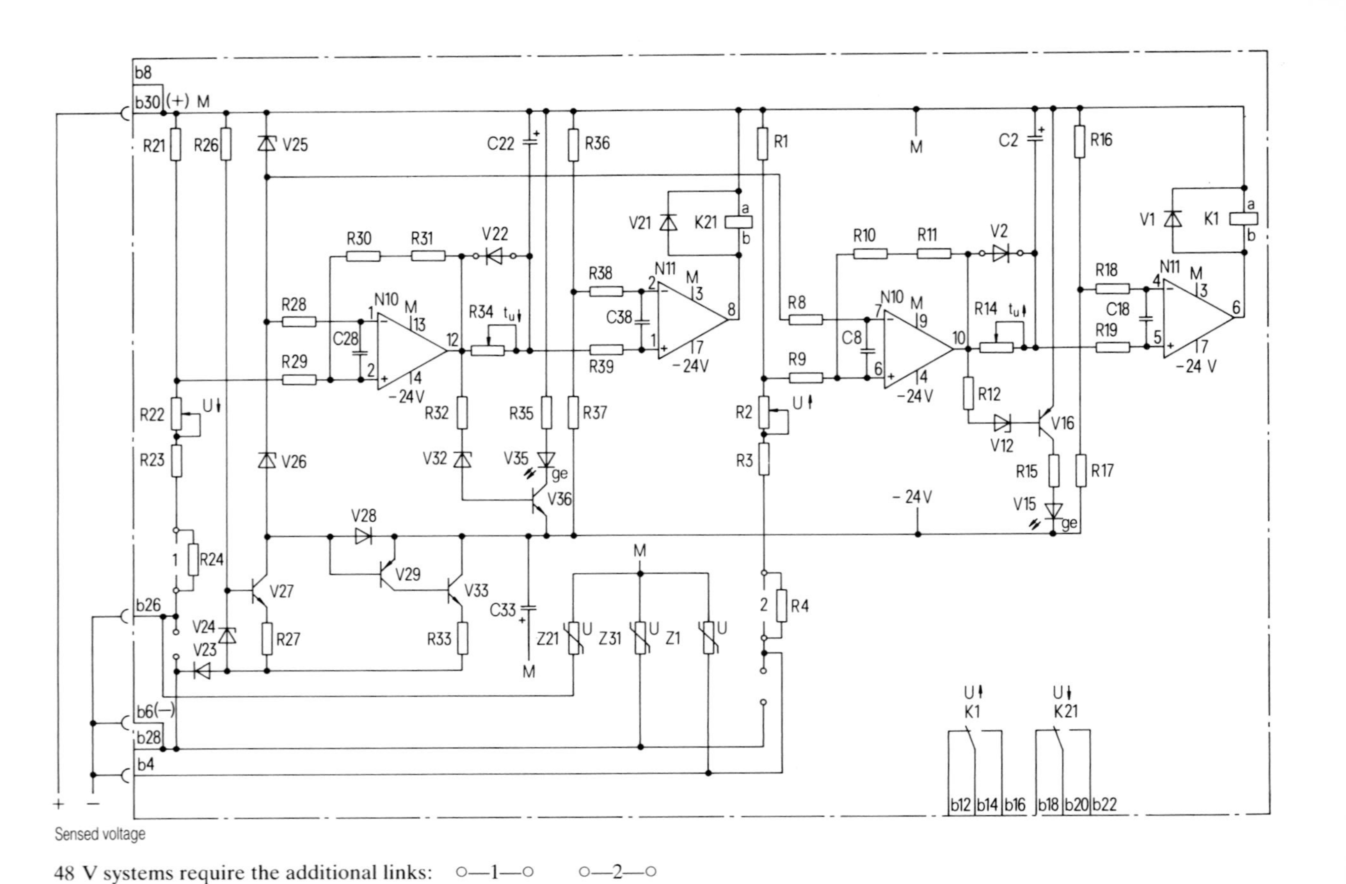

48 V systems require the additional links: ○—1—○ ○—2—○
For setting of assembly, see Table 1.23

level for the lower voltage limit can be adjusted by means of the potentiometer *R*22 and that for the upper operating limit with potentiometer *R*2 (Fig. 1.96).

The potentiometers are turned in a clockwise direction to raise the operating levels.

The voltage monitor is divided into three functional units:

▷ 24 V power supply,
▷ lower voltage-limit monitor,
▷ upper voltage-limit monitor.

24 V power supply

For the power supply to components such as operational amplifiers the voltage monitor requires a voltage of 24 V. A constant base voltage, obtained from the sensed voltage on pins b4, b6, b28 (negative) and b30 (positive) through *R*26 and zener diode V24, is applied to the transistor V27. The zener diodes V25 and V26 are consequently supplied with a constant current determined by *R*27. A stabilized voltage of –24 V (with respect to M) can thus be obtained at the collector of V27. Load currents are prevented from affecting the stabilizing circuit by the decoupling effect of V28 and V29. Voltage variations in the 24 V load circuit cause the transistor V33 to be rendered more or less conducting by V29, so that the variations are corrected (continuous longitudinal controller principle).

Lower voltage-limit monitor

The voltage monitor reference value is tapped off at the junction of the zener diodes V25 and V26 (about 11.7 V) and applied to the inverting input 1 of the operational amplifier N10.

The actual value signal is obtained from the monitored voltage through the voltage dividers *R*21 to *R*24 and applied to the non-inverting input 2 of N10. The switching point of the operational amplifier is set by the potentiometer *R*22.

So long as the actual value does not exceed the set voltage level, the reference value is more positive. The output 12 of N10 is consequently at the negative 24 V potential. Capacitor C22 is charged through the diode V22. The following operational amplifier N11 is biased by a small negative voltage at the inverting input 2. As a result of the 24 V signal at the non-inverting input 1, therefore, the output of N11 is similarly at 24 V, and the relay K21 is operated.

If the monitored voltage falls, the negative actual value similarly decreases; i.e the actual value signal becomes more positive. When the voltage falls below the set operating level, the actual value signal becomes less than the reference value of about –12 V, and the output 12 of the operational amplifier N10 switches to the M potential of the power supply.

Transistor V36 is turned on through *R*32 and V32. The light-emitting diode V35 is energized and indicates the attainment of the lower voltage limit with no delay. At the same time, the capacitor C22 is discharged through *R*34 towards M. The discharge time of C22, and thereby the response time of the circuit, is adjusted by means of *R*34.

When the voltage on the capacitor becomes more positive than the (negative) voltage at the inverting input 2 of N11, the output 8 of N11 switches to M and the relay K21 releases. If the monitored voltage increases again, the non-inverting input 2 of N10 again becomes more negative than the inverting input 1. The output 12 of N10 switches to –24 V, the light-emitting diode V35 is extinguished, C22 is recharged and N11 energizes the relay K21.

Upper voltage-limit monitor

The voltage monitor reference value is tapped off at the junction of the zener diodes V25 and V26 (about 11.7 V) and applied to the inverting input 7 of the operational amplifier N10. The actual value signal is obtained from the monitored voltage through the voltage dividers *R*1 to *R*4 and applied to the non-inverting input 6 of N10. The switching point of the operational amplifier is set on the potentiometer *R*2.

While the actual value signal does not exceed the set voltage limit, the reference potential is more negative. The output 10 of N10 is thus at the M potential. The following operational amplifier N11 is biased by a small negative voltage at its inverting input 4. As a result of the M signal at its non-inverting input, therefore, the output of N11 is similarly at the M potential and the relay K1 is released.

If the monitored voltage increases, the negative actual value signal becomes greater than the reference value of about –12 V, and the output 10 of the operational amplifier N10 switches to the 24 V potential of the power supply. Transistor V16 is turned on through *R*12 and V12. The light-emitting diode V15 is energized and indicates the attainment of the upper voltage limit with no delay. At the same time the capacitor C2 is charged through *R*14. The charging time of C2, and thus the response time of the circuit, are adjusted by means of *R*14. When the voltage on the capacitor becomes more negative than the negative voltage at the inverting input 4 of N11, the output 6 of N11 switches to 24 V and the relay K1 is operated.

If the monitored voltage falls again below the operating level, the non-inverting input 6 of N10 again becomes more positive than the inverting input 7 and the output 10 of N10 switches to M, the light-emitting diode V15 is extinguished, C2 is discharged through the diode V2 and N11 de-energizes the relay K1.

The assembly is protected against overvoltage spikes by the varistors Z1, Z21 and Z31.

Finally, the different functioning of the A36 assembly in comparison with the A11 assembly described in detail above will be considered.

In the A36 assembly both the voltage-monitoring units are responsive to under-voltage. The light-emitting diodes (V15 and V35) are illuminated at the voltage U_{Ab}. The voltage setting value U_{An} of the reducing-diode control is 64 V for K1 and 68.5 V for K21. The operating voltage level U_{Ab}, determined by the hysteresis of the circuit, is about 63.3 V for K1 and 67.8 V for K21. The response time of both monitoring units must be set to minimum – i.e. the potentiometers *R*14 and *R*34 must be fully clockwise.

If the equipment or battery voltage corresponds to a load voltage of 62 V, V15 and V35 are illuminated. K1 and K21 are released and energize the bridging contactors K13 and K14; the voltage-reducing diodes are bridged.

If the sensed voltage rises above 64 V, V15 is extinguished and K1 de-energizes K13. The reducing diodes V3/1 and V4/1 are brought into circuit and the load voltage is reduced by 5 V.

If the sensed voltage increases further to more than 68.5 V, V35 is extinguished and K21 de-energizes K14. The additional reducing diodes V3/2 and V4/2 are inserted and the load voltage is reduced by a further 3 V (now by a total of 5V + 3V = 8V).

When the sensed voltage falls below about 67.8 V, V35 lights and K21 energizes K14. The reducing diodes V3/2 and V4/2 are bridged and the load voltage is raised by 3V.

When the sensed voltage falls further to less than about 63.3 V, V15 lights and K1 energizes K13. The reducing diodes V3/1 and V4/1 are bridged and the load voltage is increased by 5 V.

1.2.5.2 Overvoltage limiter A31 and A41

The overvoltage limiter A31 is similar in design to the overvoltage limiter A41 (Fig. 1.97).

The purpose of the overvoltage limiter A31 is to protect the loads (e.g. in 60 V systems, the 62 V loads) against voltage spikes. If the preset operating level is exceeded, the trigger pulses of the relative rectifier are inhibited without any delay.

The flow of power from the supply mains is thus cut off. In addition a dr (urgent) signal is passed to the alarm unit and the appropriate indication lamp switched on.

The purpose of the *overvoltage limiter A41* is to protect the loads (e.g. in 60 V systems, the 67 V loads) against voltage spikes.

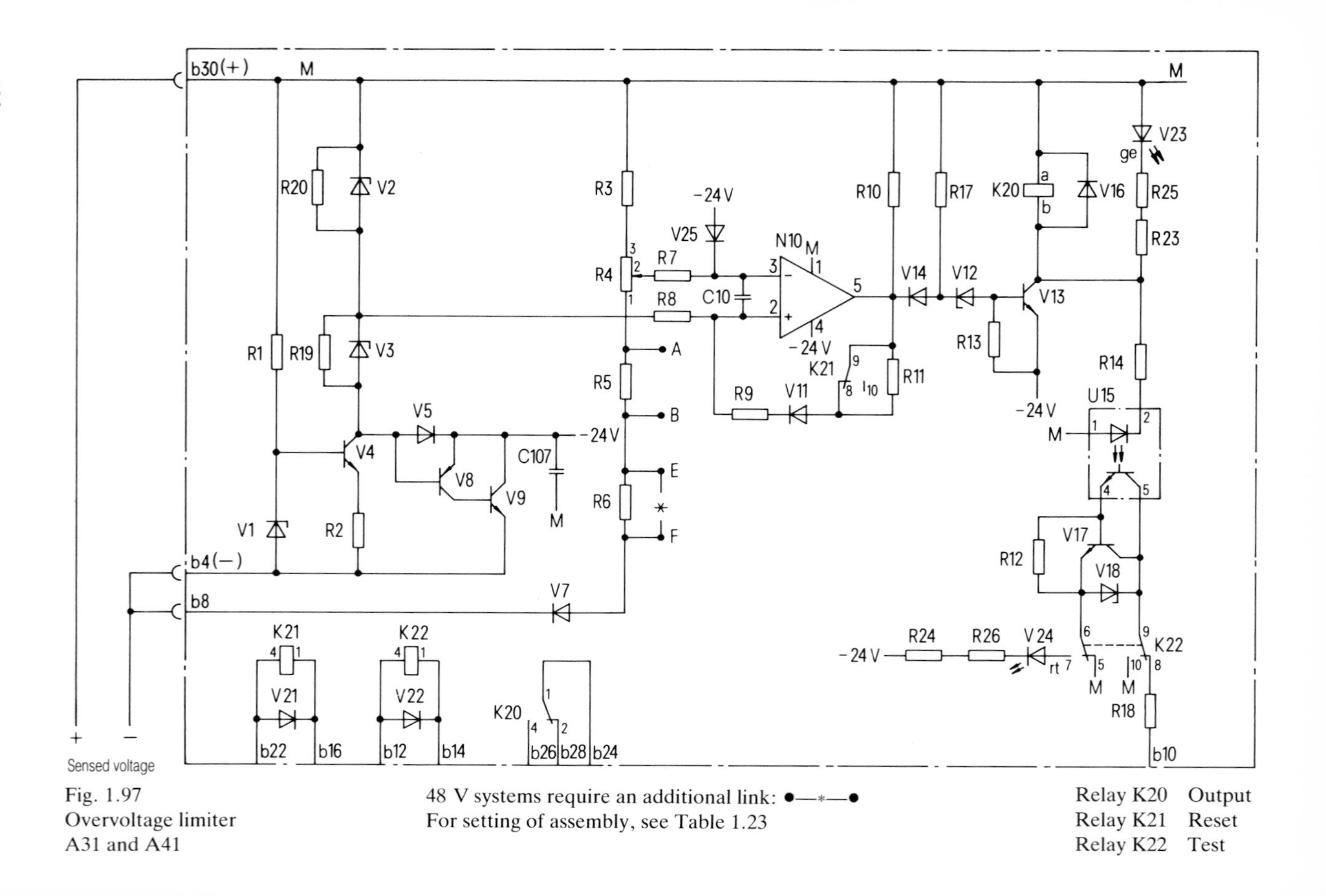

Fig. 1.97
Overvoltage limiter
A31 and A41

48 V systems require an additional link: ●—*—●
For setting of assembly, see Table 1.23

Relay K20 Output
Relay K21 Reset
Relay K22 Test

Since there is little separation between the charging voltage of 70 V and the 73 V operating level of the overvoltage limiter, supply voltage transients can cause operation of the monitoring circuit.

'Locking-out' of the 67 V rectifier in these circumstances would be inappropriate.

Following the operation of the overvoltage limiter, therefore, the 67 V rectifiers are automatically switched on again by a test circuit. If a further overvoltage occurs within a checking period of about 30 min, the possibility of a control failure is recognized and the 67 V rectifiers have to be 'locked out'; they can only be put into operation again manually (after rectification of the fault). The operating level of the circuit is adjusted by means of the potentiometer *R*4.

The A31 and A41 assemblies comprise:

▷ the 24 V power supply,
▷ the overvoltage limiter.

24 V power supply

The imformation given above in relation to the 24 V power supplies for the A11, A21 and A35 assemblies is applicable.

Overvoltage limiter

The voltage to be monitored (the sensed voltage) is applied through the inputs b30 (positive) and b4, b8 (negative) to a bridge circuit. The zener diodes V2 and V3 form the fixed arm (the reference signal), the voltage dividers *R*3 to *R*6 the variable arm (the actual value signal). The two voltage values are compared by the operational amplifier N10.

If the actual value signal at the inverting input 3 of N10 becomes more negative than the reference value at the non-inverting input 2 of N10 (the sensed voltage having reached the operating level), the output voltage of the operational amplifier N10 (output 5) changes from –24 to 0 V (M). The transistor V13 conducts, the optocoupler V15 is energized and the signal is conveyed to the base of transistor V17. This transistor is turned on and transmits a pulse inhibit signal (M) via output b10 to the closed-loop and open-loop control assembly A3 of the associated rectifier group.

Simultaneously with the operation of the optocoupler, the relay K20 is energized by V13 and the light-emitting diode V23 is switched on. The relay transmits a dr (urgent) signal, via an auxiliary relay of control section A10, to the alarm unit, and at the same time switches on the 'overvoltage' indication lamp. The operational amplifier N10 remains in the same state, by virtue of its feedback loop (contact K21, diode V11 and resistor *R*9), even though the sensed voltage falls again.

The overvoltage limiter is released with the pushbutton S81 (see Section 1.2.2), through which the relay K21 is energized. The feedback is reduced by the

Table 1.23 Adjustment of monitoring assemblies A11, A21, A31 and A41; 60 V systems for combined standby parallel and changeover mode with 30-cell lead–acid battery

Monitoring assemblies		Operating level (V)	Operating level set on poten-tiometer	Grid reference (position of potentiometer on front plate of assembly)	*Setting sequence	Range of adjustment (V)	Response time (ms) (pulse inhibit in μs region)	Response time set on poten-tiometer	Response time adjustment range (ms)
Designation	Abbreviated designation								
Voltage monitor 62 V	A11 $U\uparrow\downarrow$	$U\uparrow$ 68 ±1%	R2	G2		$U\uparrow$ approx. 67 to 75	approx. 400	R14 position 2	
		$U\downarrow$ 57 ±1%	R22	U2		$U\downarrow$ approx. 56 to 64	approx. 700	R34 position 6	
Voltage monitor 67 V	*A21 $U\uparrow\downarrow$	$U\downarrow$ 59 ±1%	R22	U2	[1])		approx. 700	R34 position 6	50 to 1200
		$U\downarrow$ 62.5 ±1%	R213 (on A10)	—	[2])		approx. 700	R34 position 6	
		$U\uparrow$ 72 ±1%	R2	G2	[3])		approx. 400	R14 position 2	
Overvoltage limiter 62 V	A31 $U\uparrow\uparrow$	$U\uparrow\uparrow$ 72 ±1%	R4	B2		approx. 65 to 80	Output b10: μs region (pulse inhibit); outputs b24 to b26: ms region	Fixed	Fixed
Overvoltage limiter 67 V	A41 $U\uparrow\uparrow$	$U\uparrow\uparrow$ 73 ±1%	R4	B2					

* The following sequences should be observed in adjusting A21:
[1]) Set the characteristic switch S61 (see Section 1.2.2) to 'continuous parallel operation' (position 1); set $U\downarrow$ to 59 V.
[2]) Set the characteristic switch S61 to 'automatic operation' (postion 2); set $U\downarrow$ to 62.5 V.
[3]) Set the characteristic switch S61 to 'automatic operation' (position 2); set $U\uparrow$ to 72 V.

insertion of the resistor *R*11 to the extent that the output of the operational amplifier reverts to –24 V.

Operation of the overvoltage limiting in the testing mode causes a changeover contact of relay K22 to switch on the light-emitting diode V24 through the transistor V17, thus checking the pulse-inhibit function without its affecting the rectifiers (i.e. without actual pulse inhibition).

1.2.5.3 Adjustment

Some examples of adjustment are given in Table 1.23 (see Section 1.2.2).

1.3 Compensators

Application

If the load voltage is to be held constant during the discharge of the battery, the battery voltage must be connected in series with a gradually increasing voltage.

This requirement is met in 60 V power supply systems by the compensator AGE (Figs 1.98 to 1.102).

While the loads are supplied from the 62 V rectifiers the compensator is by-passed. Only when the last 62 V rectifier is switched off and the load voltage falls below 62 V is the by-pass removed and the output voltage of the compensator (0 to 7 V) put in series with the battery voltage.

In this way it is possible for a lead-acid battery of 30 cells to be discharged to its final discharge voltage of 54 V while the load voltage remains constant at 61 V ± 1%.

The power required to produce the additional voltage is taken from the battery.

There are compensators for 200, 400, 600, 1000 and 2000 A.

The operating sequences in a power supply installation with a compensator have been described in Section 1.1.2.4 (see Fig. 1.23).

Basic circuit and operation

Power section

As shown in Fig. 1.101, the battery voltage is applied to the input terminals D1 (N) and C1 (P). The current consumption during operation can be read from the ammeter P2. The compensator is switched on by closing the contact K1. An input filter, consisting of L1, C2 and C4, prevents interference with other power supply units due to disturbances from the d.c. controller.

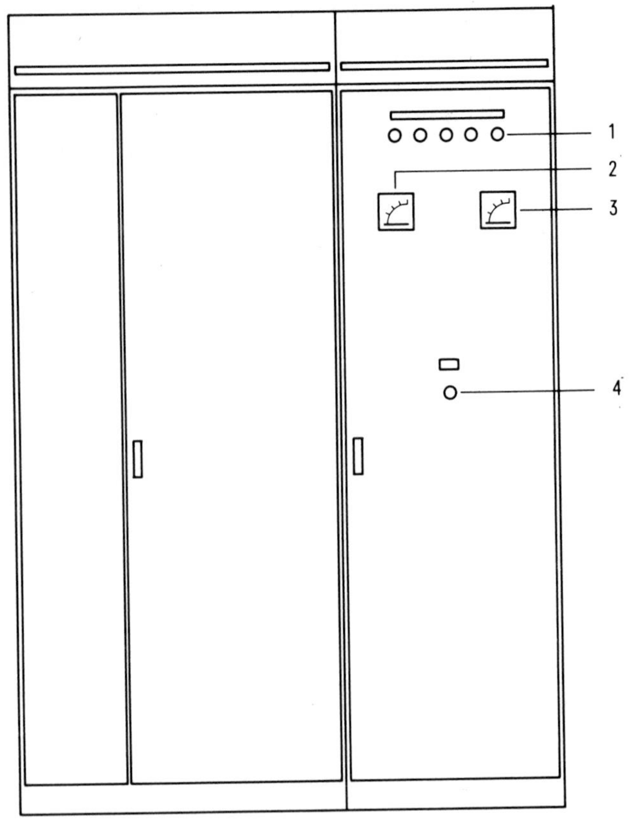

1 Indication lamps from left to right:
Operating/voltage deviation/fault/test operation/off and by-passed
2 Voltmeter
3 Ammeter
4 Lamp test button

Fig. 1.98 Compensator 7 V/1000 A UR10, exterior

The battery voltage is converted by the d.c. controller into a controlled square-wave voltage and transformed down by the main transformer T1 (T2). In the 400 and 2000 A compensators two d.c. controllers (of half the power rating) are used, operating with 180° phase displacement. The thyristors of the d.c. controllers receive their trigger pulses from the closed-loop, open-loop (regulation, control) and monitoring via the pulse transformers ⑩.

At this point the d.c. controller 1 will be described as an example.

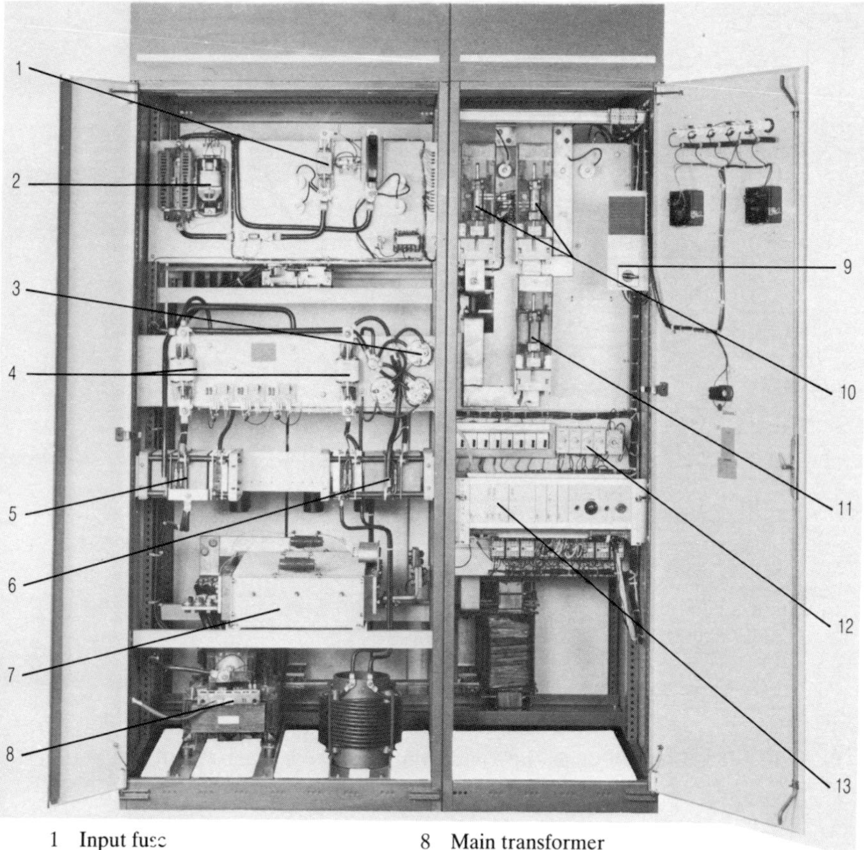

1 Input fuse
2 Input contactor
3 Commutating capacitors
4 Thyristor fuses
5 Main thyristor V11
6 Turn-off and charge-reversal thyristors V12 and V13
7 Rectifier set V1
8 Main transformer
9 Equipment switch
10 Isolating links
11 Output fuse
12 Time-delay relay
13 Closed-loop and open-loop control and monitoring

Fig. 1.99 Compensator 7 V/1000 A UR10 with doors open

When the main thyristor V11 is triggered, current flows through the path:

> $+U_{\mathrm{Batt}}$/C1 (P)/F2/L1/primary coil of main transformer T1/F11/V11/L1/K1/$R$1/F1/D1 (N)/$-U_{\mathrm{Batt}}$.

As a result of this a voltage is induced in the secondary winding of T1. Current

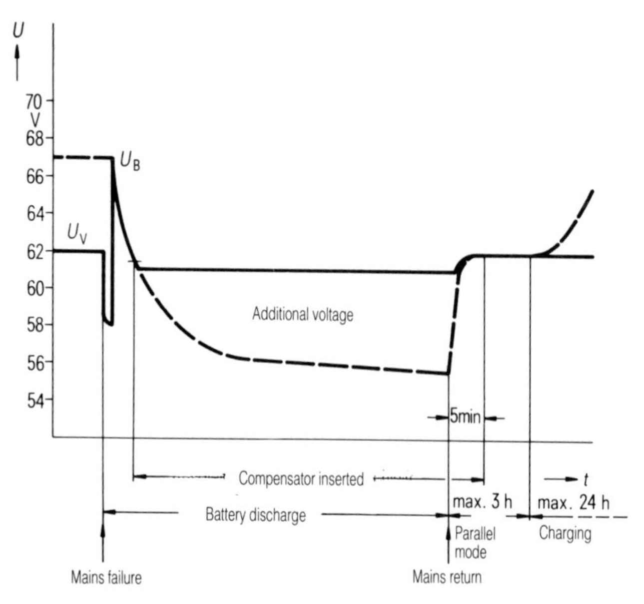

U_B Battery voltage
U_V Load voltage

Fig. 1.100 Variation of voltage with a compensator in circuit

flows to the output through the conducting diode V1 (one-way rectification circuit).

The conduction period depends upon the magnitude of the required additional voltage and is terminated by the triggering of the turn-off thyristor V13.

This initiates the discharge of the commutating capacitor C10 (polarity: positive right, negative left). The load current commutates, due to the additional voltage of C10 in series, from the main thyristor V11 to the turn-off thyristor V13. Thyristor V11 is now non-conducting and turns off. If the whole of the energy stored in the capacitor C10 is not required for the turn-off process, the surplus is dissipated in V10, L13 and L11.

With the turning-off of the main thyristor V11 the current in the primary winding of T1 falls and thereby induces a voltage in the secondary winding of the opposite polarity, so that the diode V1 blocks. During this blocking phase the energy stored in the inductors L3 and L5 is released via the free-wheel diode V3, while the capacitors C6 and C8 discharge.

The triggering of the turn-off thyristor V13 causes a current to flow in the oscillatory circuit V10, L13, L11, V13, which, together with the load current

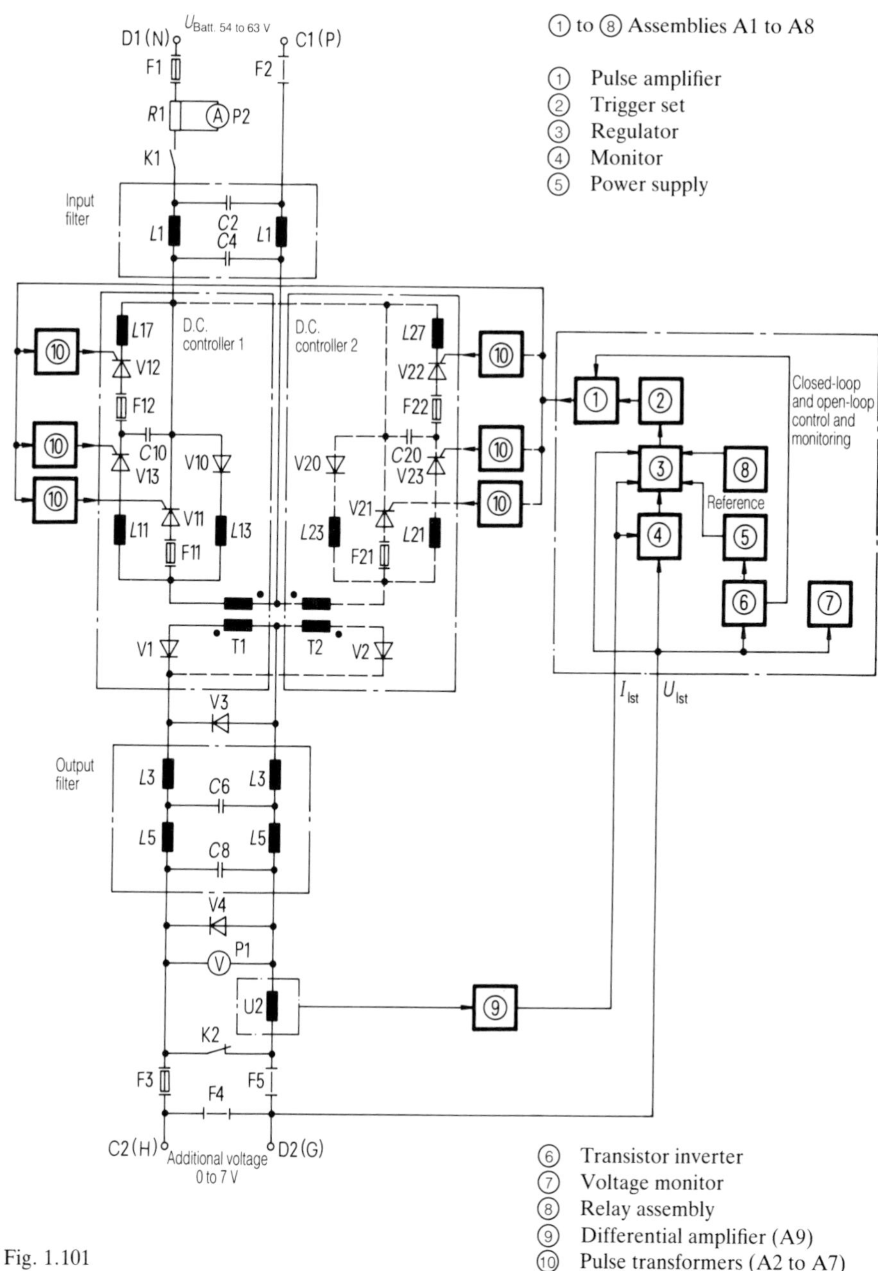

Fig. 1.101
Basic circuit of compensator UR10

flowing in the path

$$+U_{Batt}/C1\ (P)/F2/L1/T1/L11/V13/C10/L1/K1R1/F1/D1\ (N)/-U_{Batt},$$

recharges the capacitor C10 (positive left, negative right).

As the charging current falls the turn-off thyristor V13 blocks. The charge-reversal thyristor V12 is then triggered and the capacitor C10 is discharged through V12 and L17. The inductor L17 absorbs the energy from the capacitor C10. L17 then recharges the capacitor C10, but now with *reversed polarity* (positive right, negative left). V12 now blocks, and the capacitor C10 is prepared for turning off the main thyristor V11 again.

The falling voltage of the battery as it discharges and the variable loading of the compensator by the power supply system load makes it necessary to control the additional voltage. This is achieved by varying the separation between the trigger pulses of the main thyristor V11 and the turn-off thyristor V13.

The d.c. controller operates with pulse-width control (at a constant frequency of 400 Hz) to produce an additional voltage of from 0.5 to 7 V (Fig. 1.102a). At the maximum additional voltage of 7 V the conduction period of each main thyristor is at its maximum of 1.25 ms. The conduction period can be reduced to about 50 μs, which results in an additional voltage of about 0.5 V (Fig. 1.102b). To obtain a still lower additional voltage the frequency must be reduced from 400 Hz to 100 Hz (pulse-sequence–frequency control, Fig. 1.102c).

Ripple is smoothed to an acceptable level of interference voltage by the output filter, consisting of L3, L5, C6 and C8 (Fig. 1.101).

The compensator is inoperative during operation from the mains supply by virtue of a bridging contactor K2. When the compensator is switched on K2 opens, and until the compensator starts to supply power the load current flows through the by-pass diode V4. The voltage drop across this diode is shown as a negative voltage on the voltmeter P1.

When the compensator produces an additional voltage, the load current commutates from V4 to the power circuit. The additional voltage is indicated on P1.

The converter U2 provides an actual-current signal, which is amplified by the differential amplifier A9 and fed to the current regulator on the A3 assembly and also to the A4 assembly.

For maintenance purposes the compensator can be by-passed manually by means of the link F4 and isolated from the power supply system.

The output D2 (G) of the compensator is connected to assemblies A3, A4, A6 and A7 by a sensing lead.

Closed-loop, open-loop control (regulation, control) and monitoring

All the units required for closed-loop, open-loop control and monitoring are

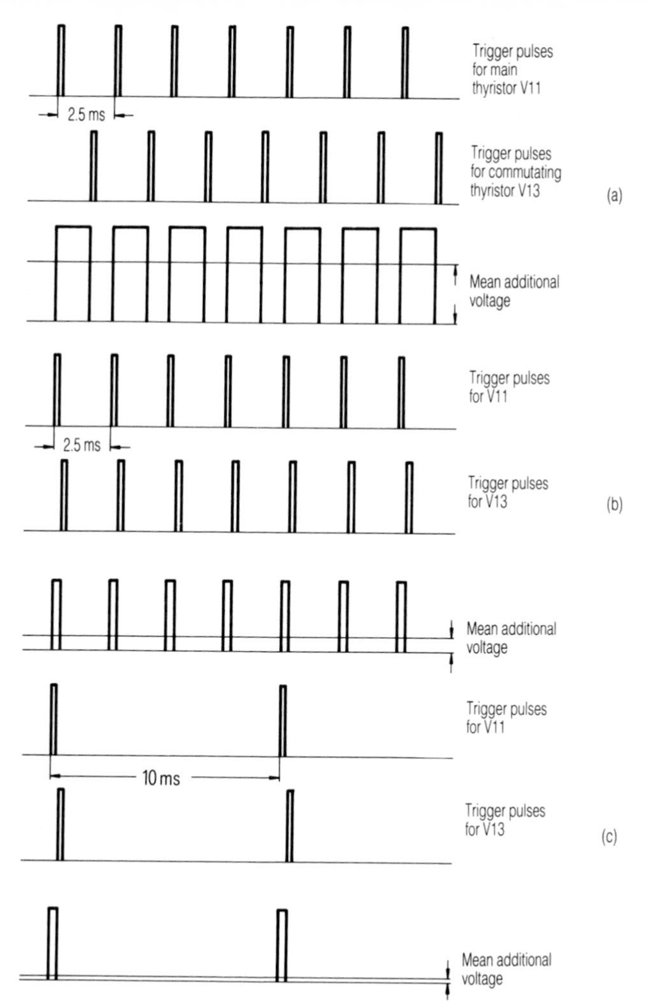

Fig. 1.102
Pulse-width control (a and b) and pulse-sequence (frequency) control (c) of a compensator

constructed as plug-in assemblies and accommodated in a hinged rack. The power supply to the assemblies is provided by the transistor inverter A6 and the power supply A5.

The monitor A4 has two purposes:

▷ to switch on the compensator when the load voltage (battery voltage) falls below 62 V during battery discharge,

▷ to switch off the compensator to protect the components against overload if the current exceeds 120% of the rated current I_{Nenn}.

The regulator A3 controls the trigger set A2 in dependence upon the d.c. output voltage and the output current. The pulses produced by the trigger set are conveyed to the thyristors of the d.c. controller via the pulse amplifier A1 and the pulse transformers ⑩.

The voltage monitor A7 monitors the 62 V power supply voltage (undervoltage and overvoltage).

The relay assembly A8 provides, for example, for pulse enable.

Technical data

The principal technical data for the compensator UR10 are listed in Table 1.24.

Table 1.24 Technical data for compensators type UR10

	Type: UR10				
	7 V/200 A	7 V/400 A	7 V/600 A	7 V/1000 A	7 V/2000 A
Number and current rating of d.c. controllers (A)	1 × 200	2 × 200	1 × 600	1 × 1000	2 × 1000
Rated current (A)	200	400	600	1000	2000
Output voltage (V)	0 to 7				
Load voltage for I = 3% to 100% (V)	58.5 to 63 ±1% (adjustable)				
Input voltage (V)	54 to 63				
Current consumption (A)	43	80	120	200	410
Interference voltage (mV)	≦1.8 (frequency-weighted with CCITT 'A' filter, measured at load output)				
Degree of radio interference	N (VDE 0875)				
Dimensions ($H \times W \times D$) (mm)	Cubicle construction:				
	2000 × 900 × 600	2000 × 1200 × 600	2000 × 1500 × 1200		2000 × 2400 × 1200
	Height with headboard: 2150				

1.4 Examples of the Application of Thyristor-controlled Rectifiers

The following section provides examples of the application of the GR10 thyristor-controlled rectifiers:

▷ power supply system for a telecommunication service building with 60 V loads <2000 A (Fig. 1.103),
▷ power supply system for a radio transmitter station with 60 V loads <2000 A (Fig. 1.104),
▷ power supply system for a telecommunication service building with 48 V loads >2000 A (Fig. 1.105).

1.5 Special Features for Transmission System Power Supplies

In considering power supplies for long-range transmission systems it is necessary to distinguish between the following:

▷ power supply systems in telecommunications towers,
▷ power supply systems in ground communications stations,
▷ power supply systems independent of mains supplies.

Radio link stations and repeater stations often have to be built in places where public mains supplies are not available. For these situations there are power supply systems which are independent of mains supplies, constructed in shelters. These can be hybrid systems, consisting, for example, of a wind generator and a solar generator together with a diesel generator set and a battery.

1.5.1 Power supply for radio link apparatus in telecommunications towers

Radio link apparatus, like other equipments, requires various supply voltages. Since telecommunications towers (Fig. 1.106) are particularly vulnerable to lightning, special measures have to be adopted in connection with the power supply systems for earthing and lightning protection (see Chapter 4). As regards the energy source, there are installations operating in the standby parallel mode and others without a battery but including an immediate standby system.

1.5.1.1 Standby parallel mode power supply systems

The power supply system is usually installed in a building adjacent to the telecommunications tower.

The basic circuit of a d.c. uninterruptible power supply system is shown in

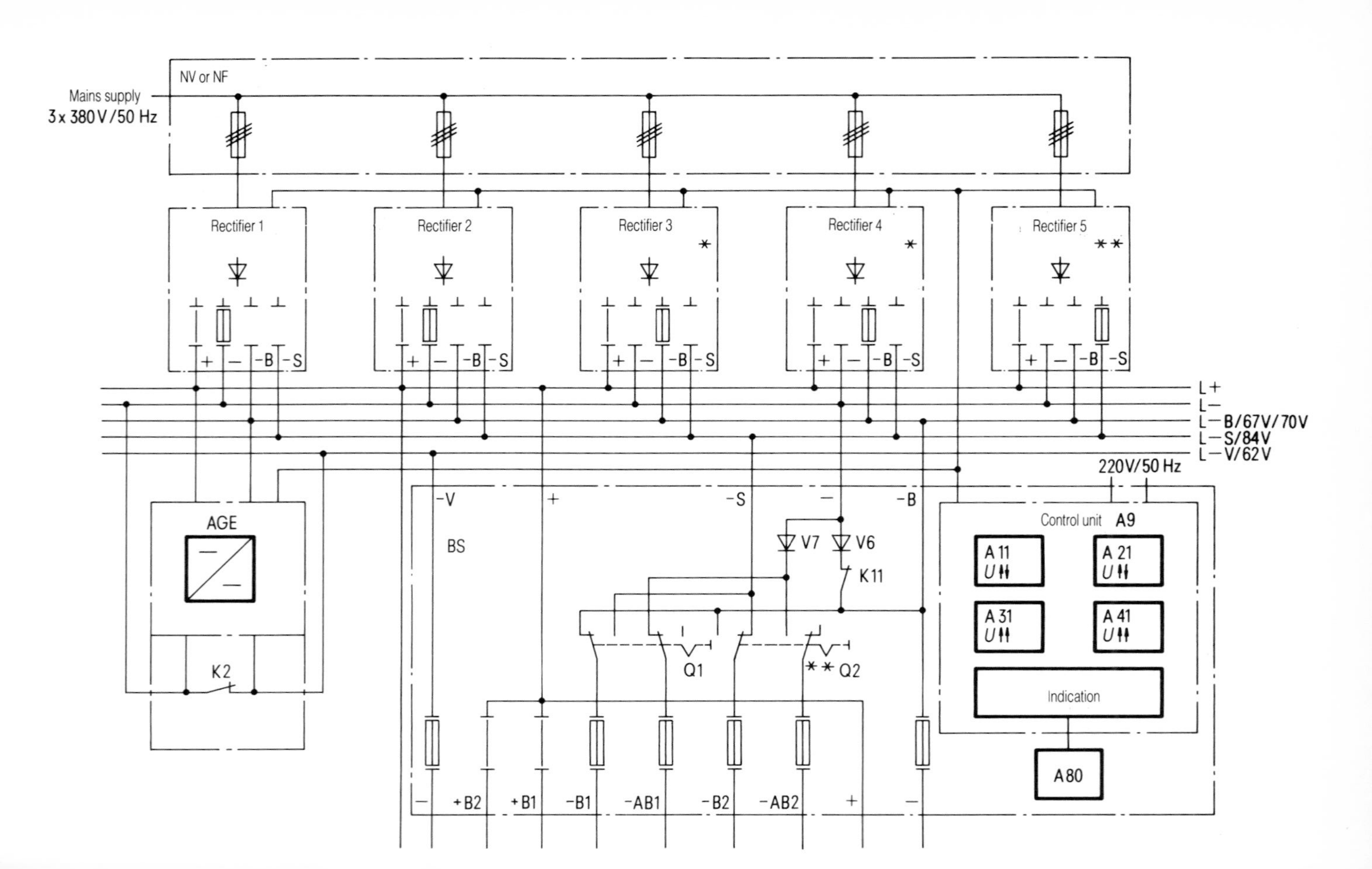

NV or NF
Mains supply
3 x 380 V / 50 Hz
Rectifier 1
Rectifier 2
Rectifier 3
Rectifier 4
Rectifier 5
L+
L−
L−B/67V/70V
L−S/84V
L−V/62V
220V/50 Hz
AGE
K2
BS
V7
V6
K11
Q1
Q2
Control unit A9
A 11
A 21
A 31
A 41
Indication
A80
+B2
+B1
−B1
−AB1
−B2
−AB2

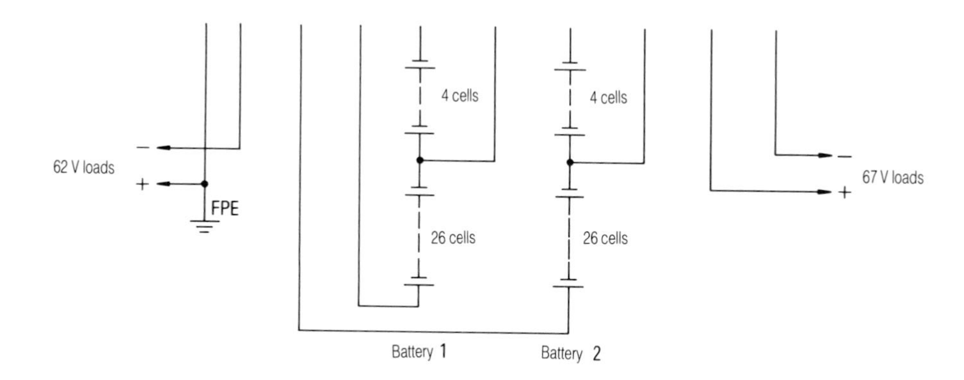

AGE Compensator
BS Battery switching panel with control unit
NF Mains switch panel
NV Mains distribution switchboard
FPE Functional and protective earth conductor
A11 Voltage monitor 62 V
A21 Voltage monitor 67 V
A31 Overvoltage limiter 62 V
A41 Overvoltage limiter 67 V
A80 Auxiliary rectifier
* These rectifiers function simultaneously as operating rectifiers for 67 V loads and as charging rectifiers
** Set for initial charging

Fig. 1.103
General circuit diagram of a power supply system for a telecommunications building with 60 V loads <2000 A; combined standby parallel and changeover mode

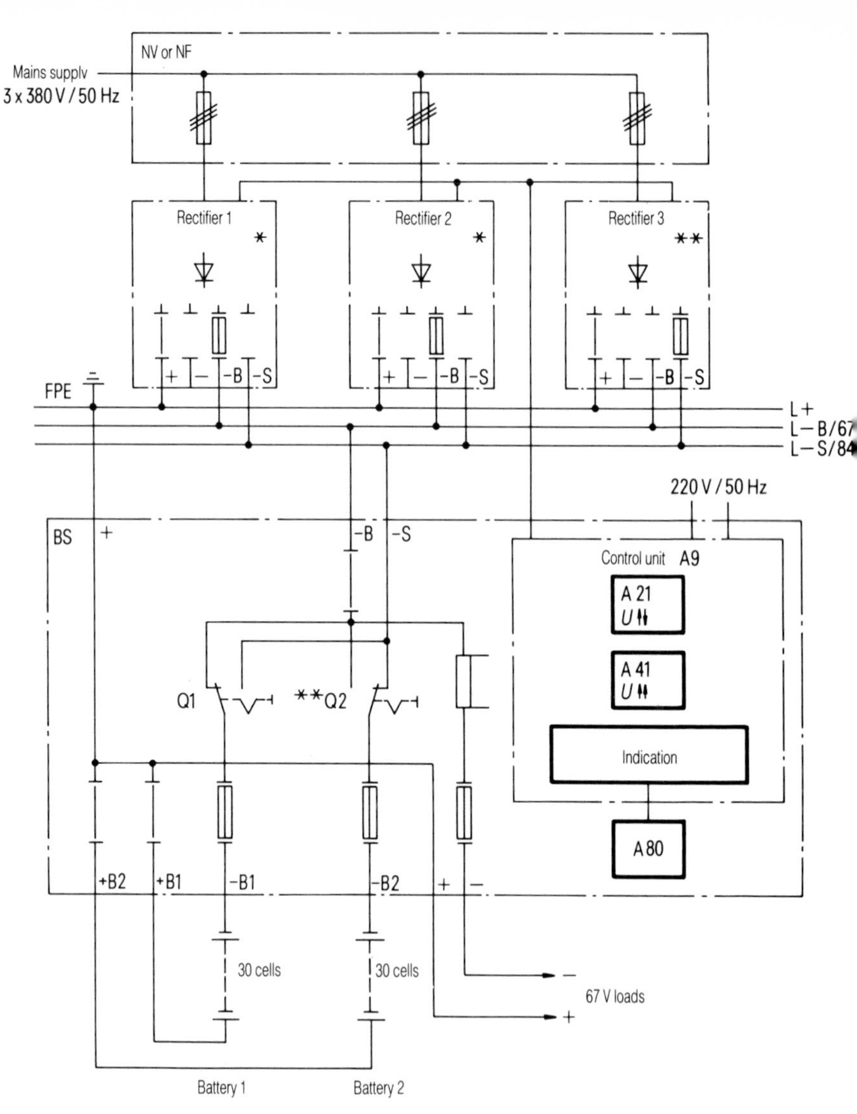

BS	Battery switching panel with control unit	FPE	Functional and protective earth conductor
NF	Mains switch panel	A21	Voltage monitor 67 V
NV	Mains distribution switchboard	A41	Overvoltage limiter 67 V
		A80	Auxiliary rectifier

* These rectifiers function simultaneously as operating rectifiers for 67 V loads and as charging rectifiers

** Set for initial charging

Fig. 1.104
General circuit diagram of a power supply system for a radio-transmission station with 60 V loads <2000 A; standby parallel mode (without reducing diodes)

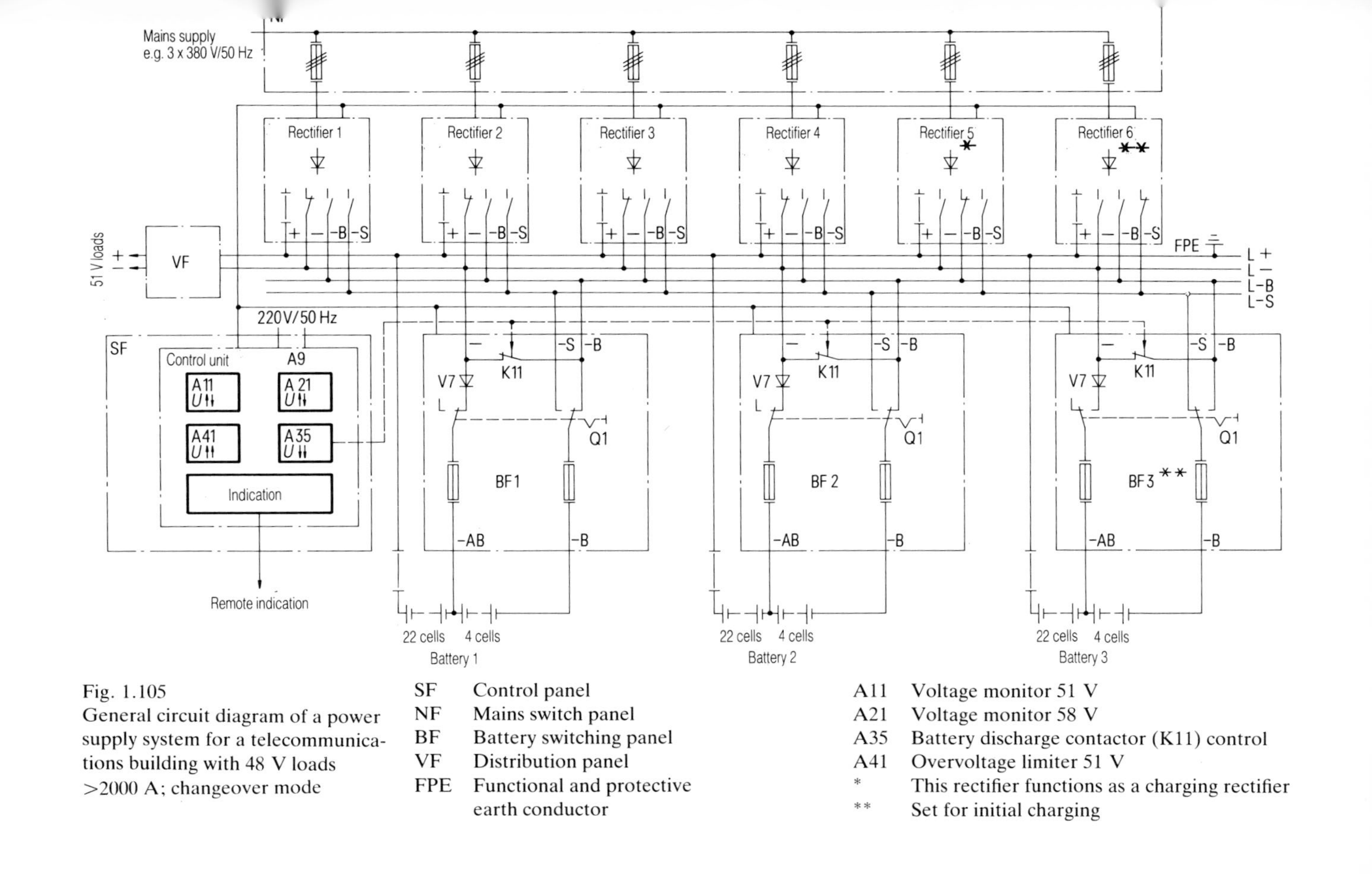

Fig. 1.105
General circuit diagram of a power supply system for a telecommunications building with 48 V loads >2000 A; changeover mode

SF	Control panel
NF	Mains switch panel
BF	Battery switching panel
VF	Distribution panel
FPE	Functional and protective earth conductor

A11	Voltage monitor 51 V
A21	Voltage monitor 58 V
A35	Battery discharge contactor (K11) control
A41	Overvoltage limiter 51 V
*	This rectifier functions as a charging rectifier
**	Set for initial charging

Fig. 1.106 Telecommunications tower

Fig. 1.107. The principal components are:

▷ standby generating set,
▷ rectifiers,
▷ lead-acid batteries (B1, B2).
▷ battery switching panel with control unit BS,
▷ d.c. distribution board with lightning protection assemblies.

For the sake of security, at least two rectifiers and two batteries are provided.

In normal operation power is obtained from the supply mains. In the event of a supply failure the rectifier switches off and the battery takes over the load without interruption. At the same time the standby generator is started automatically; it

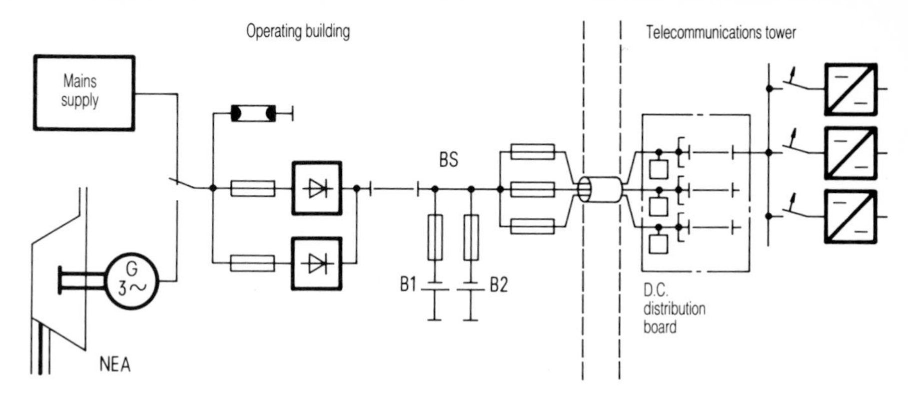

Fig. 1.107
Basic circuit of a power supply system for telecommunications towers (standby parallel mode)

provides power after a few seconds and the rectifiers switch on again. When the mains supply is restored the standby generator is shut down. The rectifiers supply the power to the loads as well as charging current for the battery.

The power is distributed by the battery switching panel BS through fuses. The battery switching panel is linked to the d.c. distribution board by a screened cable, which runs from the power installation building to the tower through a metal conduit. The distribution board, on the equipment floor of the telecommunications tower, carries lightning protection assemblies and also caters for the power distribution to the rack rows. Here the current is distributed to the individual items of transmission equipment from distributor units with automatic circuit breakers via d.c./d.c. converters.

1.5.1.2 Power supply system without battery, with immediate standby system

For telecommunication towers with a power requirement greater than 16 kW, or for those in which a.c.-supplied transmission equipment is still used, a variant of the power supply system is available (Fig. 1.108).

In this arrangement the d.c. power supply system is accommodated on the equipment floor of the telecommunications tower.

The main components are:

▷ immediate standby system (also referred to as a flywheel diesel converter set),
▷ a.c. distribution board,
▷ rectifiers,
▷ control and distribution panel,
▷ d.c. distribution board.

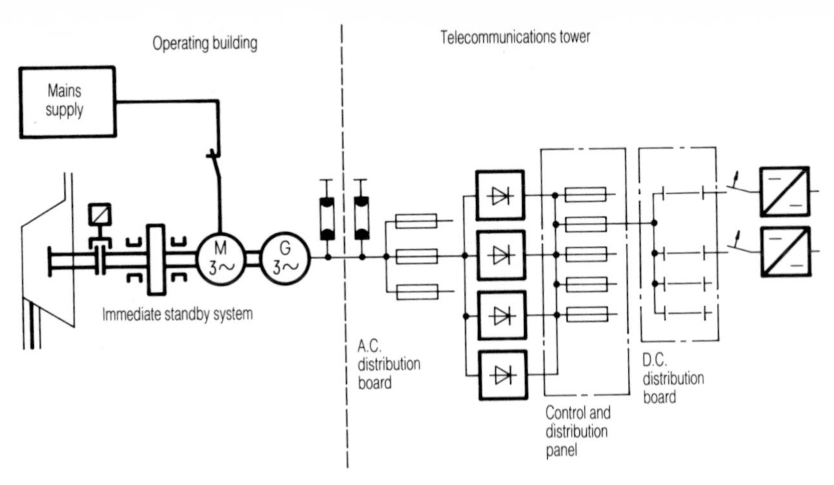

Fig. 1.108
Basic circuit of a power supply system for telecommunications towers (without battery, but with immediate standby system)

The immediate standby system serves as an a.c. power source. In normal operation a controlled three-phase a.c. supply is provided by a rotary converter set driven from the mains and provided with a flywheel. If the mains supply is interrupted the diesel engine is started; at the same time an electromagnetic clutch is actuated automatically, so that the diesel, brought up almost to rated speed by the flywheel, takes over the drive to the converter.

To guard against excessive interference voltages due to lightning strikes, cathode-drop arresters are fitted both in the a.c. switchboard (in the power equipment building) and in the a.c. distribution board (on the equipment floor of the tower). No special lightning protection measures are necessary on the d.c. side if the d.c. power installation and the d.c. loads are both on the equipment floor.

A.C. loads can be connected directly to the a.c. distribution board.

1.5.2 Power supply for ground communication stations for telecommunications satellites

Ground communication stations represent a special type of radio link installation. They are supplied with a.c. from a no-break system incorporating three-phase inverters (Figs 1.109 and 1.110). In normal operation the inverters draw their power through rectifiers from the supply mains and, in the event of a supply failure, from a battery connected to the d.c. link. To increase the reliability of the a.c. power supply the inverters are used in a half-load parallel arrangement or an $n + 1$ operation system. In addition a revert-to-mains unit (static transfer switch

1 D.C. input
2 Thyristor sets
3 Switch-on and switch-off units
4 Closed-loop and open-loop control
5 Output transformer

Fig. 1.109 125 kVA three-phase pulse inverter with doors open

STS) is provided, which on overload (a short-circuit, for example), or in the event of failure of both inverters, transfers the loads without a break to the mains.

As shown in Fig. 1.110, the a.c. mains supply is converted to d.c. by ten rectifiers connected in parallel on the d.c. side; each of the two groups of five rectifiers is fed through a 1000 A mains switching panel. The individual rectifiers are rated for an output current of 315 A at an output voltage of 245 V (trickle-charging voltage) or 256 V (charging voltage) in standby parallel mode.

The four inverter groups are supplied from a central battery switching panel. In parallel with them are connected two battery groups each of 110 cells of 6000 Ah capacity, which in total ensure four hours' operation during a mains supply failure.

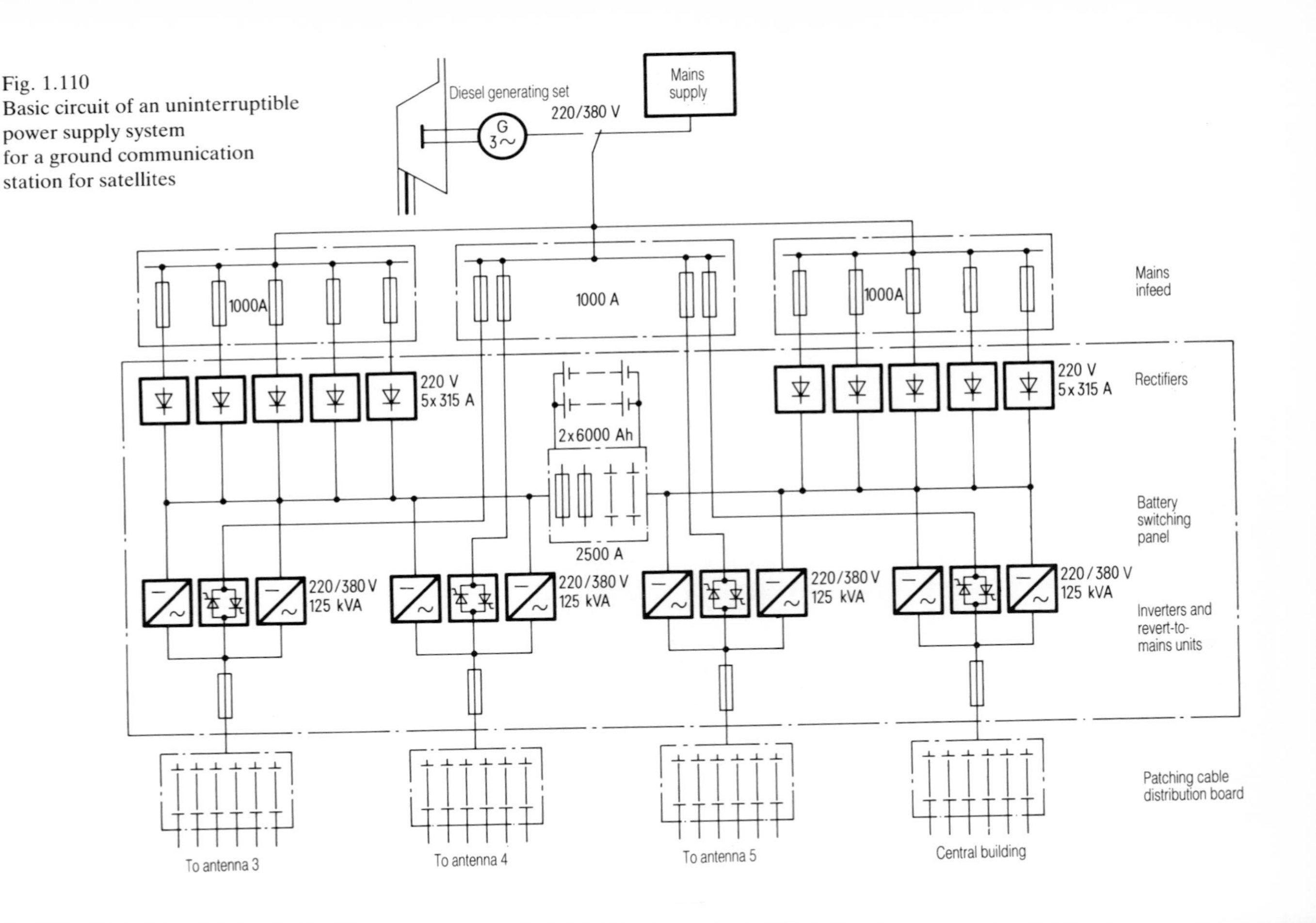

Fig. 1.110
Basic circuit of an uninterruptible power supply system for a ground communication station for satellites

Each of the load groups is fed from one inverter group; the individual inverters are rated at 125 kVA at 220/380 V. To provide passive redundancy, each pair of inverters has associated with it a revert-to-mains unit and is connected on the output side to a patching cable distribution board from which six power cables lead to the aerial installations 3, 4 and 5 and the central building. With this distribution arrangement, it is possible in the event of the failure of an inverter group to divide up its loads between the other inverters.

1.5.3 Mains-independent power supply systems

As mentioned by way of introduction, radio link stations and repeater stations, if they are located where no public mains supply is available, must be provided with mains-independent power supply systems (Fig. 1.111).

Fig. 1.111 Mains-independent power supply system in a shelter

In contrast to standby power supply systems, mains-independent power supplies in these circumstances signify small, continuously operating local 'power stations' (primary power sources).

1.5.3.1 Primary power sources

Primary power sources for use at operating locations are considered in the following paragraphs (Table 1.25).

Diesel generators

Continuously operating diesel generators

In systems employing diesel generators, two or more diesel sets operate in rotation on a continuous basis in conjunction with rectifiers and parallel-connected batteries.

Table 1.25 Technical features of primary power sources

Power source	Power range	Fuel	Availability	Maintenance cost	Other features
Diesel generator	≧2 kW	Diesel oil	Adequate	High	A.C. supply available for air-conditioning plant and lighting; rectifier and generally a battery necessary; building or shelter necessary
Small steam turbine	200 to 3000 W	Gas or diesel oil	High	Low with gas; somewhat high with diesel oil	D.C output – a.c. only through an inverter; in a dual system only 2 h max. battery standby time required; no building required
Thermogenerator	10 to 400 W	Gas, radioisotopes (diesel oil on trial)	High	Low	D.C. output; no building required; battery not always necessary
Solar generator	≦1000 W	None required	Dependent on duration and intensity of radiation	Low	D.C. output; supporting structure required; large-capacity battery necessary
Wind generator	approx. 10 kW	None required	Dependent on local wind conditions	Low	D.C. or a.c. output; mast required; battery necessary

Particular features of this mode of operation are:

- ▷ maintenance is necessary at weekly intervals, or every two or three weeks, since each generator set (without special provision) can only continue in operation for about a week,
- ▷ an a.c. supply is available.

Diesel generators with battery mode

The diesel generators charge the batteries alternately in a relatively short time. The diesels run for only a few hours a day, while the loads are supplied continuously from the battery (Fig. 1.112).

Particular aspects of this mode of operation are:

- ▷ the maintenance intervals are extended to months. The battery capacity has to be significantly larger than in the systems mentioned previously,
- ▷ the diesel fuel consumption is relatively low,
- ▷ a.c. supplies are available if required.

Steam turbines

The low-power steam turbine operates with a closed steam circuit and can be fuelled with liquefied gas, natural gas, diesel oil or kerosene.

The turbines are used either individually with batteries or in the parallel mode with or without batteries.

Thermoelectric generators

Thermoelectric generators comprise a large number of thermoelectric elements which convert heat into electrical energy; they thus contain no moving parts. The greater the temperature difference between the hot and cold sides, the higher is the efficiency. The output power of thermoelectric generators is strongly dependent upon the ambient temperature.

A number of units can be connected in parallel to form a larger system and can be used with or without a battery.

Solar generators

The solar generator converts solar energy directly into electrical energy. It consists of a number of solar panels, each of which contains a large number of solar cells; these are protected from mechanical damage on the light-sensitive side by special high-transmission glass with a low-reflection surface. The cells are encapsulated in UV-resistant PVB (polyvinylbutyral). On the back of the panel, a moisture barrier is provided by a plastic-laminated aluminium sheet, which at the same time assists heat dissipation. Optimum forms of solar panels and generators

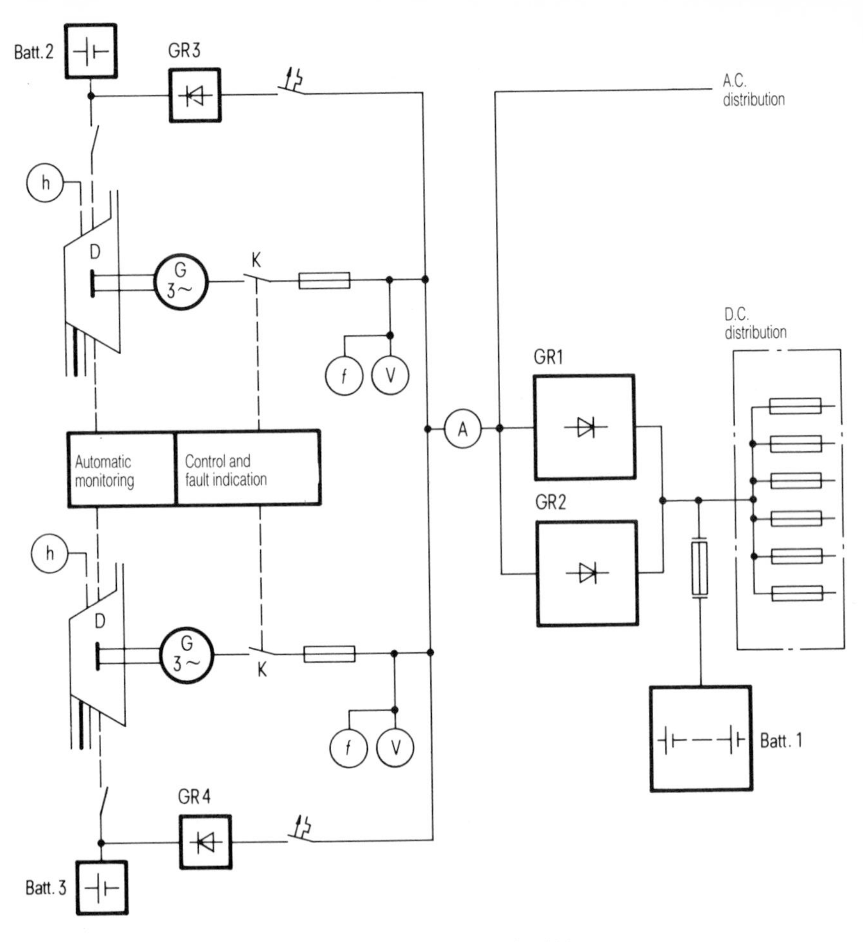

h	Operating hours counter	Batt. 1	Lead-acid battery
K	Generator contactor	Batt. 2, 3	Starter batteries
		GR1, 2	Rectifiers
		GR3, 4	Charging rectifiers for starter batteries

Fig. 1.112
Basic circuit of a mains-independent a.c. and d.c. power supply system with two diesel generators, rectifiers and battery

for particular requirements can be assembled by connecting cells in series and parallel.

Solar panels are assembled, depending upon the shape, with up to 144 mono-crystalline silicon solar cells of 100 mm diameter. With a panel area

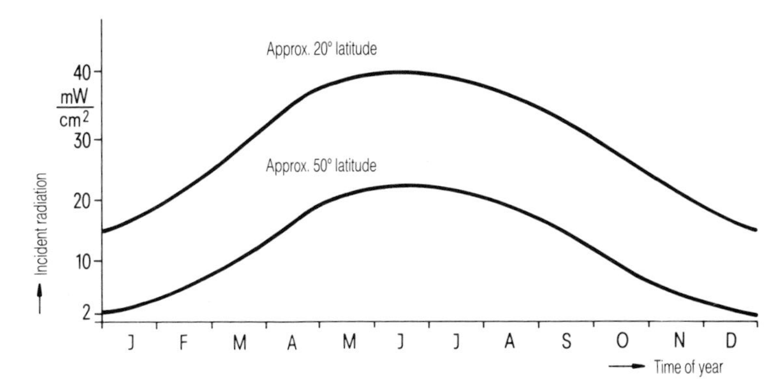

Fig. 1.113
Examples of average yearly variation of solar radiation intensity

1462 mm × 1009 mm, at the present time, a power of up to 120 W can be obtained at a solar radiation intensity of 100 mW/cm^2.

A battery must be provided to cater for the hours of darkness and for sunless days.

In the present state of the technology, it is possible to use solar generators for stations with relatively low power requirements in latitudes with average-to-good sunshine (Fig. 1.113).

Wind-generator systems

The wind generator consists of a generator coupled to a rotor or a turbine. There are several forms of construction:

- ▷ the *Savonius rotor* is notable for robust construction and easy starting. It has a vertical shaft and its operation is not dependent upon the direction of the wind; the power obtainable is small,
- ▷ the *Darrieux turbine* is simple in construction, has a vertical shaft, operates independently of wind direction and has a high efficiency. It is not self-starting,
- ▷ *horizontal-shaft turbines* are used in considerable numbers. Because of the experience gained with them, this pattern is today the most highly developed. The turbines are characterized by high efficiency.

A wind-generator installation typically consists of a rotor of 4 m (6 m) diameter mounted on the lee-side of a mast and a multi-pole synchronous generator.

Because of the large number of poles, gearing is not necessary; this is beneficial from the point of view of noise and wear in operation.

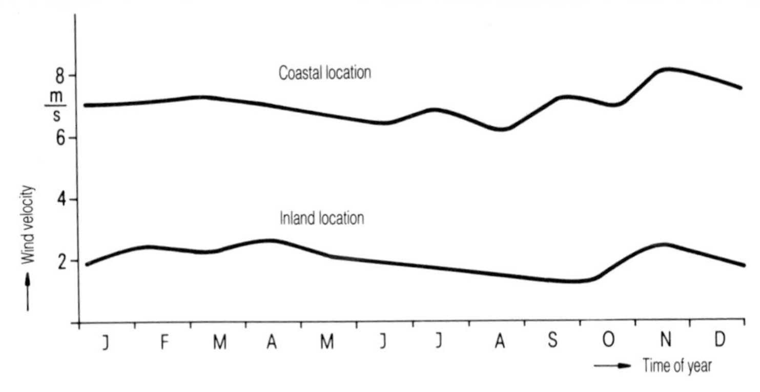

Fig. 1.114
Examples of average yearly variation of wind velocity

The a.c. power taken from the slip-rings is fed through a cable within the mast to the shelter, where it is rectified, controlled and passed to the load and the parallel-connected batteries.

The supply of power begins at a wind velocity of 3.5 m/s (13 km/h), and the maximum power of 2 kW (4 kW) is produced with a wind velocity of 8 m/s.

To protect the plant against excessive rotational speed, the speed is limited by centrifugal weights attached to the rotor blades, which modify the angle of incidence of the blades.

A vane mounted on the generator housing keeps the system in the right orientation relative to the wind.

Favourable locations for wind generators are mountain ridges, open plains and coastlines or funnel-shaped valleys (Fig. 1.114).

1.5.3.2 Hybrid power supply systems

Hybrid systems, consisting, for example, of a wind generator, solar generator, diesel generator and battery, offer exceptional reliability combined with low maintenance requirements (Figs 1.115 and 1.116).

If the battery is discharged, a small diesel generator set starts automatically, supplies the load and recharges the battery. The capacity of the battery is so chosen that the diesel generator is rarely called upon to supply the power requirements of the station. In this way long maintenance intervals can be achieved.

Existing diesel generator plants can be supplemented by wind generators and solar generators as a means of reducing the fuel consumption and maintenance costs.

Fig. 1.115
Wind-generator installation and solar generator for a mains-independent hybrid power supply system

The communications equipment, closed-loop and open-loop control (Fig. 1.117), rectifier, diesel generator and battery, are housed in shelters.

A natural-convection cooling system enables the temperature in the shelters to be retricted to a suitable level, even in hot districts.

The principle of natural-convection cooling depends upon the different densities of a liquid at different temperatures; the density of a liquid is reduced by heating and increased by cooling.

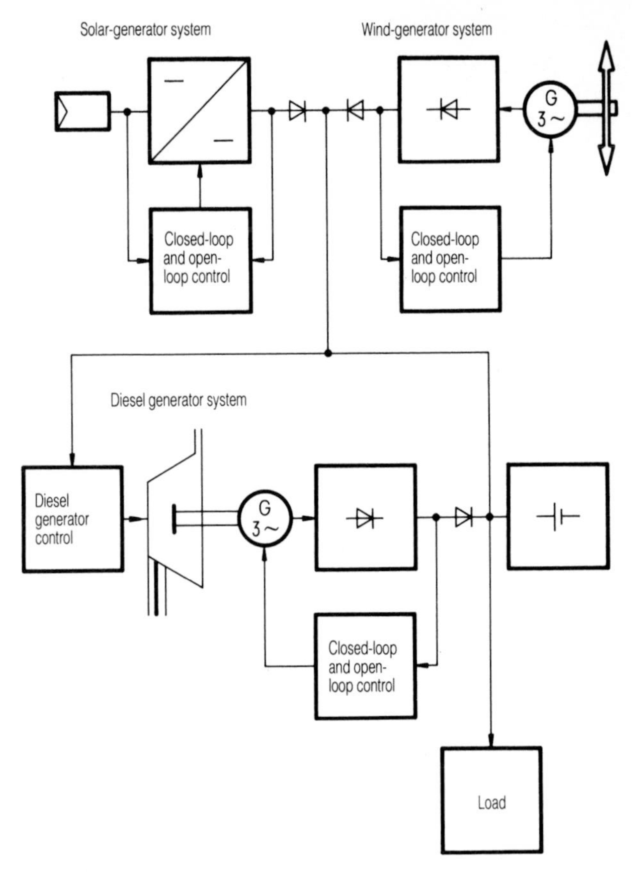

Fig. 1.116
Block diagram of a mains-independent hybrid power supply system

The cooling fluid in the system absorbs the heat generated by the radio link equipment in the *internal* heat exchanger; the heated fluid flows upwards to a tank, from which the cooler fluid descends to the heat exchanger. During the day, so long as the external temperature is higher than that of the tank, no cooling fluid can flow to the *external* heat exchanger. Only when the external temperature falls below the temperature of the tank at night does the fluid ascend to the external heat exchanger, where it is cooled and then descends again. The cooling system thus operates entirely by virtue of the losses in the shelter, and requires no additional power.

Fig. 1.117
Closed-loop and open-loop control (solar regulator) in a shelter

2 Decentralized Power Supply Systems

At this point examples of decentralized power supply equipment will be described. All equipment are proof against overload and short-circuits.

2.1 Rectifiers with Transistor Power Sections

2.1.1 Equipments with longitudinal controllers

2.1.1.1 Type 24 V/6 A GR1

The 24 V/6 A GR1 rectifier (Fig. 2.1) embodies the principle of the linear longitudinal controller (series regulator).

Application

The GR1 equipment (wall-mounting) is used to supply 24 V communications systems such as:

▷ branch (PABX) systems,
▷ control installations,
▷ monitoring installations,
▷ clock installations.

Fig. 2.1
Rectifier type 24 V/6 A GR1, door open

Three equipments can be connected in parallel, so that a current demand for the communications system and the battery of up to 18 A can be catered for.

Operating modes

The possible operating modes are the rectifier mode and the parallel mode (usually the standby parallel mode) without voltage-reducing diodes.

Basic circuit and operation

The *basic circuit* (Fig. 2.2) can be divided as follows:

▷ power section and
▷ closed-loop and open-loop control A1.

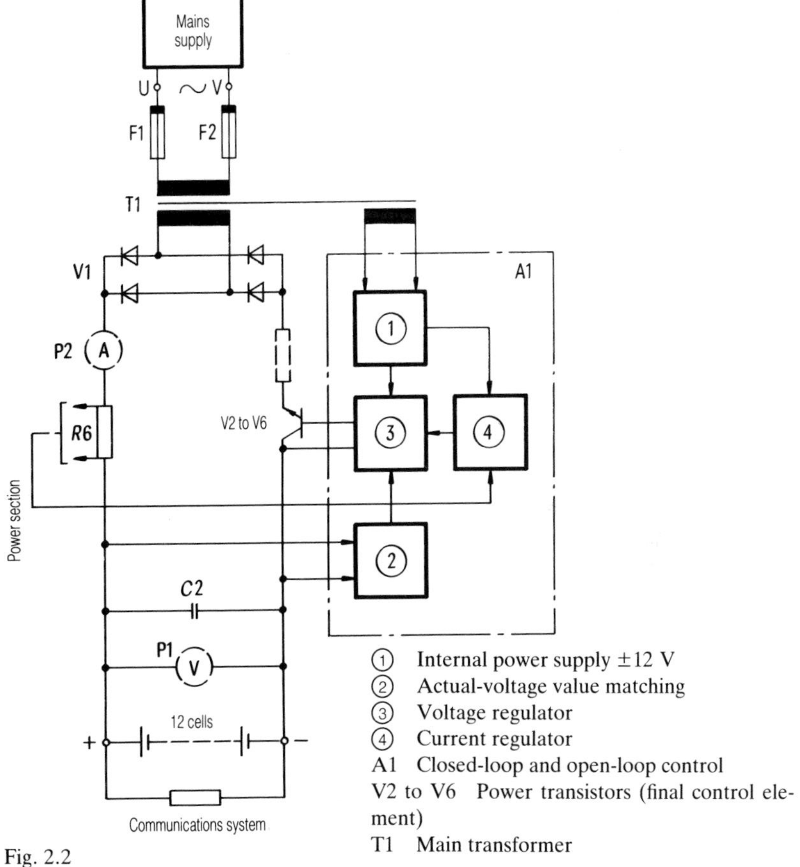

Fig. 2.2
Basic circuit of rectifier type 24 V/6 A GR1

Power section

The mains supply voltage is transformed down by the transformer T1 and rectified by the bridge V1. To this is connected the transistor final control element block (power transistors V2 to V6). These transistors are controlled by the closed-loop and open-loop control assembly A1; the d.c. output voltage U_A is held constant despite disturbing factors by virtue of the variable voltage drop across the transistor V2 to V6.

Closed-loop and open-loop control A1

The closed-loop and open-loop control assembly A1 embraces the functional units shown in Fig. 2.2.

The *internal power supply* +12 V ① supplies the constant set-point voltage value (reference U_{Soll}) to the voltage regulator (controller). The actual voltage value (U_{Ist}) is sensed at the output of the rectifier and passed to the voltage regulator via the *actual value matching* ② (which also includes the voltage characteristic changeover). The *voltage regulator (controller)* ③, consisting of the operational amplifier V10 with associated circuits (see Fig. 2.3), compares the reference voltage value with the actual voltage value, and from the difference between the two quantities the control signal for the final control element is produced.

The *current regulator (controller)* ④ provides for current limiting, intervening in the control circuit to avoid overloading of the rectifier.

The current regulator–operational amplifier V9 with associated circuits (see Fig. 2.3) – compares the actual current signal (I_{Ist}) from the measurement shunt R6 in the power section with the constant current reference (I_{Soll}) from the power supply. If the first value is greater than the second, the current regulator reduces the d.c. output voltage through the voltage regulator and the final control element.

In the *circuit diagram* (Fig. 2.3), the main circuit is through the fuses F1 and F2, the main switch S1, transformer T1, bridge circuit V1, smoothing capacitor C1, transistors V2 to V6, measurement shunt R6 to the output terminals N and P.

▷

① Internal power supply ±12 V
② Actual-voltage value matching
③ Voltage regulator
④ Current regulator
A1 Closed-loop and open-loop control
V2 to V6 Power transistors (final control element)
T1 Main transformer

Fig. 2.3 Circuit of rectifier type 24 V/6 A GR1

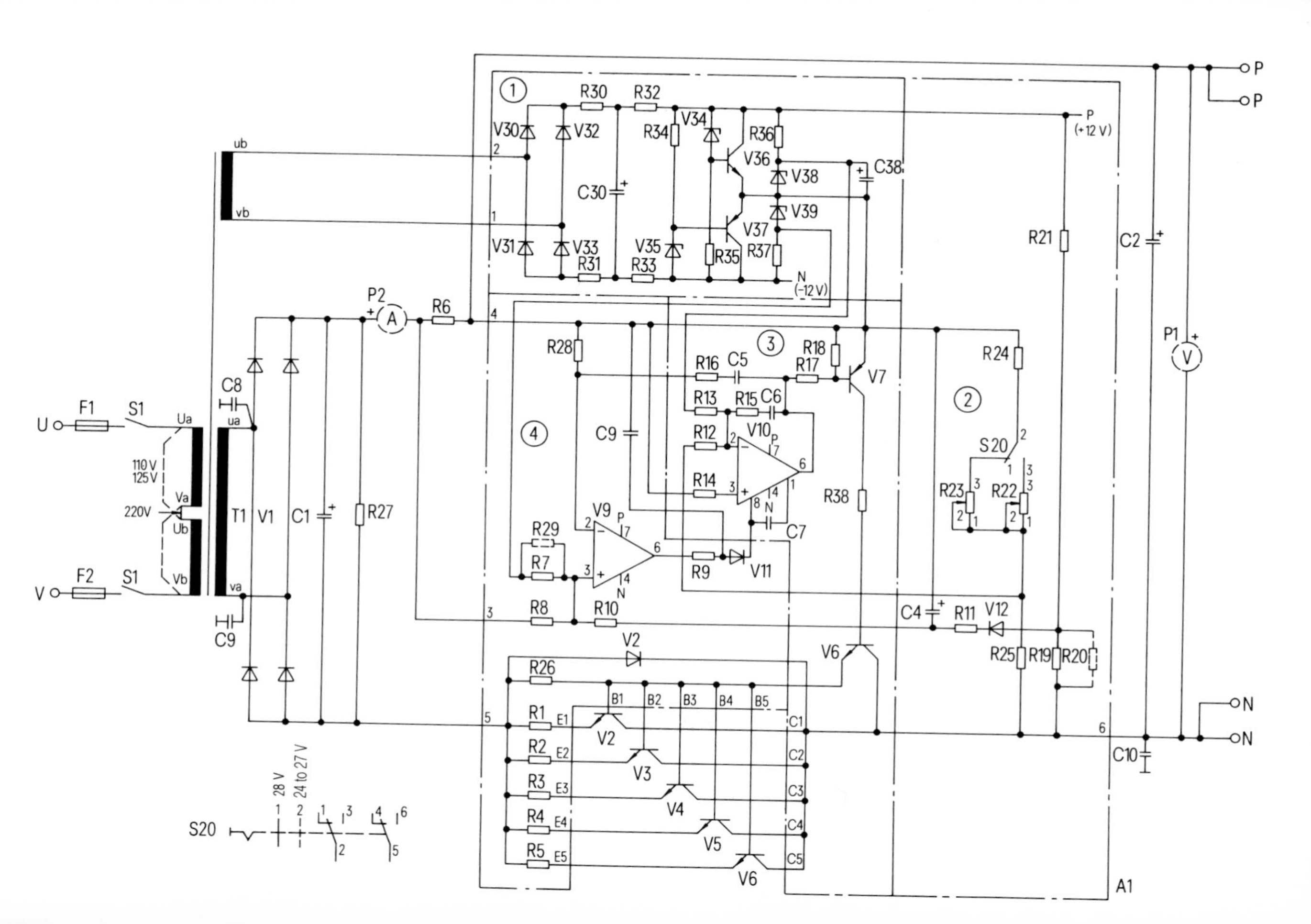
P
P
P (+12 V)
N (−12 V)
N
N
R30
R32
V30
V32
R34
V34
R36
V36
V38
C38
ub
vb
C30
V39
V37
V31
V33
V35
R35
R37
R31
R33
R21
C2
P2
A
R6
R28
R16
C5
R18
R17
V7
R24
P1
V
R13
R15
C6
U
F1
S1
Ua
ua
C8
V10
C9
R12
S20
110 V
125 V
Va
220V
Ub
T1
V1
C1
R27
R14
R23
R22
V9
R29
R7
R9
V11
R38
C7
V
F2
S1
Vb
va
R8
R10
C4
R11
V12
C9
V2
V6
R25
R19
R20
R26
B1
B2
B3
B4
B5
R1
E1
V2
C1
R2
E2
V3
C2
C10
R3
E3
V4
C3
R4
E4
V5
C4
R5
E5
V6
C5
A1
28 V
24 to 27 V
S20

① consists of an auxiliary winding of transformer T1, a bridge circuit (V30 to V33), a smoothing circuit (R30, R31, C30) and a voltage stabilizer circuit (R32, R33, R34, R35, V34, V35, V36 and V37).

The positive voltage reference value is obtained from resistor R36 and the zener diode V38; it increases slowly at switch-on due to the effect of the capacitor C38.

The negative actual voltage value is matched by means of the voltage dividers R24, R23 (interchangeable with R22 through the switch S20) and R25 across the output terminals ②.

The voltage at the inverting input 2 of the operational amplifier V10 ③ is positive when U_{Ist} (on R12) is less than U_{Soll} (on R13). The output voltage of the operational amplifier is then driven in a negative direction, and the transistor V7, and similarly V6 (on A1), conduct to a greater extent (i.e. offer a lower resistance). When the voltage drop across the transistors V2 to V6 (in the power section) decreases, the output voltage of the equipment increases until the level set on the potentiometer R22 or R23 is reached.

The negative current reference value is fed from resistor R37 and the zener diode V39 through resistors R7 and R29 to the input of the operational amplifier V9. The value of resistor R29 is so selected that current-limiting occurs at the required level.

The positive actual current signal is taken from the measurement shunt R6 and passes through R8 to the operational amplifier V9 ④. The output of V9 is connected to the limiting input of the operational amplifier V10, resulting in an 'overriding' control.

If the load current is less than the value set with R29, the voltage at the non-inverting input 3 of the operational amplifier V9 is negative. The output voltage of V9 is similarly negative, diode V11 is on reverse and the current regulator is isolated from the control circuit.

If the load current exceeds the level corresponding to the adjustment of R29, the voltage at the non-inverting input of the operational amplifier V9 becomes positive and its output rises. When it equals the output voltage of the voltage regulator, the diode V11 conducts and the output voltages of the operational amplifiers V9 and V10 become more positive than before. The transistors V7 and V6 (A1) conduct less (offer a higher resistance), the voltage drop across transistors V2 to V6 (power section) increases and the output voltage of the equipment falls, so that the load current is restricted to the present level.

If the equipment output voltage falls below 21 V due to current-limiting, an additional variable positive actual value is added to the fixed negative reference current value. The additional actual value is produced by the arrangement comprising R19, R20, R21 and diode V12, and is applied through R11 and R10 to the non-inverting input of the operational amplifier V9.

The voltage divider R21, R19 is so adjusted, by means of R20, that the voltage on the anode of the diode V12 reaches 0.5 V when the equipment output voltage is about +21 V. When the output voltage drops below 21 V, the diode V12 conducts and the negative current reference value is reduced as a result of the positive actual value addition. This reduces the output voltage of the equipment. With constant loading the output current falls.

With the output short-circuited the output voltage of the equipment is approximately 0 V (the current about 1 A). The purpose of resistor R18 is to protect the transistor V7.

For the purposes of radio interference suppression, capacitors C8 and C9 are provided on the secondary side of the transformer T1 and C10 at the output of the equipment.

Technical data

The principal technical data for the 24 V/6 A rectifier are listed in Table 2.1.

Table 2.1 Technical data for rectifier type 24 V/6 A GR1

Mains input supply		
Voltage	(V)	220 +10 to −15% or 110 +25 to −15%
Frequency	(Hz)	50 or 60 ±2%
Equipment fuse protection	(A)	6
Degree of radio interference		N (VDE 0875)
D.C. output		
Operating mode or condition		Rated direct voltage (V) (tolerance for 1.5%–100% I_{Nenn}) Equipment voltage for lead–acid battery of 12 cells
Rectifier operation		24 ±1%
Parallel operation/trickle charging (2.23 V/cell)		26.8 ±1%
Parallel operation/charging (2.33 V/cell)		28 ±1%
Initial charging		32.5
Rated direct current	(A)	6
Interference voltage	(mV)	≦1 (frequency-weighted with CCITT 'A' filter)
Dimensions ($H \times W \times D$)	(mm)	Wall-mounting: 446 × 338 × 230

2.1.2 Equipments with switching controllers

2.1.2.1 Type 150 W

The rectifier unit type 150 W (Fig. 2.4) employs the switching controller principle (30 kHz, pulse-width control).

Application

The equipment (in assemblies construction) is used as the power supply for the EMS 20 and EMS 30 telephone systems.

If the telephone system is equipped with additional power items the power requirement is increased to 220 W. For this purpose an extension of the rectifier unit is possible (with rectifier 70 W). There are also special equipments and plug-in units for particular purposes, such as floppy-disk drives.

Operating mode

The 150 W rectifier is used in the rectifier mode.

In the event of a mains supply failure, one or more exchange lines in EMS telephone system are switched to selected branches (exchange line transfer), so that a limited telephone service is provided. Communication from branch to branch – without exchange facilities – is, however, not possible.

If it is required to keep the telephone service in operation without restriction during a supply interruption or other disruption, a reserve of energy – preferably a lead–acid battery – can be kept in readiness. The standby parallel mode is then instituted. For this purpose the switching-mode power supply described in

Fig. 2.4 Rectifier type 150 W

Section 2.1.3.2, for example, may be used in conjunction with the 80 W d.c./d.c. converter (Section 2.2.3). The d.c./d.c. converter is similar in construction to the rectifier discussed in this section; the difference is that the mains input assembly NEBG is replaced by a battery input assembly BEBG.

Block diagrams, basic circuit diagrams and operation

The 150 W rectifier consists of a *mains input assembly* NEBG (150 W) and a number of *output assemblies* ABG (Fig. 2.5; see Fig. 2.6 and 2.8).

Mains input assembly

The 150 W rectifier is arranged as an equipment with protectively insulated parts (see Chapter 5) in accordance with VDE 0804. Those parts which are connected to mains voltage have, in addition to functional insulation, protective insulation from the secondary circuit (the communications system). Varistors and surge arresters are provided at the mains input as a protection against surge voltages

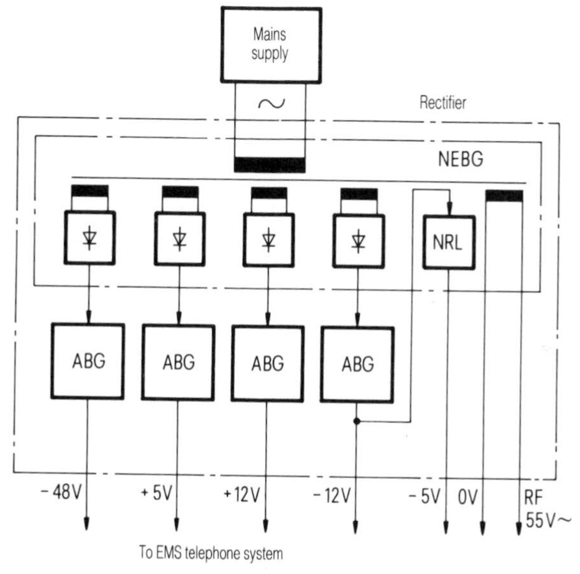

NEBG Mains input assembly
ABG Output assembly
RF A.C. ringing supply
NRL Adjustment controller

Fig. 2.5 Block diagram of rectifier type 150 W

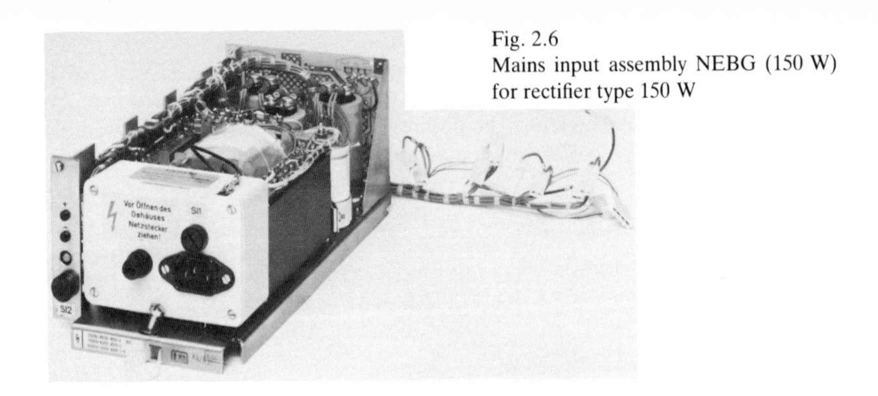

Fig. 2.6
Mains input assembly NEBG (150 W) for rectifier type 150 W

(e.g. atmospheric discharges). This necessitates a connection from the equipment to the earth conductor (PE) of the mains supply.

The mains voltage is transformed down in the mains input assembly NEBG (150 W) by the mains transformer (50 Hz) (Fig. 2.6). The transformer can be matched to the available mains voltage (110/115/120/127/**220**/240/250 V). The equipment is supplied as for 220 V input.

On the secondary side the transformer supplies, from isolated windings, four alternating voltages, which, after rectification by bridge circuits, constitute the d.c. input supplies for the associated output assemblies ABG. The transformer also provides auxiliary supplies of 2 × 25 V centre-tapped for the output assemblies and the a.c. ringing supply RF of 55 V.

The –5 V/10 mA *'adjustment controller'* NRL is accommodated in the mains input assembly. It produces from the constant –12 V supply obtained from the output assembly ABG a controlled supply of –5 V (Fig. 2.7). The –5 V supply is used as one of the supplies for the telephone system. The adjustment controller functions as a continuous longitudinal controller and consists of an integrated –5 V negative voltage regulator I1, a zener diode, resistors and electrolytic capacitors. A light-emitting diode (LED) indicates the 'on' condition.

Output assemblies

From the smoothed direct voltage supplied by the mains input assembly NEBG each output assembly ABG (Fig. 2.8) provides a separately controlled and monitored d.c. output voltage.

Four output assemblies are used in the rectifier:

▷ ABG –12 V/1.5 A,
▷ ABG +5 V/5.0 A,
▷ ABG +12 V/3.5 A,
▷ ABG –48 V/1.5 A.

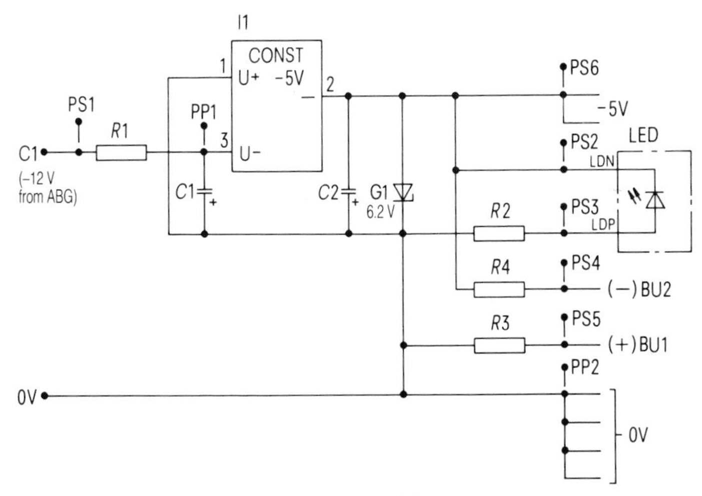

Fig. 2.7 Adjustment controller NRL – 5 V/10 mA

These assemblies (with the exception of ABG –48 V) function on the 30 kHz switching controller principle.

The –48 V output assembly has two operating ranges. It functions either as an electronic filter or as a continuous longitudinal controller, depending on the d.c. input voltage. If the d.c. input voltage is less than 60 V (with a low supply voltage to the mains input assembly), the d.c. output voltage varies in proportion to the mean value of the input voltage. This means that the output assembly, operating under this condition in an *uncontrolled* mode, functions only to keep the *interference voltage* to *less than 0.5 mV* at the output (i.e. as an electronic filter). In this mode the losses are low.

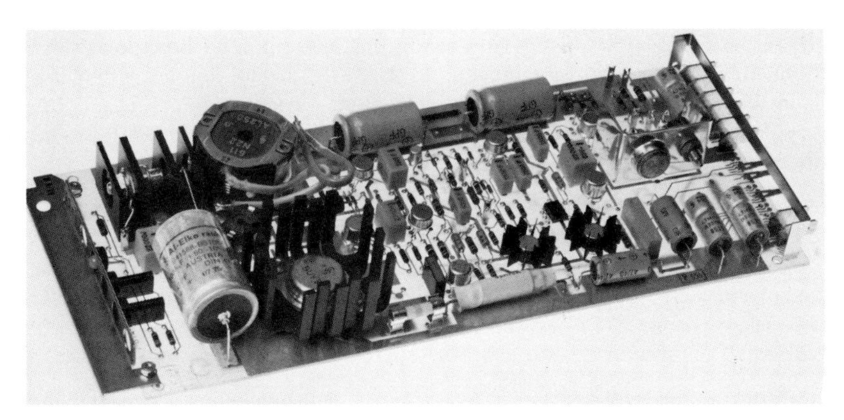

Fig. 2.8 Output assembly ABG

If the d.c. input voltage rises above 60 V (due to a higher supply voltage at the input to the mains input assembly) the output assembly changes over to the 'continuous longitudinal controller' mode; it then controls the d.c. output voltage to a constant level of about 54 V regardless of d.c. input voltage variations.

The output assemblies are provided with reverse battery protection. They are also short-circuit-proof by virtue of a fold-back current-limiting characteristic.

Figure 2.9 shows the *block diagram* of the *output assembly* ABG –12 V/1.5 A, which may be taken as an example of the operation of the 30 kHz switching controller assemblies.

The d.c. supply voltage of 21 to 40 V is fed from the mains input assembly to the power section of the output assembly, and there applied via radio electromagnetic interference (EMI) suppression components to the 30 kHz switching transistor T4 (final control element), which receives its control signal from the '30 kHz switching transistor T4 control', located within the 'closed-loop, open-loop control and monitoring' section. The current measurement shunt downstream of the switching transistor provides the actual-current signal, which is used in the current limiter.

In the event of an overvoltage the output voltage monitor triggers the limiting thyristor, which short-circuits the output and thereby prevents a further unacceptable increase in the d.c. output voltage. A light-emitting diode (LED) indicates operation. A constant part supply voltage of –12 V is provided at the output for the telephone system EMS.

Figure 2.10 shows the *basic circuit* of the power section of the output assembly ABG. The switching transistor T4 is switched on and off at a frequency of 30 kHz by the '30 kHz switching transistor T4 control' via the coupling transformer winding 3/4. This results in a 30 kHz rectangular-wave voltage at the emitter of transistor T4.

When the switching transistor T4 is conducting, current flows from point CPO through the EMI suppression inductor L1/fuse Si1/collector-emitter path of T4/ inductor L2/load and back through the current-measuring resistors R53 and R52 to point CNE. During this phase a magnetic field builds up in the inductor L2 and the smoothing capacitor C17 is charged.

When the switching transistor T4 is non-conducting, the energy in the inductor L2 decreases and effects a flow of current in the same direction through the load as during the conducting phase of the switching transistor T4, in the path consisting of inductor L2/load/resistors R53 and R52/free-wheel diode G8/back to inductor L2. The capacitor C17 is now partly discharging, so that it supports the flow of current. The inductor L2 also serves to suppress electromagnetic interference.

If the d.c. output voltage changes (e.g. due to a changing current demand in the telephone system), the conduction period in each cycle of operation of the

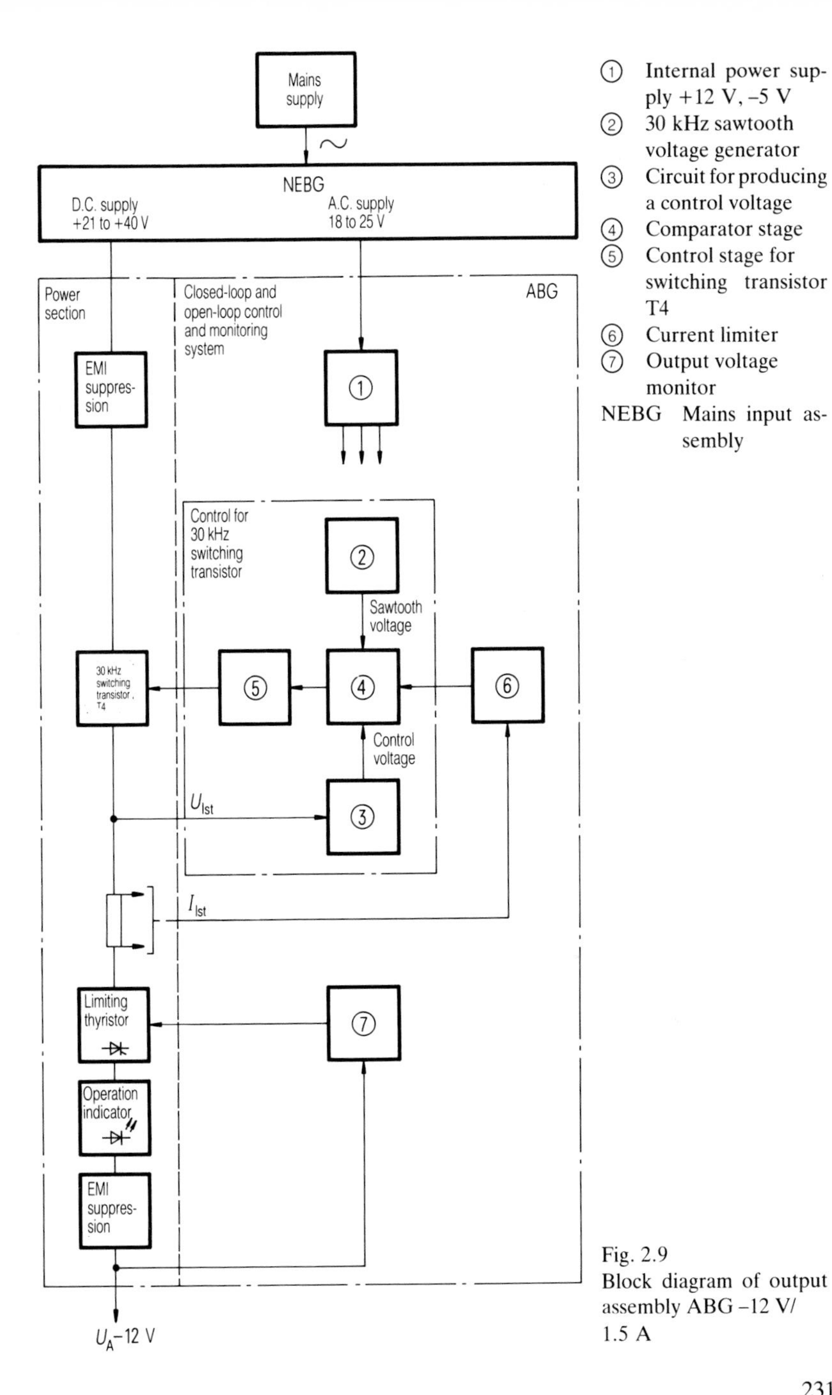

① Internal power supply +12 V, –5 V
② 30 kHz sawtooth voltage generator
③ Circuit for producing a control voltage
④ Comparator stage
⑤ Control stage for switching transistor T4
⑥ Current limiter
⑦ Output voltage monitor

NEBG Mains input assembly

Fig. 2.9
Block diagram of output assembly ABG –12 V/ 1.5 A

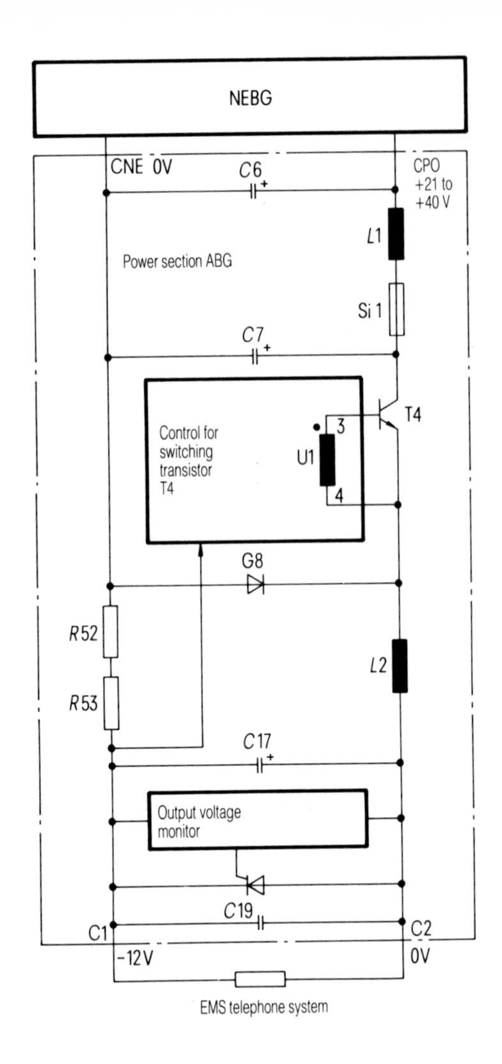

Fig. 2.10
Basic circuit of power section of output assembly ABG–12 V/1.5 A

switching transistor T4 (mark/space ratio) is changed by the '30 kHz switching transistor T4 control' to such effect that the d.c. output voltage regains its original value.

The voltage waveforms at the emitter of the 30 kHz switching transistor T4 and on capacitor C19 are shown in Fig. 2.11. The mark/space ratio is variable (pulse-width control) and the cycle period T (frequency) constant.

Capacitors C6, C17 and C19 are included for EMI suppression (Fig. 2.10).

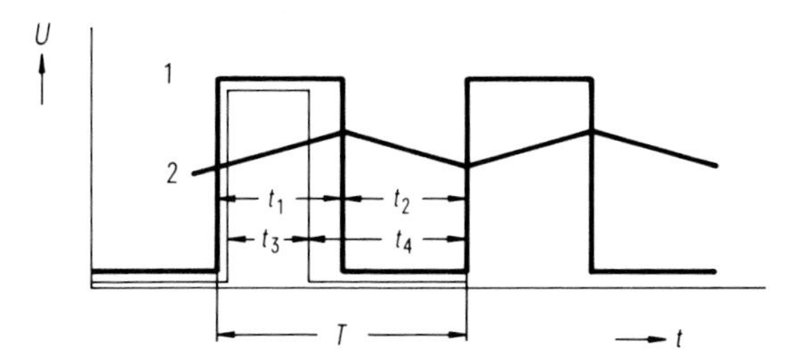

T Cycle period approximately 33 μs (30 kHz clock frequency)
$T = t_1 + t_2$
$T = t_3 + t_4$
1 Voltage on emitter of switching transistor T4
2 Voltage on output capacitor C19

Fig. 2.11
Waveforms of voltage on switching transistor T4 and capacitor C19

If a fault occurs in the output assembly which causes an unacceptable increase in the output voltage, the thyristor is triggered and short-circuits the output.

The current-measuring resistors R52 and R53 are for the purpose of providing the actual current signal.

The operation of the closed-loop, open-loop control (regulation, control) and monitoring sections of the output assembly will be explained with reference to the block diagram of Fig. 2.9. The a.c. supply voltage of 18 to 25 V is obtained from the mains input assembly NEBG and fed to the *internal power supply* ①.

The *30 kHz switching transistor T4 control* can be divided into the following functional sections:

② 30 kHz sawtooth voltage generator,
③ a circuit producing a control voltage, from the difference between the output voltage feedback (actual voltage value) and the set-point (reference) voltage value,
④ comparator stage (comparing the 30 kHz sawtooth voltage with the control voltage and including auxiliary voltage monitoring),
⑤ control stage for the switching transistor T4.

The *current limiter* ⑥ is provided to prevent the load current from exceeding the maximum permissible level. If the voltage drop across the current-measuring resistor, proportional to the load current, reaches a predetermined value, the conduction period of the switching transistor T4 is reduced by the comparator and control stages, so that the d.c. output voltage is reduced, with the result that the output current is limited. The output assembly ABG has a 'fold-back'' current-limiting characteristic; i.e., the current that flows into a short-circuit (the short-circuit current) is less than the current at the onset of current limiting (see Fig. 2.20).

The *output voltage monitor* ⑦ has been described previously in connection with the voltage-limiting thyristor (power section).

Described below are the details of the functional units indicated in the block diagram (Fig. 2.9) as closed-loop and open-loop control and monitoring, and their combined operation.

① *Internal power supply +12 V and –5 V* The auxiliary supply of +12 V is provided by a continuous longitudinal controller consisting of the integrated positive voltage regulator I1, a transistor T1, resistors R1, R2, R3 and R41 and a capacitor C3 (Fig. 2.12). The direct voltage at the test point PP63 (25 to 35 V) is applied to pin 8 of the regulator I1. Pin 5 of I1 is connected to input CH2 (reference potential 0 V). A voltage tapped off from the voltage divider R3/R41 is applied to I1 at pin 2 as an actual-voltage signal. I1 compares this with an internal reference voltage (set-point value) and controls the transistor T1 via pin 7. As a result, a constant direct voltage of +12 V, measured with respect to point CH2, is maintained at the collector of transistor T1 (point PP55) regardless of variations in input voltage and loading.

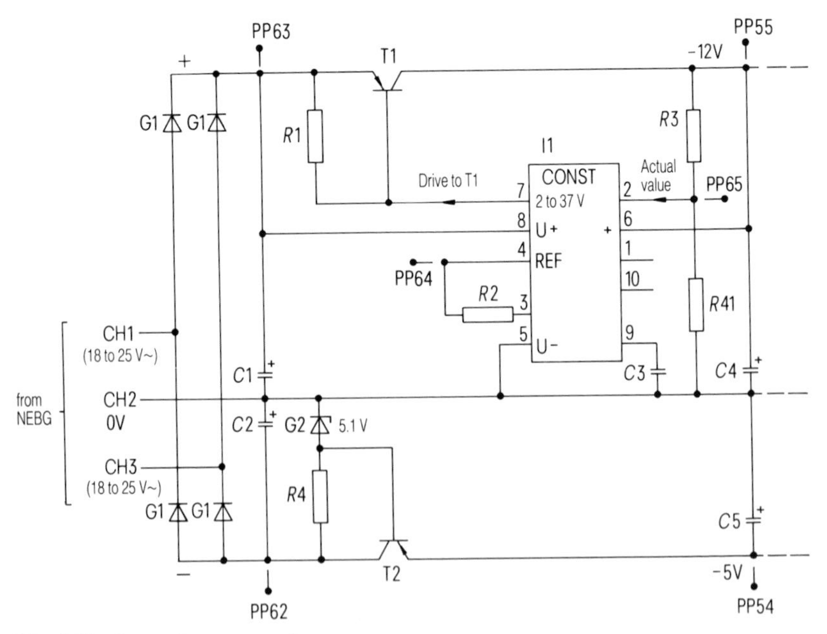

Fig. 2.12 Internal power supply

A stabilized voltage of approximately –5 V relative to point CH2 is available at point PP54.

② *30 kHz sawtooth voltage generator* A 30 kHz square-wave voltage (about +12 V high, –5 V low) is first produced by means of the integrated switching circuit I3 and the voltage of +12 V on R15, and applied through resistor R50 to the diode G5 (Fig. 2.13).

When there is no low signal (–5 V) at pin 7 of I3, a high signal (+12 V) is applied through resistors R15 and R50 to the cathode of diode G5, which is consequently non-conducting. A high signal can then pass through resistor R13 to the base of transistor T6. Since the base of T6 is now positive with respect to its emitter, the transistor conducts (Fig. 2.14, instant t_1) and the capacitor C8 receives a rapid initial charge through resistors R13 and R12 to a fixed level (interval t_1 to t_2). When this level is reached (instant t_2), transistor T6 turns off and the capacitor C8 is charged with a longer time constant through R10 towards +12 V (up to about +7 V) (sawtooth ramp: interval t_2 to t_3).

During the low period (–5 V) of the square-wave voltage on pin 7 (I3), the capacitor C8 discharges through diode G6, resistor R50 and the internal circuit of I3 to about –3 V.

Transistor T6 remains non-conducting during this period (t_3 to t_4), since its base is negative relative to its emitter. The alternation of the high and low signals, the conducting and non-conducting states of transistor T6 and the charging and discharging of capacitor C8 result in a 30 kHz sawtooth voltage waveform fed to the comparator stage I2 (test point PP24).

③ *Control voltage source* The actual-voltage feedback signal taken from the output voltage via the voltage divider R38/R39 is applied to the integrated circuit

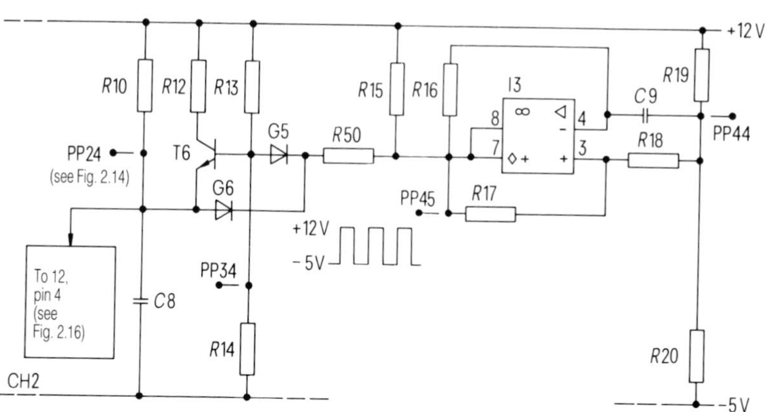

Fig. 2.13 30 kHz sawtooth voltage generator

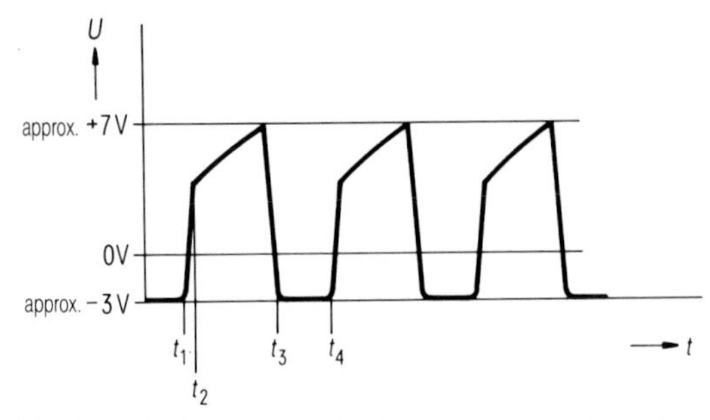

Fig. 2.14 30 kHz sawtooth voltage at emitter of transistor T6

I5 at pin 3 (Fig. 2.15). The amplified difference between the actual-voltage value and a reference voltage (set-point voltage value) is supplied from pin 6 of the integrated circuit I5 through R22 as a control voltage to the comparator stage.

④ *Comparator stage* The comparator stage (Fig. 2.16) compares the control voltage with the 30 kHz sawtooth voltage by means of I2. When the ramp of the sawtooth voltage reaches a positive value relative to the control voltage (Fig. 2.17a, instant t_2), pin 7 of I2 is switched to a low signal (–5 V).

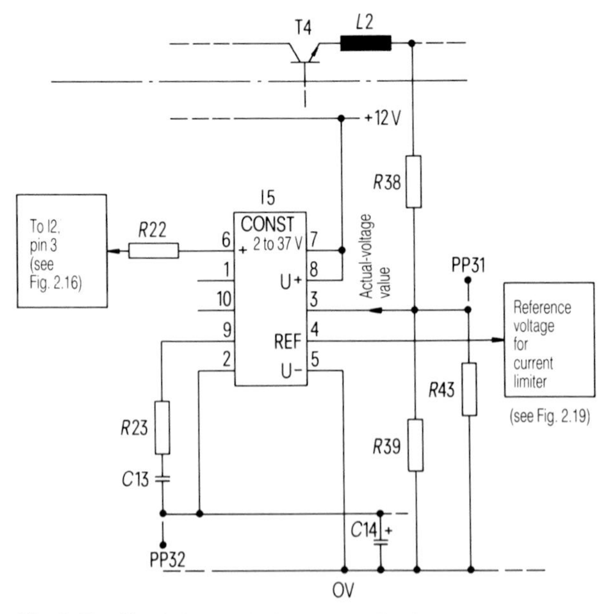

Fig. 2.15 Circuit for producing a control voltage

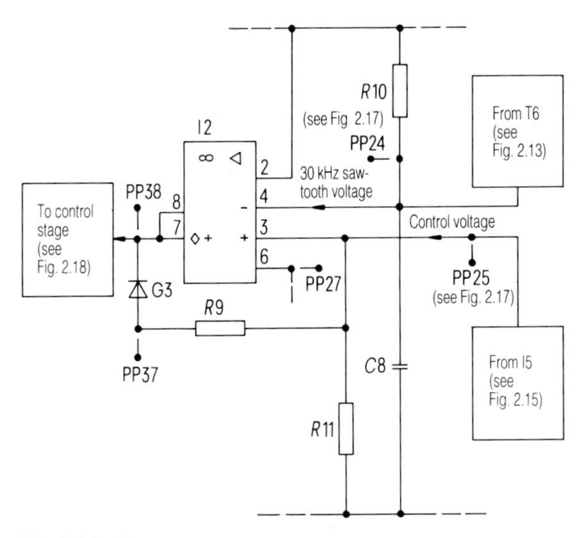

Fig. 2.16 Comparator stage

As soon as the instantaneous sawtooth voltage falls below the control voltage (instant t_3), pin 7 of I2 becomes high (+12 V) again. As a result, with a constant control voltage, the state of pin 7 alternates periodically between low and high.

If the input voltage to the output assembly increases or the loading on it is reduced, the actual voltage feedback signal tapped off from the voltage divider R38/R39, and in consequence the control voltage generated in I5, increases (see Fig. 2.15). The ramp of the sawtooth voltage passes through the level of the control voltage at a later instant (t_5, Fig. 2.17b), so that the low period at pin 7 of I2 is shortened and the high period lengthened.

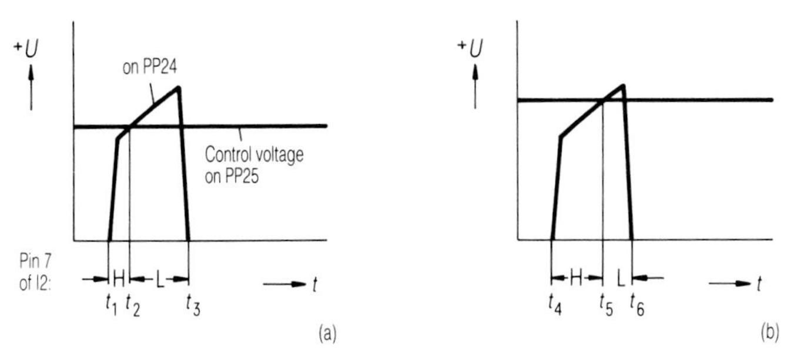

Fig. 2.17 Control voltage and 30 kHz sawtooth voltage

Similarly, a reduction in input voltage or an increase in loading on the output of the output assembly leads to a longer low period at pin 7 and a shorter high period.

⑤ *Control stage for switching transistor T4* If pin 7 of the integrated switching circuit I2 (see Fig. 2.16) is high (+12 V), diode G9 is on reverse and transistor T3 (Fig. 2.18) is driven by a positive base-emitter voltage through resistor R6. The transistor therefore conducts. The resulting rise of current in the primary windings of the coupling transformer U1 induces in the secondary winding a voltage whose polarity (negative base, positive emitter) turns off the switching transistor T4 (power section).

When the signal at pin 7 of I2 changes again to low (–5 V), the diode G9 conducts and the transistor T3 is turned off (base negative relative to emitter). The flow of current through the primary windings of U1 is thus interrupted. A voltage is generated in the secondary winding in the opposite direction to that induced previously (base of T4 positive, emitter negative). The switching transistor T4 is now turned on.

These switching processes are repeated, so that at the emitter of T4 a pulsed direct voltage is produced with a repetition frequency of 30 kHz. Since the conduction period of T4 is always proportional to the duration of the low period at pin 7 of I2,

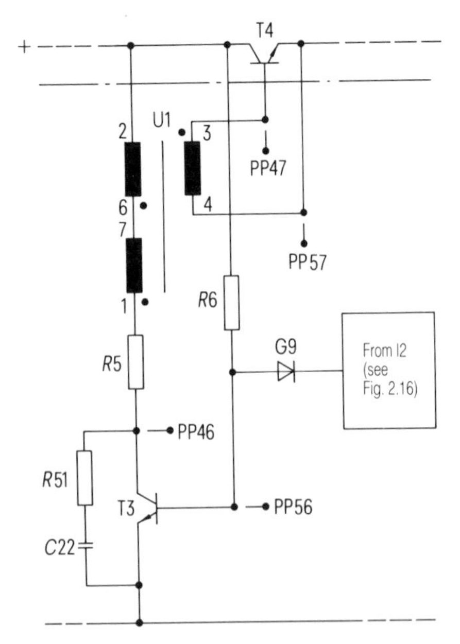

Fig. 2.18
Control stage for switching transistor T4

the output voltage of the output assembly is held constant irrespective of the input voltage or the output loading.

⑥ *Current limiter* The current limiter (Fig. 2.19) is for the purpose of preventing the load current from exceeding the maximum permissible level. I5 (see Fig. 2.15) produces at pin 4 a reference voltage which, via the voltage divider formed with R25/R27, appears across R29 as a current set-point value (a threshold of 80 to 90 mV).

If the voltage drop proportional to current across the current-measuring resistors R52 and R53 reaches the threshold level, the voltage on pin 3 of I4 becomes more positive relative to that on pin 2 (for the function of the current-measuring resistors R52 and R53 see the power section, Fig. 2.10).

The signal from pin 6 of I4, through diode G7, then causes the control voltage to the comparator stage (I2) to become more positive (see Fig. 2.16), and the conduction period of the switching transistor T4 is shortened. This results in a reduction of the output voltage and the limitation of the output current. A current whose magnitude is dependent upon the output voltage flows in the resistor R30. As the output voltage falls due to the current-limiting process, the threshold voltage across R29 decreases. This results in a 'fold-back' current-limiting

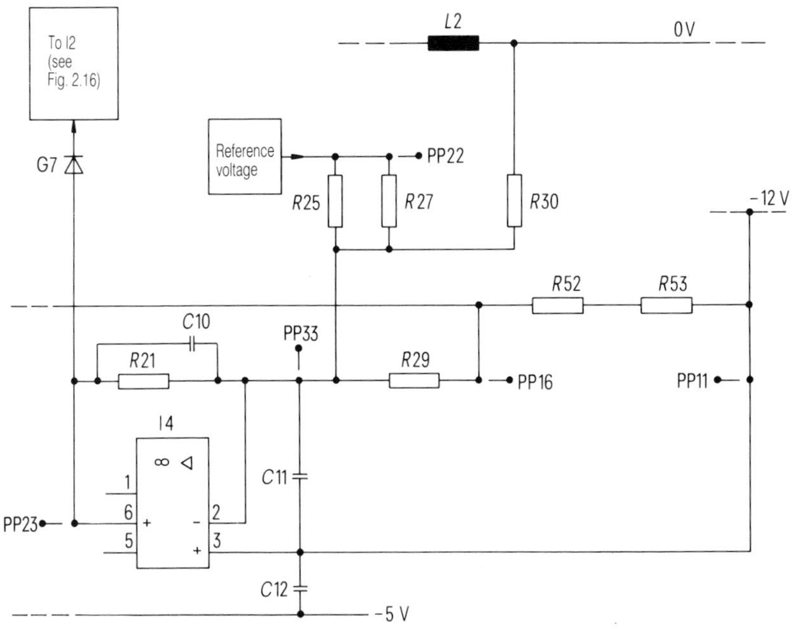

Fig. 2.19 Current limiter

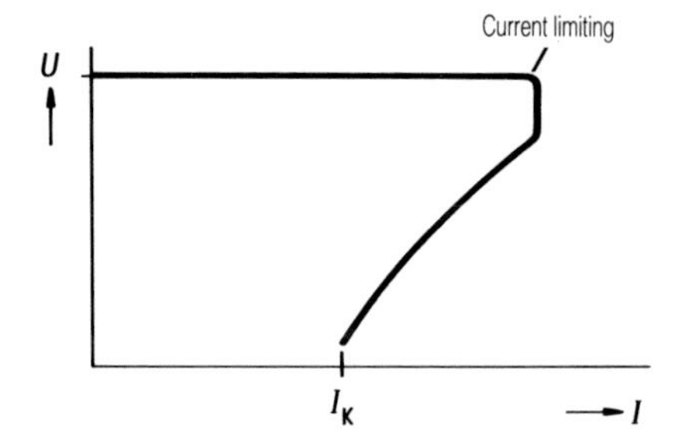

I_K Short-circuit current

Fig. 2.20
Current-limiting characteristic

characteristic, i.e. the current that flows into a short-circuit is less than that at the onset of current limiting (Fig. 2.20).

⑦ *Output voltage monitor* Through the resistor network R31, R32, R42 and R33, there is applied to pin 2 of the integrated switching circuit (threshold switch) I6 a voltage which, so long as the output assembly is functioning correctly, is below a threshold voltage produced in I6 (Fig. 2.21). If the output voltage rises – e.g. through loss of the control voltage generated by the switching circuit I5 – the

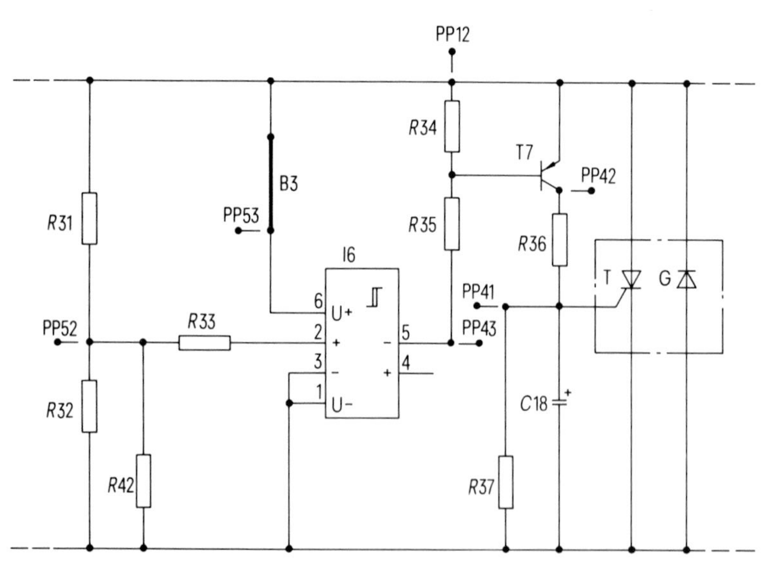

Fig. 2.21 Output voltage monitor

Table 2.2 Technical data for rectifier type 150 W

Mains input supply							
Voltage	(V)	110/115/120/127/220/240/250 +10 to –20%					
Frequency	(Hz)	50 or 60 ±1%					
Equipment fuse protection	(A)	4					
Degree of radio interference		Limit class B (VDE 0871)					
D.C. outputs		ABG	ABG	ABG	ABG	Adjustment controller in NEBG 150 W	
Rated direct voltage	(V)	–48 (–44 to –56)	+12 ±3%	–12 ±3%	+5 ±3%	–5 ±3%	
Rated direct current	(A)	1.5	3.5	1.5	5	0.01	
Interference voltage	(mV)	≦0.5 (frequency-weighted with CCITT 'A' filter)	–	–	–	–	
A.C. ringing supply output							NEBG 150 W
Ringing supply voltage RF	(V)						55 (40 to 65)
Frequency	(Hz)						50
Rated apparent power	(VA)						3.2
Ringing supply current	(A)						0.06
Dimensions (W)	(mm)	SIVAPAC® assembly chasis with a width of 342 mm (1/2 width) or 447 mm (2/3 width)					
		40	40	35	40	–	134

voltage on pin 5 of I6 becomes negative; transistor T7 conducts and thyristor T is triggered, short-circuiting the output. The resulting increase in current causes the current limiter to operate; this reduces the output voltage through the switching transistor T4 towards 0 V. If the current limiting does not function, the fuse Si1 blows (see Fig. 2.10). The thyristor can be turned off by withdrawing the mains plug.

Technical data

The principal technical data of the 150 W rectifier are listed in Table 2.2.

2.1.3 Switching-mode power supplies

The switching-mode power supplies described in the following sections employ the single-ended flow converter principle (20 kHz pulse-width control).

2.1.3.1 Type 650 W

Application

The switching-mode power supply 650 W (Fig. 2.22, modular construction) is used to supply the telephone systems EMS 150 and EMS 180. The maximum

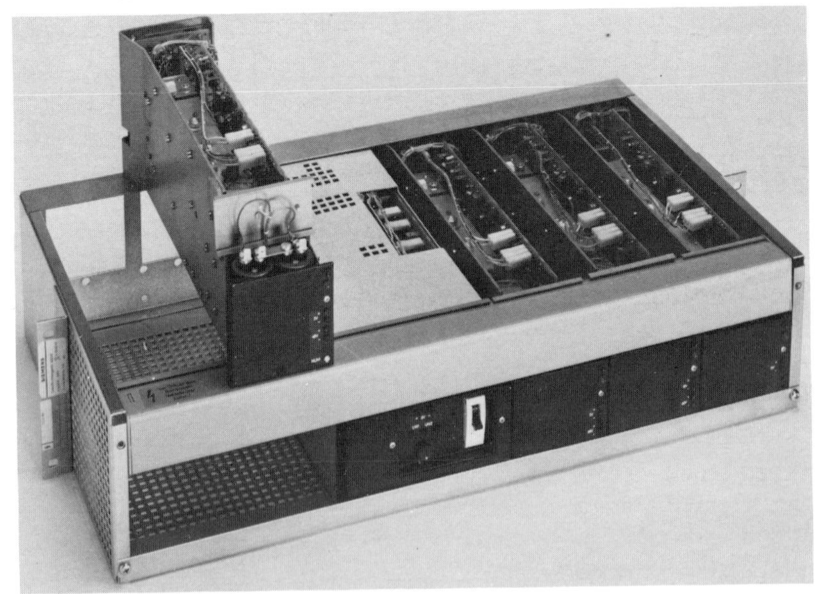

Fig. 2.22 Switching-mode power supply, 650 W

power requirement of these systems amounts to 650 W, so one rectifier is sufficient. As well as the switching-mode power supply 650 W, there are also, for lower power demands, the 300 W and 300 W + 100 W equipments (EMS 80 telephone system).

Operating mode

The 650 W switching-mode power supply is used in the rectifier-operation mode.

Since in the EMS 150 and EMS 180 telephone systems, in the event of a mains supply failure, one or more exchange lines are switched to selected branches (exchange line transfer), a limited telephone service is then available. Communication from branch to branch – without exchange facilities – is, however, not possible. If it is required to maintain the communications traffic without restriction, a reserve of energy (preferably in the form of lead–acid battery) can be provided. Standby parallel mode operation is then instituted. To this end the switching-mode power supply described in Section 2.1.3.2 may, for example, be used in conjunction with the 300 W or the 450 W d.c./d.c. converter (Section 2.2.5). These d.c./d.c. converters are similar in construction to the switching-mode power supply discussed in this section; the differences are that the mains input module NETE is replaced by a battery input module BETE and the mains converter module NUM by a battery converter module BUM.

Block diagrams, basic circuit diagrams and operation

Figure 2.23 shows the equipment of the 650 W switching-mode power supply with the associated modules and the subassemblies incorporated into them. This contains a mains input module NETE 650 W, three 'dual-voltage' mains converter modules NUM and one 'single-voltage' mains converter module NUM.

Figure 2.24 is the block diagram of the 650 W switching-mode power supply. It consists of one mains input module NETE and a number of mains converter modules NUM.

The mains input module NETE is a common input unit for the operation of the mains converter modules NUM. These embody single-ended flow converters (20 kHz) with pulse-width control.

Mains input module

The mains input module NETE consists essentially of a *power section* with a mains filter NEFI, mains input rectifier and smoothing, an auxiliary transformer and a *mains circuit board* NEPL3.

The *mains filter* is provided for EMI suppression; it contains mains overvoltage protection networks and includes a changeover facility for the two supply voltage

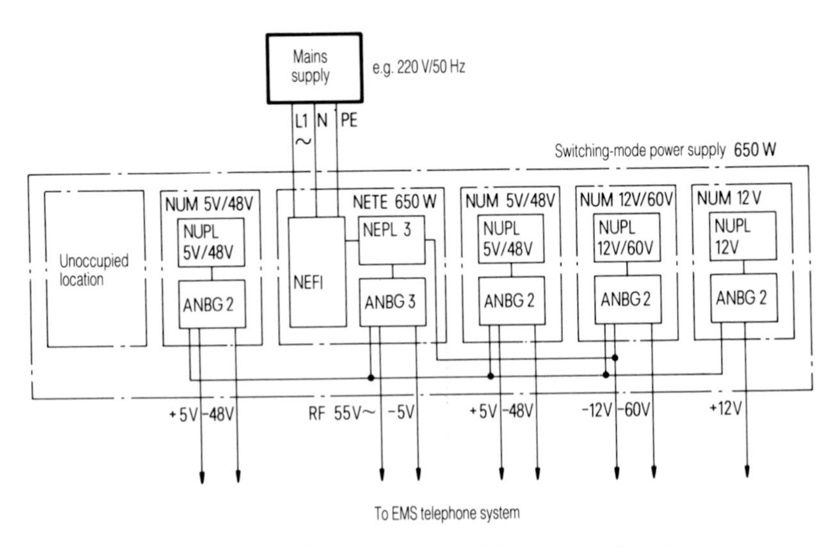

NUM	Mains converter module	NUPL	Mains converter board
ANBG	Connection assembly	NEPL	Mains circuit board
NETE	Mains input module	NEFI	Mains filter
RF	A.C. ringing supply		

Fig. 2.23
Arrangement of switching-mode power supply type 650 W and component supply voltages

variants. It can be supplied in two designs:

▷ NEFI 14A (supply voltage variant W/1) for a.c. supply voltages of 110, 115 and 127 V – the following mains input rectifier is operated as a voltage doubler,

▷ NEFI 7A (supply voltage variant W/2) for a.c. supply voltages of 220, 240 and 250 V – the following mains input rectifier operates as a bridge.

---

①	Internal +10 V power supply	U_{A2}	Jointly controlled output voltage of a mains converter module
②	Stabilized –5 V power supply	HS	Auxiliary supply to control circuit
③	Central monitoring unit	FS	Remote indication ‘overvoltage’
④	Voltage control	FA	Remote switch-off signal
⑤	Control circuit	RF	A.C. ringing supply
⑥	Current limiting circuit	NETE	Mains input module
⑦	Overvoltage-monitoring detector circuit	NEFI	Mains filter
U_E	Rectified mains voltage	NEPL	Mains circuit board
U_{A1}	Controlled output voltage of a mains converter module	NUM	Mains converter module
		NUPL	Mains converter board

Fig. 2.24 Block diagram of switching-mode power supply type 650 W ▷

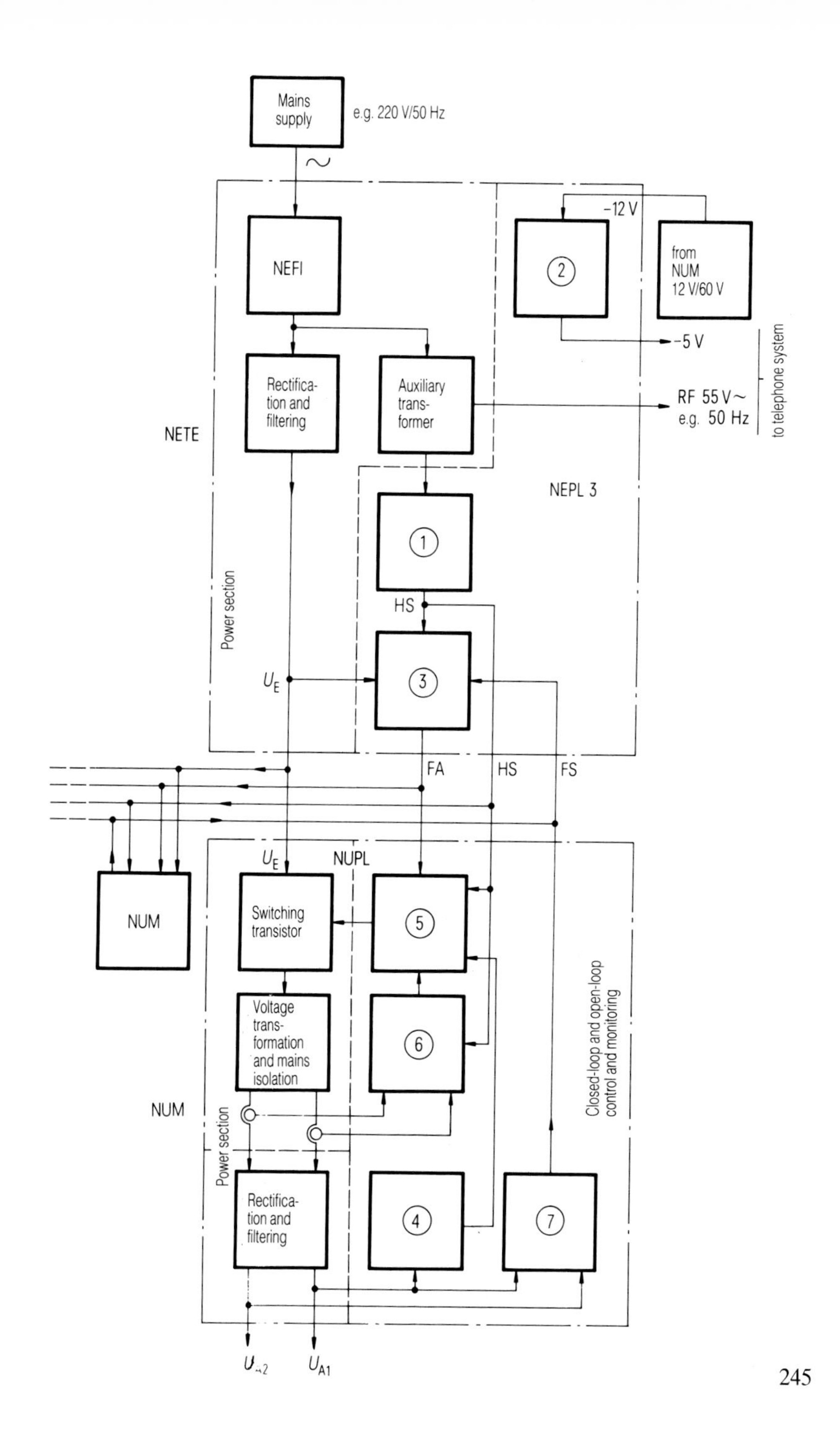
Mains supply
e.g. 220 V/50 Hz
~
−12 V
NEFI
2
from NUM 12 V/60 V
−5 V
Rectifica-tion and filtering
Auxiliary trans-former
RF 55 V~ e.g. 50 Hz
to telephone system
NETE
NEPL 3
1
Power section
HS
U_E
3
FA
HS
FS
U_E
NUPL
NUM
Switching transistor
5
Voltage trans-formation and mains isolation
6
Closed-loop and open-loop control and monitoring
NUM
Power section
Rectifica-tion and filtering
4
7
U_{A2}
U_{A1}

In the *rectification and filtering* the supply voltage is rectified and smoothed. With either supply voltage variant a variable d.c. input voltage U_E (d.c. link voltage) of 175 to 350 V is provided for the mains converter modules NUM. This is checked by the 'central monitoring unit'.

An *auxiliary transformer* supplies the alternating voltage for the +10 V *internal power supply* ① and the a.c. ringing supply of 55 V at, for example, 50 Hz for the EMS telephone system. The internal power supply, with a stabilized auxiliary voltage of +10 V, is supplied via a continuous longitudinal controller to the mains converter modules NUM and to the common monitoring unit (HS). The auxiliary supply is required in the mains converter module NUM for the control and current-limiting circuit in the closed-loop, open-loop control and monitoring section.

The –5 V *stabilized power supply* ② functions similarly with a continuous longitudinal controller. It is supplied at –12 V from the 12 V/60 V mains converter module NUM, and delivers the controlled d.c. supply of –5 V for the smaller loads in the EMS telephone system.

The *central monitoring unit* ③ contains:

▷ a central circuit for the evaluation of fault signals due to overvoltages at the outputs of the mains converter modules NUM.

  On overvoltage the overvoltage detection circuit in the mains converter module NUM transmits the remote signal FS to the central monitoring unit and thence to a *test circuit* (permanent fault recognition); this transmits without delay the remote switch-off signal FA to all the mains converter modules NUM, and thereby turns off all the switching transistors. The test circuit then attempts to switch on the transistors at intervals of about 0.3 to 0.7 s. If an overvoltage is still present, a 'lock-out' results.

▷ monitoring circuit for the d.c. input voltage U_E to the mains converter modules NUM.

  If the d.c. input voltage rises above 360 V, due, for example, to atmospheric discharges or switching overvoltages, the central monitoring unit transmits the signal FA to all the mains converter modules. This turns off all the switching transistors. If the d.c. input voltage U_E returns to its normal level, the switching transistors are allowed to turn on again. The test circuit is *not* operative in the case of d.c. input voltage deviations.

▷ +10 V auxiliary voltage monitoring circuit.

  If the auxiliary voltage drops to less than about 9 V, the remote switch-off signal FA is fed to the mains converter modules NUM, turning off the switching transistors. The test circuit is unaffected by this voltage deviation also.

  The operation of the mains input module NETE (650 W) is explained below with reference to the *basic circuit diagram* of Fig. 2.25. A supply voltage of

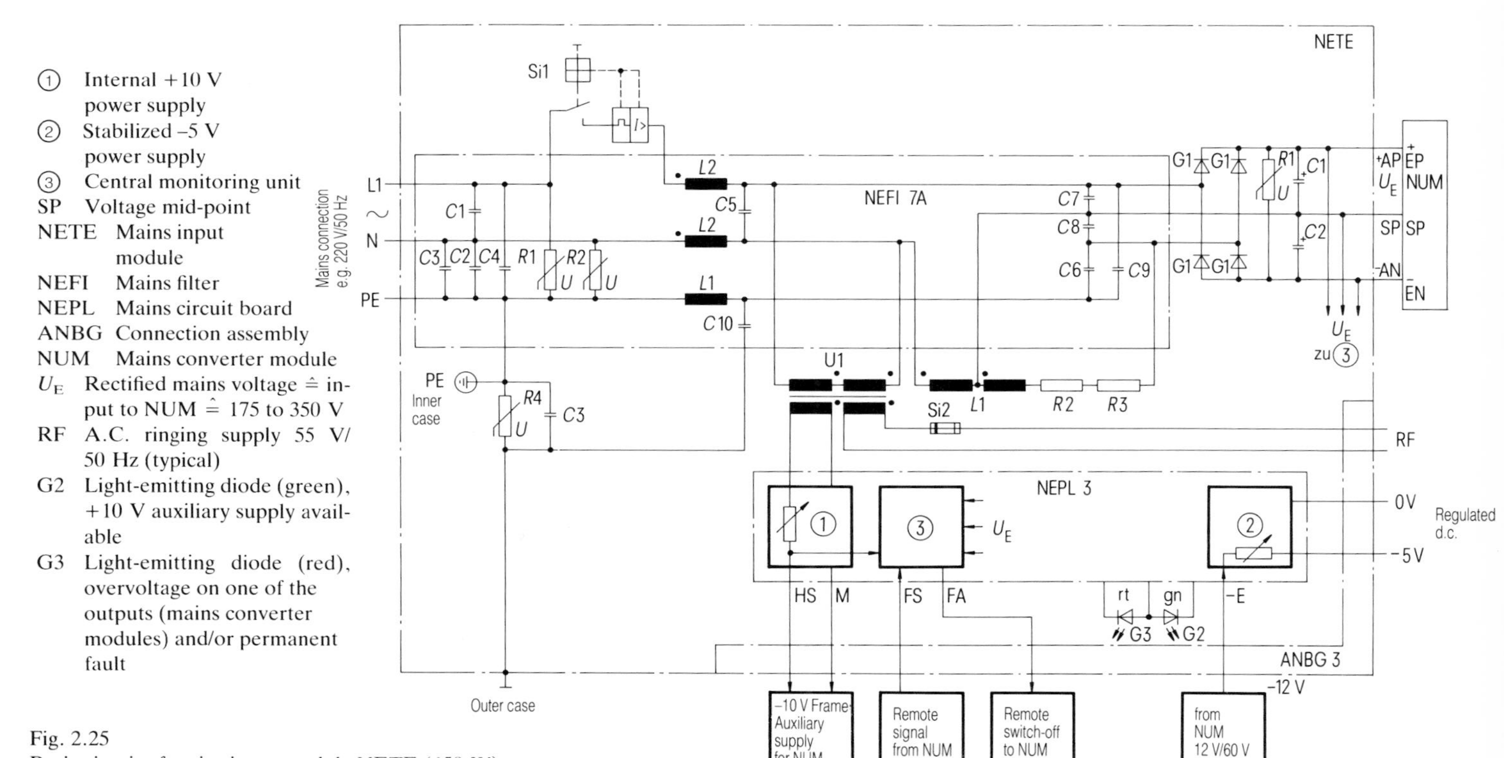

① Internal +10 V power supply
② Stabilized –5 V power supply
③ Central monitoring unit
SP Voltage mid-point
NETE Mains input module
NEFI Mains filter
NEPL Mains circuit board
ANBG Connection assembly
NUM Mains converter module
U_E Rectified mains voltage $\hat{=}$ input to NUM $\hat{=}$ 175 to 350 V
RF A.C. ringing supply 55 V/50 Hz (typical)
G2 Light-emitting diode (green), +10 V auxiliary supply available
G3 Light-emitting diode (red), overvoltage on one of the outputs (mains converter modules) and/or permanent fault

Fig. 2.25
Basic circuit of mains input module NETE (650 W)

220 V (mains voltage variant W/2) is assumed for the purposes of an example.

The mains input module NETE contains the subassemblies:

- ▷ mains filter NEFI,
- ▷ mains circuit board NEPL 3,
- ▷ connection assembly ANBG (not described further).

The circuit board of the *mains filter* NEFI also carries EMI suppression components. The bridge rectifier in the mains input module, and those in the following mains converter modules, generate interference, which has to be attenuated at the mains inputs L1, N and PE. EMI suppression is effected in the mains filter by the capacitors C1 to C10 and the inductors L1 and L2. The varistors R1 and R2, in conjunction with the varistors R1 and R4 in the mains input module, the inductor L1 and the resistors R2 and R3, provide protection against mains overvoltages. The inductor L1 (NETE) and resistors R2 and R3 (NETE) also serve the purpose of inrush current limiting at switch-on.

In the supply voltage variant 2 the bridge rectifier G1 (NETE) is connected to the mains filter. The capacitors C1 and C2 store energy and effect smoothing. In the supply voltage variant 1 they serve the purpose of voltage doubling (G1 is then used as a voltage doubler). The varistor R1 limits the d.c. input voltage to the mains converter module to 700 V.

The rectified voltage (175 to 350 V d.c.–link voltage) is taken from the outputs +AP and –AN (the d.c. input voltage U_E to the mains converter modules).

The a.c. ringing voltage is taken from a secondary winding of the auxiliary transformer U1 and fed to the connection assembly ANBG 3.

The *mains circuit board* NEPL 3 has the following functions:

- ▷ monitoring of the d.c. input voltage U_E to the mains converter modules,
- ▷ central evaluation of fault signals associated with overvoltages at the outputs of the mains converter modules,
- ▷ provision of a controlled +10V auxiliary supply for the mains converter modules,
- ▷ monitoring of the +10 V auxiliary supply,
- ▷ provision of the –5 V stabilized d.c. supply for small loads in the EMS telephone system.

A further secondary winding of U1 provides an a.c. supply for the +10 V *internal power supply* ① (NEPL 3), which after rectification and smoothing is stabilized by a continuous longitudinal controller. The +10 V auxiliary supply is fed through the connection assembly ANBG 3 to the mains converter modules and is checked in the central monitoring unit ③ (NEPL 3). A green light-emitting diode (G2) indicates that the auxiliary supply is available, while a red LED (G3) indicates an overvoltage at an output of a mains converter module and/or a permanent fault.

The remaining essential functions of the mains circuit board NEPL 3, such as the central monitoring unit ③ and the –5 V stabilized power supply ②, have previously been described with reference to Fig. 2.24.

Mains converter module

The mains converter module NUM (Fig. 2.24) consists essentially of a *power section* – i.e. a part of the mains converter board NUPL (20 kHz switching transistor/20 kHz transformer/current transformers) – the secondary rectification and output smoothing and, also mounted on the mains converter board, the *closed-loop, open-loop control and monitoring*.

The mains converter modules NUM provide, from the d.c. input voltage U_E of between 175 and 350 V obtained from the mains input module NETE, a d.c. supply, or two d.c. supplies, according to whether the module is of the single-voltage or dual-voltage type, for the loads in the EMS telephone system (U_{A1}, U_{A2}).

The d.c. input voltage U_E is chopped by a switching transistor, transformed down by the following 'voltage transformation and mains isolation' and rectified and filtered by the 'secondary rectification and filtering'. A constant direct voltage (with close tolerance) is produced at the master (main) output U_{A1}. It will be seen from Fig. 2.24 that for the purpose of voltage control only the actual-voltage value at the output U_{A1} is sensed. Only this voltage, therefore, is stabilized by the control circuit.

A 'jointly controlled' d.c. supply (with a wider voltage tolerance) can be taken from the output U_{A2}. The master-controlled voltage from the output of the mains converter module is the lower of the two voltages in all cases.

In the mains converter module NUM 5 V/48 V, for example, the +5 V master-controlled output voltage U_{A1} has a tolerance of ±4%, while the –48 V jointly controlled output voltage U_{A2} varies over a range from –44 to –56 V.

All the outputs of the mains converter module are potential-free and isolated from the input; they are also overload- and short-circuit-proof.

The polarities of the output voltages can be changed by connecting the appropriate poles (positive or negative) to the frame of the power supply (0 V connection).

The actual-voltage feedback signal is taken from the output U_{A1} and fed to the *voltage control* ④ in the *closed-loop, open-loop control (regulation, control) and monitoring*. The signal from the voltage control, in which the actual voltage is compared with the reference voltage, governs the control circuit ⑤ (pulse-width control).

The actual-current feedback signal is produced by the two current transformers. One measures the current at output U_{A1}, the other that at output U_{A2}. The alternating measurement voltage is fed to the *current-limiting circuit* ⑥. If the load current exceeds the present level, the actual current feedback signal becomes greater than the current reference value, so that current limiting takes effect and the d.c. output voltage is reduced. The mains converter module is thus protected against overload.

The *overvoltage-monitoring detector circuit* ⑦ registers unacceptable overvoltages at the outputs U_{A1} and U_{A2} and actuates the central monitoring unit in the mains input module. Reverse battery protection is provided by a built-in shunt diode (not shown in the block diagram).

The 650 W switching-mode power supply incorporates protectively insulated parts in accordance with VDE 0804 (see Chapter 5).

The parts of the equipment connected to the mains have, in addition to functional insulation, protective insulation from the secondary circuit (the communications system).

To suppress overvoltages, varistors are connected between the live side of the mains and the earth lead (PE), which is insulated from the equipment case, and also between PE and the case.

The operation of a mains converter module NUM is explained below in greater detail with reference to the *basic circuit diagram* of Fig. 2.26.

The mains converter module NUM contains the sub-assemblies:

▷ mains converter board NUPL and
▷ connection assembly ANBG 2.

The input supply voltage U_E from the mains input module NETE is brought to terminals +EP and –EN of the mains converter module.

The filter circuit at the input of the mains converter module consists of the capacitors C1 and C2.

The mains converter board consists of a power section and the closed-loop, open-loop control (regulation, control) and monitoring. The functions of the control and monitoring have been explained with reference to Fig. 2.24.

The following components are mounted directly in the mains converter module, not on the mains converter board:

▷ resistor R2,
▷ power-switching transistor T1,
▷ rectifier diodes G2 and G4,
▷ free-wheel diodes G3 and G5,
▷ output filter capacitors C5 and C6.

The d.c. input voltage (175 to 350 V) obtained from the mains input module is chopped by the switching transistor T1 at a repetition frequency of 20 kHz and

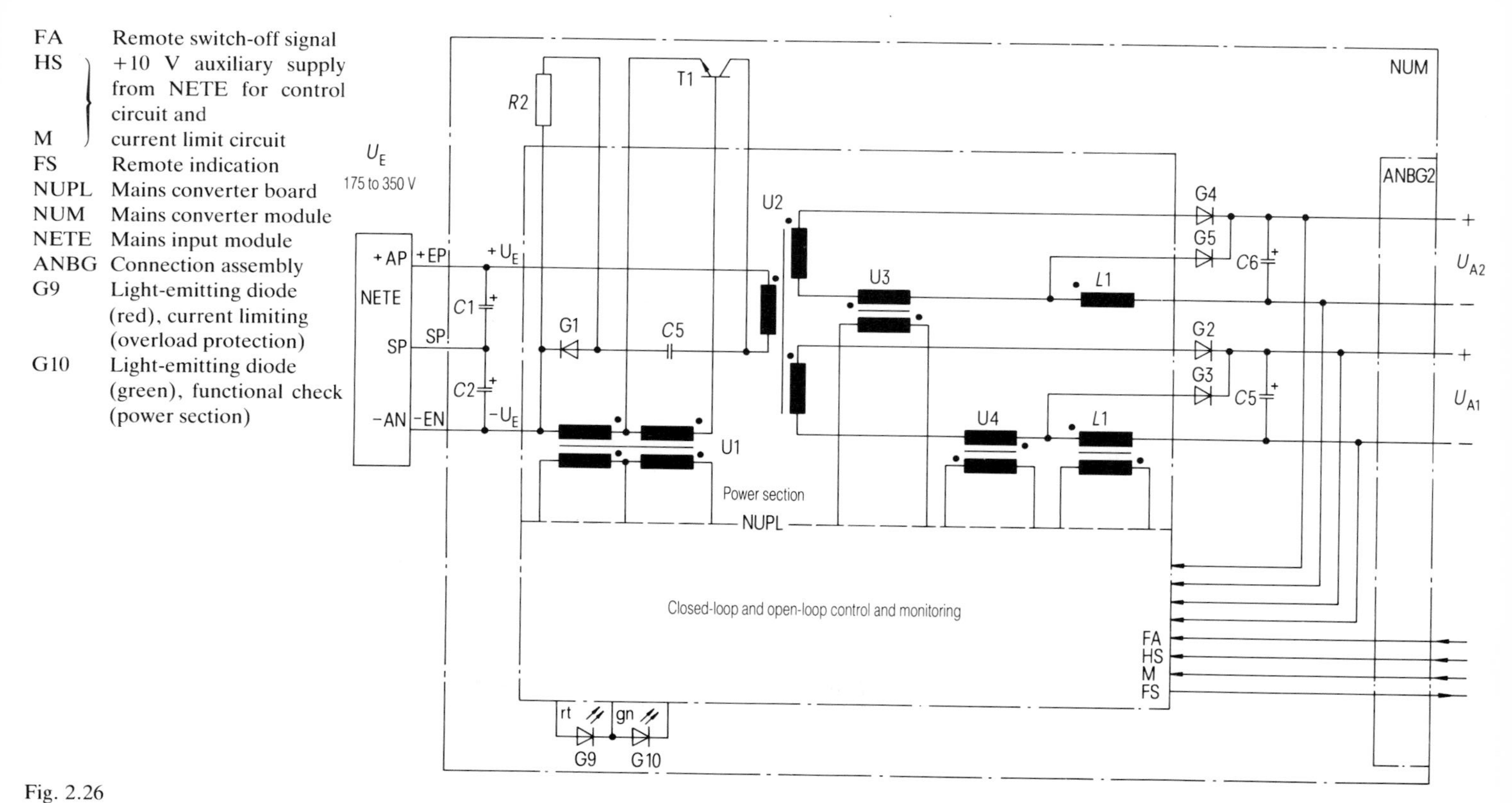

Fig. 2.26
Basic circuit of a mains converter module NUM

applied to the primary winding of the transformer U2. The drive for T1 is generated by the control circuit in the control and monitoring (NUPL) and coupled into the base-emitter circuit of the switching transistor T1 by the coupling transformer U1 (NUPL). When the transistor conducts, the input voltage U_E is applied to the primary winding of the transformer U2; the rectifier diodes in the two output circuits, G2 and G4, conduct. Energy is stored in the inductor L1, and the two filter capacitors C5 and C6 are charged.

When the transistor turns off, the currents in the two windings of the inductor L1 are maintained through the free-wheel diodes G3 and G5. Charge is released from the filter capacitors C5 and C6.

The voltage regulator and, when necessary, the current-limiting circuit – both in the closed-loop, open-loop control and monitoring – control the switching ratio V_T ($V_{Tmax} = 0.5$) at the constant frequency of 20 kHz.

The components G1, C5 (NUPL) and R2 (NUM) limit the turn-on current and reduce the turn-off loss in the switching transistor T1.

The direct current in the inductor L1 has an a.c. component superimposed upon it. This component is absorbed by the filter capacitors C5 and C6.

The current limiter operates on the occurrence of an overload, or large current steps within the load limit, and when the equipment is switched on (charging the capacitors at the output). The measured quantity is the pulse current on the secondary side of U2.

A voltage U_{mess} proportional to the load currents is obtained from the two current transformers (U4, U3) connected respectively in the master-controlled and jointly controlled output circuits and added to a rectified and filtered negative voltage obtained from the auxiliary winding on L1, which is proportional to the output voltage U_{A1} of the mains converter module.

The green light-emitting diode (LED) G10 lights when the power section of the mains converter module is operating. The red LED G9 indicates when current limiting occurs on overload. U_{A1} is the master-controlled and U_{A2} the jointly controlled output voltage.

Technical data

The principal technical data for the switching-mode power supply type 650 W are listed in Table 2.3.

2.1.3.2 Type 48 V/30 A GR3 (GR31), GR12

Application

The switching-mode power supply 48 V/30 A GR3 (GR31), GR12 (Fig. 2.27, built-in construction) is used as a power supply for the EMS telephone system, the KN system and the EWSD system; it can also feed a parallel-connected battery.

Table 2.3 Technical data for switching-mode power supply type 650 W

Mains input supply							
Voltage	(V)	Supply-voltage variant 1 110, 115, 127 +10 to −20%				Supply-voltage variant 2 220, 240, 250 +10 to −20%	
Frequency	(Hz)	50 or 60 ±1%					
Equipment fuse protection	(A)	16				10	
Degree of radio interference		Limit class B (VDE 0871)					
D.C. outputs		NUM		NUM		NUM	Stabilized power supply in NETE 650 W
Rated direct voltage	(V)	−12 ±4%	−60 (−55 to −70)	+5 ±4%	−48 (−44 to −56)	+12 ±4%	−5 ±4%
Rated direct current	(A)	7.5	1.5	10	2.7	12.5	0.03
Interference voltage (frequency weighted with CCITT 'A' filter)	(mV)	–	≦0.5	–	≦0.5	–	–
A.C. ringing supply output							NETE 650 W
Ringing supply voltage RF	(V)						55 (40 to 65)
Frequency	(Hz)						50
Rated apparent power	(VA)						6
Ringing supply current	(A)						0.11
Dimensions ($H \times W \times D$)	(mm)	The power supply chassis (single-width) matches the system cubicles. The mounting dimensions correspond to those of the single-height SIVAPAC® assembly chassis, except for the somewhat greater depth. 148 × 658 × 358					

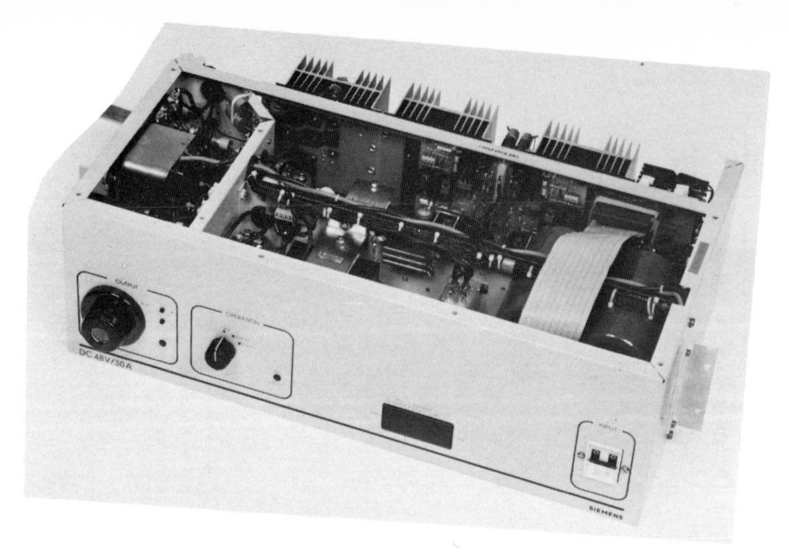

Fig. 2.27 Switching-mode power supply 48 V/30 A GR3 (GR31), GR12

Up to four equipments can be connected in parallel. In this way it is possible to cater for a current demand for the communications system and the battery of up to 120 A.

Individual equipments are intended for incorporation into system cubicles. Two or more equipments can be accommodated in the connection panel ASF (in a separate cubicle).

The GR3 and GR31 versions are similar in construction and intended as power supplies for the EMS telephone system. The battery disconnection unit (battery undervoltage monitoring and battery disconnection contactor) is included only in the GR31; only one is required for each power supply installation. If the connection panel ASF is available the unit can be accommodated centrally in it. The GR12 version can supply the KN and ESWD systems. GR12 differs from GR3 (GR31) in that it is supplied with higher output voltages – for 25 lead–acid battery cells instead of 24. If required, the GR12 equipment can also be furnished with a battery disconnection unit.

As well as the switching-mode power supplies 48 V/30 A GR3 (GR31), GR12, there are also smaller equipments, similar in form, which need not be described here.

Operating modes

The switching-mode power supplies 48 V/30 A GR3 (GR31), GR12 can be used in the rectifier-operation mode or in the parallel-operation mode (usually the standby parallel mode) without voltage-reducing diodes.

The additional d.c./d.c. converters incorporated into the systems are described in Section 2.2.

Basic circuit and operation

The basic circuit of the switching-mode power supply 48 V/30 A GR31 (GR12) is shown in Fig. 2.28. This is drawn on the basis of a 220 V/50 Hz mains input.

Power section

The single-phase mains supply is applied through the mains filter Z1 (high-frequency input filter), the automatic circuit breaker Q1, resistor R31 and the mains relay contact K1 to the power assembly A2. The input resistor R31 limits the switch-on inrush current; it is bridged by the relay K1 as soon as the switch-on enable signal is received from the A1 closed-loop and open-loop control (regulation and control) assembly.

Rectification (V1) of the a.c. mains voltage is followed by smoothing by means of the capacitors C1 and C2.

For an input voltage of 110 to 127 V the input filter capacitors form a voltage-doubler circuit in conjunction with the rectifier V1 (not illustrated). For input voltages between 220 and 250 V the input rectifier diodes V1 form a bridge circuit.

The d.c.-link voltage after the input smoothing amounts to about 310 V at the rated supply voltage. The ratio of the conducting to the non-conducting periods of the power switching transistors V11 and V51 (20 kHz switch) is determined by the closed-loop and open-loop control assembly A1 at a constant frequency (mark-space ratio, pulse-width control).

The switching transistors V11 and V51 periodically supply pulses corresponding to the d.c.-link voltage to the 20 kHz transformer. The magnitude of the d.c. output voltage depends upon the primary voltage and the conduction period.

The safe turning-on of the switching transistors by the leading edges of the control pulses from the closed-loop and open-loop control assembly A1 is ensured by the drive circuit A3, with which the two current transformers T401 and T451 are also associated.

When the switching transistors are switched off, the transformer current commutates initially from the transistors V11 and V51 to C111, V111, C151 and V151, so that as the collector-emitter voltage rises only a small current is flowing through the transistors. The capacitors (C111 and C151) are charged to the d.c.-link voltage. From this point the transformer current is taken over by the resetting diodes V13 and V53, so that the blocking voltage to which the transistors are subjected is limited to approximately the level of the d.c.-link voltage. When the power transistors turn on, the discharge current from the capacitors C111 and C151 is limited by the resistors R111 and R151, so that the transistors are not overstressed.

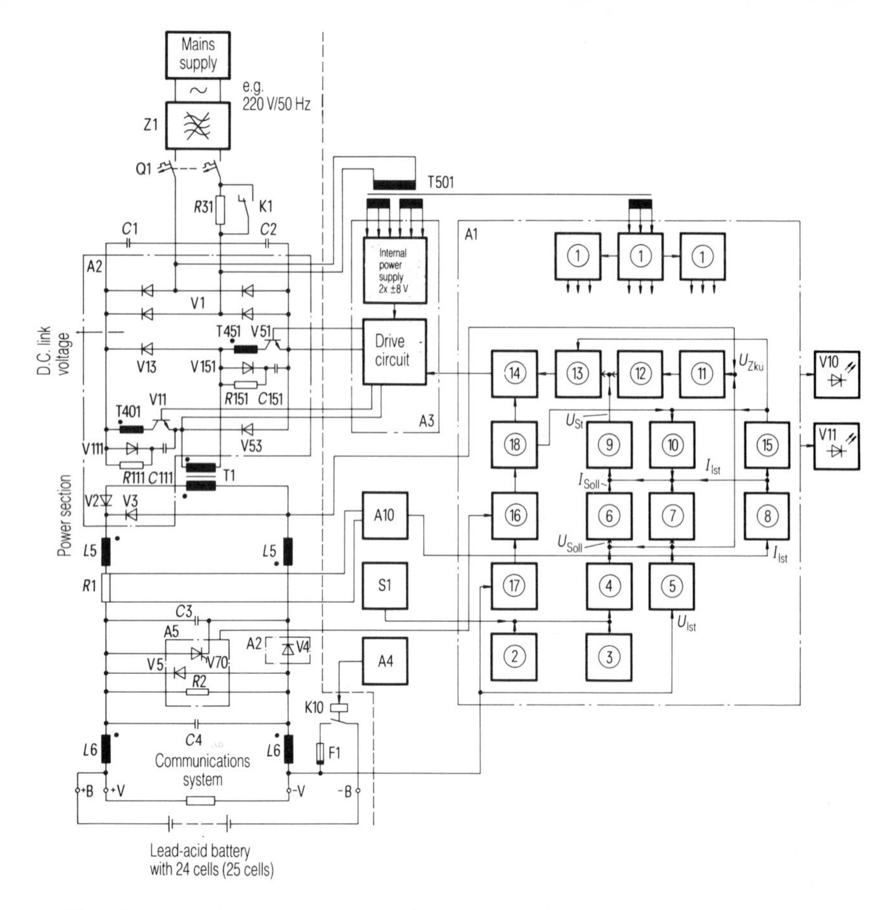

① Internal ±10 V power supply (reference voltage)
Internal ±15 V power supply
Internal +5 V power supply
② Switch S300: 1. Operation 2.2.7 V/cell
③ Electronic time-delay relay 2 to 10 h
④ Set-point device for voltage reference
⑤ Actual-voltage value matching
⑥ Voltage regulator
⑦ A.C. voltage control
⑧ Actual-current value amplification
⑨ Current regulator
⑩ Automatic start-up delay
⑪ Feedforward control

Fig. 2.28
Basic circuit of switching-mode power supply 48 V/30 A GR31 (GR12)

To translate energy the 20 kHz transformer T1 requires a time-varying voltage on the primary side. This is provided by means of the switching transistors V11 and V51. The transformer T1 translates the pulses in a 'potential-free' manner to the secondary circuit. Apart from power translation and voltage transformation it provides positive isolation between the supply mains and the communications circuits. The secondary voltage is rectified by a one-way rectifier (diode V2).

Since the transformer operates in a unipolar mode, it has to be reset in order to prevent saturation of the core. The magnetic energy stored during the conducting period is returned as a current through the diodes V13 and V53 to the capacitors C1 and C2. To ensure demagnetization the switching ratio V_T (the ratio of the conduction period of the power transistor to the total cycle period) must be less than 0.5. The closed-loop and open-loop control A1 permits a switching ratio $V_T = 0.4$; because of the turn-off time of the power transistor T1 sees a maximum ratio $V_T = 0.45$. During the resetting period, the transformer generates a negative voltage equal to the d.c.-link voltage across its primary terminals. As a result, the secondary current (forced by the inductor L5) commutates after the zero-crossing of the secondary voltage from the rectifier diode V2 to the free-wheel diode V3.

The secondary rectification produces direct voltage pulses whose magnitude differs from the primary d.c.-link voltage according to the transformation ratio of the transformer T1. Their duration is determined by the switching ratio of the switching transistors. The pulses are smoothed by the output filter L5, C3 and C4.

◁ ⑫ 20 kHz clock generator
⑬ Trigger set
⑭ Pulse amplification
⑮ Overcurrent switch-off circuit
⑯ Instantaneous overvoltage monitor with test circuit
⑰ Delayed overvoltage monitor
⑱ Pulse cancelling and switch-on enable

A1 Closed-loop and open-loop control
A2 Power assembly
A3 Drive assembly
A4 Battery undervoltage monitoring
A5 Voltage limiter and indication
A10 Actual-current value matching
T1 20 kHz main transformer
T501 Auxiliary transformer
S1 Switch: 1. Automatic
2. 2.23 V/cell
3. 2.33 V/cell
V10 Light-emitting diode (operation indication)
V11 Light-emitting diode (automatic 2.33 V/cell)
$U_{Zkü}$ Transformed d.c.-link voltage

This is so dimensioned that it is effective only for the switching frequency of 20 kHz and its harmonics. The low-frequency components of the direct voltage are smoothed by an a.c. voltage regulator (in the closed-loop and open-loop control assembly A1).

The current-measurement shunt R1 provides the actual-current signal, which is used in the closed-loop and open-loop control assembly A1.

The decoupling diode V4 (on the power assembly A2) prevents current from the parallel-connected battery or another power supply equipment from flowing through the voltage-limiting thyristor V70; it also prevents the voltage limiting from responding to overvoltages which arise outside the equipment. The direct voltage passes to the output terminals of the rectifier via filters (EMI suppression elements consisting of the suppression inductor L6 and suppression capacitors not shown in the diagram).

The communications system is connected to the positive and negative load terminals (+V and –V) and the battery to +B and –B. F1 is the battery fuse.

Battery undervoltage monitoring A4

The output to the battery, in the GR31 equipment, for example, is taken through the battery disconnection contactor K10, which is controlled by the battery undervoltage monitoring A4. K10 and A4 represent protection against heavy discharge of the battery.

After a successful ‘start-up’, the battery is switched in parallel with the rectifier and the communications system through K10. If the communications system is supplied by the battery – due to a supply failure, for example – and the voltage drops below 42 V, the battery undervoltage monitoring A4 causes the contactor K10 to release, so that the battery is disconnected from the communications system. The lead–acid battery is thus protected against excessive discharge.

Voltage limiting and indication A5

The voltage-limiting and indication assembly A5 serves the purpose, among others, of protecting the communications system from overvoltage spikes. Other functions of the A5 assembly are:

▷ measurement and adjustment of the instantaneous overvoltage monitor,
▷ limitation of overvoltages,
▷ indication, via a potential-free contact (operated in normal operation), of loss of rectifier output,
▷ reverse battery protection.

If the equipment output voltage U_A exceeds the preset limiting value, the measuring circuit in the A5 assembly triggers the voltage-limiting thyristor V70, which short-circuits the output of the rectifier; the equipment output voltage is thus prevented from rising further.

In addition, the measuring circuit actuates the instantaneous overvoltage monitor with test circuit (in the closed-loop and open-loop control assembly A1), inhibiting the triggering pulses. The residual load resistor *R*2 prevents an undesirable rise in voltage on no-load.

Reverse-battery protection is provided by the diode V5 and the fuse F1. An incorrectly poled battery is short-circuited by V5, so that the load is disconnected from the battery by the rupture of F1.

Switch S1

The switch S1 selects the operating mode:

1 automatic,
2 2.23 V/cell,
3 2.33 V/cell.

In the 'automatic' position, all the operating sequences proceed automatically; in normal operation the rectifier supplies a voltage of 2.23 V/cell (trickle-charging voltage).

Assuming a lead–acid battery of 24 (25) cells, this implies an output voltage of 53.5 V (56 V). If the mains supply is lost, operation is maintained by the battery, and on restoration of the supply the equipment is automatically switched to charging at 2.33 V/cell. The rectifier output voltage, for an adjustable period (between two and ten hours) is about 56 V (58.5 V). At the end of the timing period of the electronic time-delay relay (in assembly A1) the characteristic is automatically switched back to 2.23 V/cell and the normal operating condition is reinstated. During the timing period the light-emitting diode V11 indicates the 'automatic' condition.

The light-emitting diode V10 indicates operation of the rectifier equipment. In position 2 of the switch (2.23 V/cell) the output voltage is 53.5 V (56 V) and in position 3 (2.33 V/cell) 56 V (58.5 V).

S1 selects the characteristic by modifying the set-point device (reference value selector) which supplies the voltage reference.

The auxiliary transformer T501 provides a.c. power supplies to the drive assembly A3 and the closed-loop and open-loop control assembly A1, which derive stabilized auxiliary power supplies from them by means of rectifiers and continuous longitudinal controllers.

Drive assembly A3

The drive assembly A3 consists of an auxiliary power supply of $2 \times \pm 8$ V and the drive circuit.

Trigger pulses pass from the pulse amplifier in the closed-loop and open-loop control assembly A1 via the drive assembly A3 to the power transistors.

The drive assembly A3 ensures observance of the following conditions:

▷ the base current of the switching transistors at turn-on and turn-off must follow a defined waveform, in order to minimize the switching losses,
▷ at turn-on the transistors require a steeply rising drive current, so that the collector-emitter voltage collapses rapidly and the rising collector current does not produce an excessive turn-on loss,
▷ the negative base current, which is necessary at turn-off, should not rise at more than a defined rate, in order that the turn-off loss should be as low as possible.

Closed-loop and open-loop control A1

The closed-loop and open-loop control (regulation and control) assembly A1 determines the switching frequency and the conduction period of the power transistors, to maintain the output voltage of the equipment at the required level with any permitted loading and to prevent the current from exceeding the rated current of the equipment in the event of unacceptably high loading. The switching frequency is set at about 20 kHz. The conduction period is controlled according to the mains supply voltage, the equipment output voltage, the a.c. component of the direct voltage and the output current. As well as the regulator and the clock generator, the ±15, ±10 and +5 V auxiliary power supplies, the overvoltage monitoring and the electronic time-delay relay are also accommodated in the A1 assembly.

① *Internal power supply* The A1 assembly obtains from the auxiliary transformer T501 an a.c. supply from which stabilized direct voltages of ±15 V are produced. Reference voltages of –10 and +10 V are produced from the –15 V stabilized supply. The +15 V stabilized supply is used among other purposes to produce the +5 V auxiliary supply.

② *Switch S300* The switch S300 has two positions: 1, 'operation' and 2, '2.7 V/cell'. Position 1 is the normal operating position. Position 2 may be selected only with the load disconnected, for the initial charge of the lead–acid battery of 24 (25) cells. This produces a voltage of 64.8 V (67.5 V). The switch S300 adjusts the set-point device to produce the appropriate voltage reference values.

③ *Electronic time-delay relay 2 to 10 h* Every time the rectifier is switched on after an inoperative period, and on the restoration of the mains supply after an interruption, assuming the switch S1 to be in the 'automatic' position, the electronic time-delay relay automatically switches the output voltage from 2.23 V/cell (trickle charging) to 2.33 V/cell (charging) for a period adjustable from about 2 to 10 h. The electronic time-delay relay adjusts the reference circuit of the voltage regulator through the relay K300.

When the set time has expired, the 2.23 V/cell characteristic is restored automatically.

④ *Set-point device for voltage reference values* The set-point device (reference value selector) for the voltage references is adjusted by the switches S1 and S300, and by the electronic time-delay relay. It supplies the appropriate reference voltage values to the voltage regulator.

⑤ *Actual-voltage value matching* The actual-voltage feedback value from the equipment output is fed to the actual-voltage value matching, where it is filtered and inverted.

⑥ *Voltage regulator* The purpose of the voltage regulator is to hold the output voltage of the equipment constant at the required level.

The control principle used embodies a voltage control system with an inner current loop. Thus the voltage regulator does not influence the trigger set directly but produces a current reference for the current regulator.

The voltage regulator compares the constant set-point voltage (reference value) with the variable, matched actual-voltage feedback signal from the actual-voltage feedback matching. The result of this comparison is fed to the following current regulator as a set-point current (reference value).

⑦ *A.C. voltage control* The smoothing circuit for the direct voltage (output filter), L5, C3, C4, as previously explained, attenuates the high-frequency components in the output voltage. The harmonics of the supply frequency arising from the input rectification have therefore to be removed from the output in another way in order to keep the frequency-weighted interference voltage to not more than 0.5 mV. This is effected by the a.c. voltage regulator, which, like the d.c. voltage regulator, acts upon the current regulator.

⑧ *Actual-current value amplification* The voltage developed across the measuring shunt $R1$ by the equipment output current is applied to the actual-current feedback value matching assembly A10; at the rated current this supplies a voltage of about 50 mV to the actual-current feedback amplifier, which is part of the A1 assembly. The A10 assembly makes it possible to exchange the A1 assembly without having to readjust the current limiter.

The actual-current feedback amplifier has to raise the actual-current feedback signal to the level of the output voltage of the voltage regulator. The maximum current of the equipment (current limiting) can be adjusted by means of a potentiometer in the actual-current feedback amplifier.

⑨ *Current regulator* The current regulator (controller) receives from the voltage regulator a set-point current (reference value), which is modulated by the effect of the a.c. voltage regulator. The current regulator compares the current reference with the actual-current feedback signal supplied by the actual-current feedback amplifier, and thereby produces the control voltage U_{St}; the control voltage is applied to the trigger set, and hence determines the conduction period of the power transistors.

The control holds the output voltage of the equipment constant up to the rated current: beyond that point, the voltage is reduced to the extent necessary to limit the current to the rated value.

⑩ *Automatic start-up delay* The automatic start-up (soft start) circuit (capacitor) ensures, when the rectifier is switched on, that the output voltage rises only slowly, so that no overshoot occurs. This takes effect at the input to the current regulator.

⑪ *Feedforward control* The object of the feedforward control is to vary the conduction period of the power transistors V11 and V51 in accordance with the d.c.-link voltage so that approximately constant voltage-time integrals are applied to the 20 kHz transformer T1. It also determines the maximum voltage-time integral for the single-ended flow converter. The voltage and current control thus has only to compensate for deviations due to loading and temperature. For this purpose a voltage proportional to the d.c.-link voltage, and hence also to the mains supply voltage, is tapped off from the secondary circuit of the 20 kHz transformer and applied to the feedforward control ($U_{Zkü}$).

⑫ *20 kHz clock generator* The clock (sawtooth) generator determines the switching frequency (about 20 kHz) and the maximum switching ratio of the power transistors V11 and V51.

⑬ *Trigger set* The trigger set determines the conduction period of the power transistors by comparing the sawtooth voltage from the clock generator with the variable control voltage from the current regulator. Switching off on overcurrent is also effected through the trigger set.

The switching ratio V_T can be varied between zero and a maximum of 0.4 (Fig. 2.29).

The control voltage is smaller in Fig. 2.29(a) than in Fig. 2.29(b). The conduction period of the power transistors V11 and V51 is therefore shorter in example (a) than in example (b).

⑭ *Pulse amplification* The pulse amplifier receives trigger pulses from the trigger set and raises them to the power level required to drive the power transistors V11 and V51. It is also controlled by the pulse inhibition and switch-on

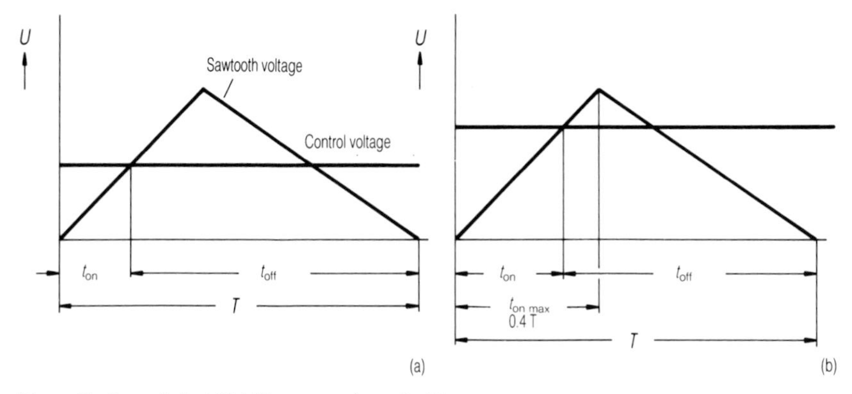

T	Cycle period at 20 kHz, approximately 50 μs
t_{on}	'On' period
t_{off}	'Off' period

Fig. 2.29 Sawtooth voltage and control voltage

enable. The trigger pulses can be short-circuited, and the power transistors thereby turned off, in the following circumstances:

▷ overcurrent (the overcurrent switch-off circuit controls the trigger set directly),
▷ overvoltage,
▷ pending the switch-on enable.

⑮ *Overcurrent switch-off circuit* If the current regulator fails to limit the output current to the maximum level set in the actual-current feedback amplifier, the overcurrent switch-off circuit turns off the power transistors V11 and V51 via the trigger set, and also the current regulator, instantaneously when the limit is exceeded by about 20%.

⑯ *Instantaneous overvoltage monitor with test circuit* The sensing circuit and the voltage-limiting unit are located in the voltage-limiting and indication assembly A5, the logic and the test circuit in assembly A1.

If the output voltage exceeds a maximum value set in the A5 assembly, the voltage-limiting thyristor V70 is triggered, the pulse-inhibit command is transmitted by the instantaneous overvoltage monitor with test circuit to the pulse inhibition and switch-on enable ⑱ and the equipment switched off.

After a disconnect time (inoperative period) of about 3 min the rectifier is automatically switched on again (pulse enable). If a second overvoltage occurs within a further 3 min (the second half of the 6 min total testing time), the monitoring circuit 'locks out' the rectifier. The voltage-limiting thyristor V70 is again triggered – as on all overvoltages – and the pulse inhibit is applied. Thereafter, the equipment can only be restored to operation manually – i.e. by switching off and on again by means of Q1.

If no further overvoltage occurs during the last 3 min of the testing time, the rectifier remains in operation.

⑰ *Delayed overvoltage monitor* The measuring circuit of the delayed overvoltage monitor checks the equipment output voltage. The rectifier is switched off if the product of overvoltage and time exceeds a predetermined limit. The response level of this circuit is different from that of the instantaneous overvoltage monitor.

In this case also the rectifier is switched off by the pulse inhibition; this can be released only by switching off and on again by means of Q1.

⑱ *Pulse cancelling and switch-on enable* The pulse cancelling (inhibit) and switch-on enable prevents the application of drive to the power transistors until the input filter capacitors C1 and C2 are properly charged.

Technical data

The principal technical data for the 48 V/30 A GR3 (GR31), GR12 rectifiers are listed in Table 2.4.

2.2 D.C./D.C. Converters

D.C./d.c. converters (GUR) are designed either as single-height or double-height power supply units or as complete built-in equipments; they all deliver constant controlled component supply voltages.

2.2.1 Type 20 to 30 W

The single-height d.c./d.c. converters 20 to 30 W (Figs 2.30 and 2.31), in SIVAPAC® construction, operate as single-ended blocking converters (20 kHz, pulse width control); they produce, from a 48 or 60 V input (40 to 75 V), an output voltage of 5, 12, 24 or 60 V (of either polarity) or two outputs of +12 V and –5 V. To provide the second controlled output in the latter case, the circuit board is provided with an additional plug-in board (see Fig. 2.31). All d.c./d.c. converters have the same basic circuit and the same circuit board design. Their grip boards carry sockets for voltage measurement and a light-emitting diode to indicate operation.

Table 2.4 Technical data for rectifier type 48 V/30 A GR3 (GR31), GR12

Mains input supply			
Voltage	(V)	110 to 127 +10 to −20% or * 220 to 250 +10 to −20% **	
Frequency	(Hz)	50 or 60 ±5%	
Equipment fuse protection	(A)	40 * or 25 **	
Degree of radio interference		Limit class B (VDE 0871)	
D.C. output		GR3 (GR31)	GR12
Operating mode or condition		Rated direct voltage (V) (tolerances for 1.5 to 100% rated current) Equipment voltage ≙ load voltage (except for initial charging with communications system disconnected)	
		Lead–acid battery with 24 cells	Lead–acid battery with 25 cells
Rectifier or parallel operation/ trickle charging (2.23 V/cell)		53.5 ±1%	56 ±1%
Parallel operation/charging (2.33 V/cell)		56 ±1%	58.5 ±1%
Initial charging (at rated mains supply voltage with communications system disconnected)		65	67.5
Rated direct current	(A)	30	
Interference voltage	(mV)	≦0.5 (frequency-weighted with CCITT 'A' filter)	
Dimensions ($H \times W \times D$)	(mm)	208 × 650 × 413	

Application

The d.c./d.c. converters are used in local power supplies for loads in 48 and 60 V communications systems, such as:

▷ telephone systems EMS 600 and 12 000,
▷ system EWSA.

Operating modes

The d.c./d.c. converters can be used for the rectifier operation mode or the parallel standby mode (Fig. 2.32). The example relates to the telephone systems EMS 600 and 12 000.

Fig. 2.30 D.C./D.C. converter 5 V/4 A

Fig. 2.31
D.C./D.C. converter +12 V/2 A, –5 V/1 A (with additional circuit board)

Basic circuit and operation

As shown in Fig. 2.33, the input voltage is applied periodically through the input filter and the switching transistor to the energy-storage transformer. Thus the transformer absorbs energy during the conducting period of the power transistor. In the non-conducting period this energy is delivered to the output through the rectifier diode and the output filter. The switching frequency is 20 kHz. Current and voltage control are effected by shifting the periodic switching instant of the transistor T3. The output is controlled, short-circuit-proofed and monitored with regard to overvoltage and undervoltage. Operation is indicated by a light-emitting diode.

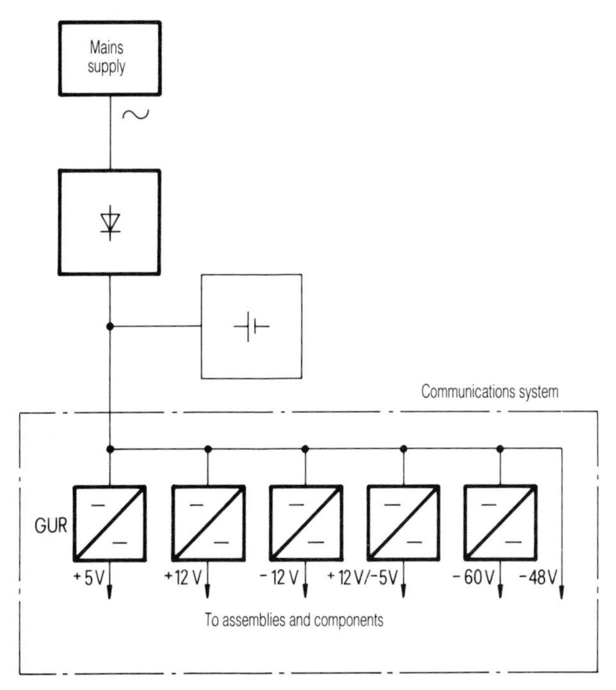

Fig. 2.32 Rectifier mode or standby parallel mode

Power section

The power section of the d.c./d.c. converter consists of:

▷ input filter L1, L2, C2, C3, C7 and C8,
▷ switching transistor T3,
▷ storage transformer Tr2,
▷ rectifier diode G9,
▷ output filter L3, C16, C17, C19 and C21.

The primary and secondary circuits are coupled through the storage transformer Tr2. The switching transistor T3 periodically switches the input voltage onto the primary winding. During the conducting period of T3, only primary current flows to magnetized the transformer so that it stores energy. In the ensuing non-conducting period of T3, secondary current flows in the process of releasing the stored energy to the load.

The transformer Tr2 provides 'galvanic separation'. It transforms the rectangular-wave voltage to the level required in the output circuit, where it is rectified by the diode G9 and smoothed by the capacitors C16, C17, C19 and C21 and the

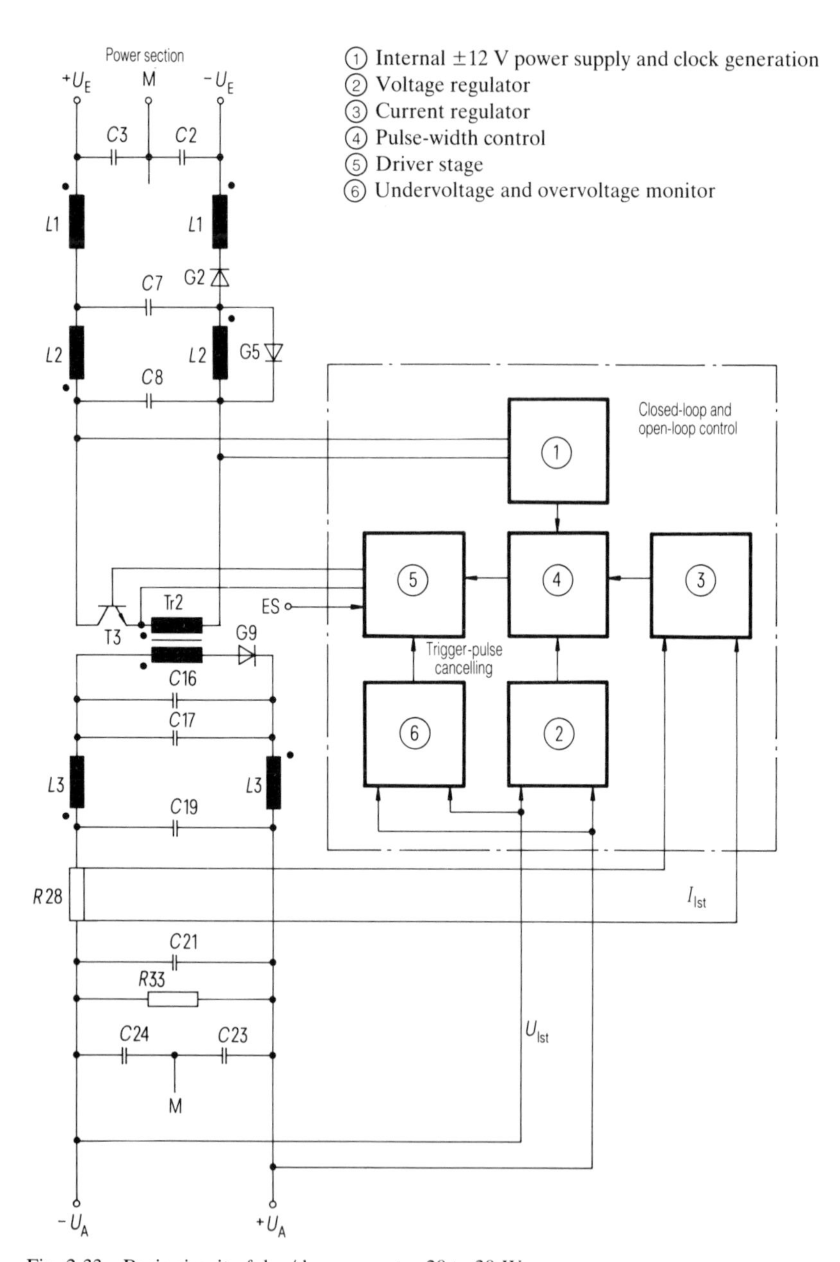

Fig. 2.33 Basic circuit of d.c./d.c. converter 20 to 30 W

inductor L3. Capacitors C23 and C24 are provided for EMI suppression. High- and low-frequency effects on the input supply due to the 20 kHz switching processes are limited to acceptable levels by the inductors L1 and L2 and the capacitors C2, C3, C7 and C8. Diode G5 suppresses overshoot when the supply is switched on, the residual load resistor *R*33 assists control stability, diode G2 blocks reverse current and provides reverse polarity protection and resistor *R*28 provides the actual-current feedback signal.

Closed-loop and open-loop control

The auxiliary supplies are produced from the input by the ±12 V *internal power supply* ① and stabilized by an e.m.f. (electromagnetically coupled) control system and open-loop correction. The switching frequency from the *clock generation* ① is about 20 kHz and the maximum switching ratio V_T is approximately 0.5. The switching ratio necessary for control is determined by a sawtooth voltage, which is superimposed on the actual-voltage feedback signal and compared with the set-point (reference) voltage.

The actual-value feedback signal for the e.m.f. control is obtained through a capacitor; with variations in input voltage or load this produces a continuous opposing effect on the mark-space ratio. The comparison between the reference voltage and the controlled output voltage is made by the *voltage regulator (controller)* ②; the difference is amplified. The output produces a sawtooth voltage, which determines the duration of the control pulses for the driver stage through the input threshold of a logic circuit. In parallel with this, the *current regulator (controller)* ③ (current limiter) reduces the pulse width of the control signals if a preset current limit is exceeded. The direct current in the output circuit is measured using a defined section of circuit-board track as a measurement shunt (*R*28). The difference between the measured voltage and a predetermined opposing voltage is converted by an operational amplifier into a corresponding output signal, which modifies the switching ratio. Since the reference is constant, the onset of current limiting is not affected by input voltage variations.

A logic circuit in the *pulse-width control* ④ generates a drive signal for the driver stage which can be shortened in comparison with the basic pulse, depending upon the required output power. The power transistor T3 is turned on or off by the *driver stage* ⑤.

The output voltage is monitored by the *undervoltage and overvoltage monitor* ⑥. In operation within the rating limits the green light-emitting diode on the board of the assembly is lit, the 'potential-free' contact is closed and a high signal is produced. If an undervoltage is detected at the output, it is indicated through the contact (relay contact opened) and by a low signal. If the rated voltage is exceeded by up to 40%, the output is switched off by suppression of the trigger pulses. The switched-off condition is maintained, since the auxiliary supply continues to be produced. The power section can be restored to operation by a brief application

Table 2.5 Technical data for d.c./d.c. converters 20 to 30 W

D.C./D.C. converter		Type					
		5 V/4 A	12 V/2 A	+12 V/2 A –5 V/1 A	60 V/0.4 A	24 V/1 A	+12 V/2 A –5 V/0.01 A
Abbreviated designation							
As used in switching system EWSA (with alarm relay)		EWA	EWB	EWC		EWF	EWG
As used in telephone system EMS (without alarm relay)		GAA	GAB	GAC	GAE	GAF	GAG
Input							
Rated direct voltage	(V)	–48 or –60 (–40 to –75)					
Rated direct current							
at 48 V	(A)	0.62	0.68	0.92	0.63	0.63	0.68
at 60 V	(A)	0.49	0.54	0.73	–	0.52	0.55
Degree of radio interference		Limit class B (VDE 0871)					
Output							
Rated direct voltage	(V)	5 ±3%	12 ±3%	+12 ±3%/ –5 ±3%	60 ±3%	24 ±3%	+12 ±3%/ –5 ±3%
Rated direct current	(A)	4	2	2/1	0.4	1	2/0.01
Polarity		Either	Either	Prescribed (rel. to 0 V)	Either	Either	Prescribed (rel. to 0 V)
Dimensions ($L \times W$)	(mm)	SIVAPAC® construction, single-height: 277 × 110					

of earth potential (>50 ms) to the remote control input ES, or by switching the input supply off and on.

Technical data

The principal technical data for the 20 to 30 W d.c./d.c. converters are listed in Table 2.5; this refers to the application of the assemblies to the EMS 600 and 12 000 telephone systems and the EWSA system.

2.2.2 Type 11 to 125 W

The double-height d.c./d.c. converters 11 to 125 W (Fig. 2.34) in SIVAPAC® construction operate, in general, on the single-ended flow converter principle (40 kHz, with pulse-width control). From an input supply of 48 or 60 V (40 to 75 V), they produce an output voltage of 5, 12 or 24 V (of either polarity).

Assemblies with two or four outputs employ the single-ended blocking-converter principle. The grip boards carry sockets for voltage measurement, a light-emitting diode (indicating operation) and a switch.

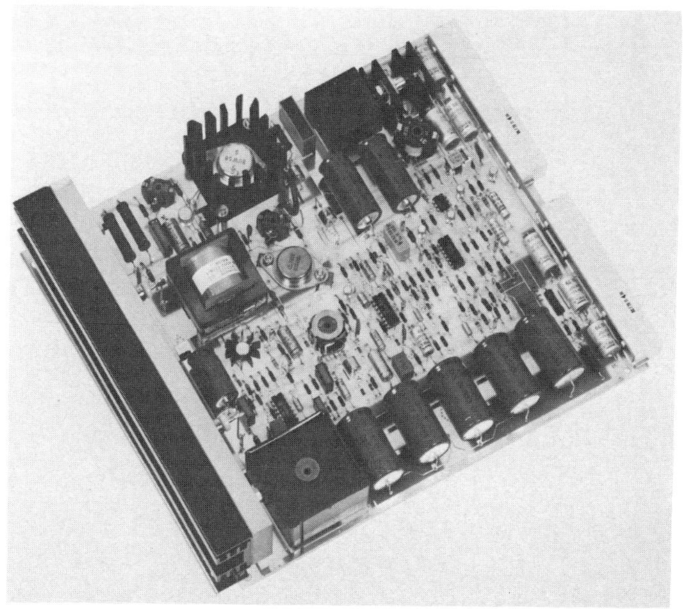

Fig. 2.34 Double-height d.c./d.c. converter

Application

The d.c./d.c. converters can be used as decentralized power supplies to loads in 48 and 60 V PCM communications systems – for example:

▷ the EWSD system,
▷ the KN system.

Operating modes

The d.c./d.c. converters are used for the rectifier mode or the standby parallel mode (see Fig. 2.32).

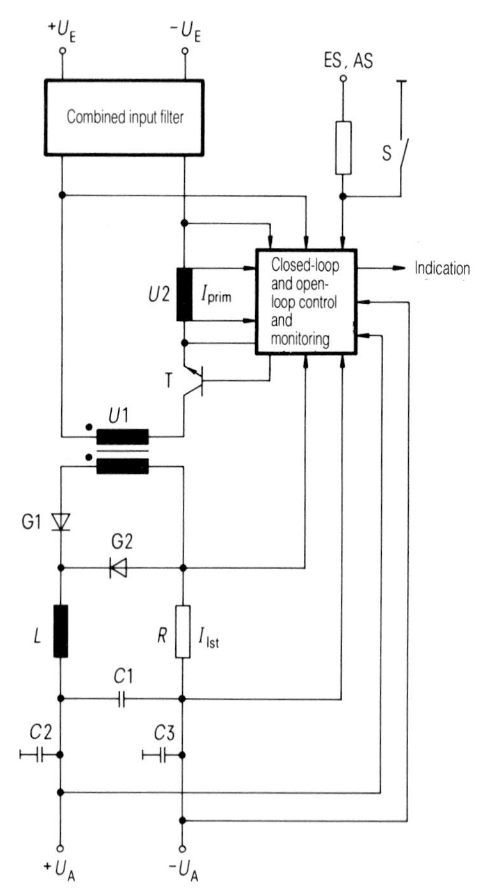

Fig. 2.35
Basic circuit of a d.c./d.c. converter (single-ended flow converter)

Basic circuit and operation

The basic circuit of the double-height d.c./d.c. converters is shown in Fig. 2.35.

The input is applied periodically to the transformer U1 through the input filter and the switching transistor T, transformed to the required output level, rectified by the diode G1 and smoothed by the output filter L and C1. The switching frequency is 40 kHz. The conduction period of the transistor T is determined by the closed-loop and open-loop control.

The output is controlled, short-circuit-proofed and monitored for undervoltage and overvoltage.

Operation is indictated by a green light-emitting diode, which lights when the output voltage is at its rated value.

The *power section* of the d.c./d.c. converter consists of

▷ combined input filter (h.f. and l.f.),
▷ switching transistor T,
▷ primary current transformer U2,
▷ power transformer U1,
▷ rectifier diode G1,
▷ free-wheel diode G2,
▷ output filter L and C1,
▷ current-measuring resistor *R* for current limiting.

The transistor T applies the voltage periodically to the primary winding of the transformer U1. During the conduction period of the transistor T, current flows through the rectifier diode G1. The transformer U1 provides 'galvanic separation' and transforms the input voltage to the level required in the output circuit. In the non-conducting period the energy stored in the inductor L flows through the diode G2. The output voltage is filtered by the inductor L and the capacitor C1. Capacitors C2 and C3 are for EMI suppression.

The resistor *R* is used to obtain the actual-current feedback signal in the output circuit. The input current of the converter is measured with transformer U2, and the transistor T is thereby protected from overcurrent spikes.

The principles of *closed-loop, open-loop control (regulation, control) and monitoring* are illustrated in Fig. 2.35. The basic mode of operation is generally as in Fig. 2.33.

The auxiliary supply of ±12 V is obtained from the input and stabilized by e.m.f. (electromagnetically coupled) control.

The switching frequency of 40 kHz is produced by an integrated control circuit. The maximum switching ratio is 0.5.

An integrated control circuit is similarly used to control the output voltage, synchronized to the frequency of the auxiliary power supply. The voltage-regulator unit performs the set-point-actual comparison, amplifies the control signal and effects the conversion into a corresponding pulse-width-controlled trigger pulse. If the voltage across the resistor R exceeds a predetermined value due to overload, the regulator unit reduces the conduction period.

The functions of the undervoltage and overvoltage monitor and the remote control input ES are as explained in Section 2.2.1. A particular converter can be switched on and off by means of the switch S, while the control point AS enables a number of converters to be switched off simultaneously.

Technical data

The principal technical data for the 11 to 125 W d.c./d.c. converters are listed in Table 2.6, which refers to the application of the assemblies in the ESWD and KN systems.

2.2.3 Type 80 W

The 80 W d.c./d.c. converter (Fig. 2.36, assemblies construction) employs the single-ended flow-converter principle (25 kHz with pulse-width control).

Application

The equipment is used in power supplies for the EMS 20 and EMS 30 telephone systems.

Mode of operation

The 80 W d.c./d.c. converter is used in the standby parallel mode, in conjunction, for example, with the rectifiers described in Section 2.1.3.2.

In the 80 W d.c./d.c. converter, the mains input assembly of the 150 W rectifier (see Section 2.1.2.1) is replaced by a battery input assembly.

Block diagram and operation

The block diagram of the 80 W d.c./d.c. converter is shown in Fig. 2.37. Besides the battery input assembly BEBG it contains the output assemblies ABG used in the 150 W rectifier and also a 25 Hz ringing generator G 60 V/8 VA. The battery input assembly consists of two SIVAPAC® printed circuit boards – the battery input assembly control BEBS and the battery input assembly power section BEBL. On the front board of the battery input assembly are mounted the protective devices SI1 to SI3, two test sockets and a light-emitting diode. SI1 and

Table 2.6 Technical data for d.c./d.c. converters 11 to 125 W

D.C./D.C. converter		Type						
		5 V/14 A	12 V/6 A	12 V/2.5 A 5 V/4 A	+12 V/0.8 A −5 V/0.3 A	+12 V/1.1 A −12 V/0.5 A +5 V/4 A −5 V/1.2 A	24 V/3 A	5 V/20 A
Abbreviated designation		EZA	EZB	EZC	EZG	EZH	EZF	EZL
Input								
Rated direct voltage	(V)	−48 or −60 (−40 to −75)						
Rated direct current at 48 V at 60 V	 (A) (A)	 2.02 1.64	 1.92 1.54	 1.53 1.23	 0.45 0.37	 1.61 1.29	 1.86 1.50	 3.72 3.00
Degree of radio interference		Limit class B (VDE 0871)						
Output								
Rated direct voltage	(V)	5 ±3%	12 ±3%	12 ±3%/ 5 ±3%	+12 ±3%/ −5 ±3%	+12 ±3%/ −12 ±3% +5 ±3%/ −5 ±3%	24 ±3%	5 ±3%
Rated direct current	(A)	14	6	2.5/4	0.8/0.3	1.1/0.5 4/1.2	3	20
Polarity		Either	Either	Either	Prescribed (rel. to 0 V)	Prescribed (rel. to 0 V)	Either	Either
Dimensions ($L \times W$)	(mm)	SIVAPAC® construction, double-height: 277 × 220						

Fig. 2.36 D.C./D.C. converter 80 W

SI2 are remote indication cutouts for the protection of the –48 V loads. The fuse SI3 provides short-circuit and reverse-battery protection for the battery input assembly.

The battery input assembly comprises a single-ended flow converter and a –5 V adjustment controller. Its functions are:

▷ to provide from the –48 V battery supply the d.c. inputs and auxiliary supplies required for the operation of the following output assemblies,
▷ to supply a controlled direct voltage of –5 V to the telephone system.

The battery input assembly control BEBS produces from the –48 V battery supply an alternating voltage with a frequency of 25 kHz and applies it, via the battery input assembly power section BEBL, to a power transformer. From the secondary windings are taken alternating voltages which, after rectification and smoothing, are available as input and auxiliary supplies for the individual output assemblies ABG.

The battery input assembly control embodies the following functional units:

▷ control (25 kHz switching-regulator integrated circuit with peripheral components),
▷ driver stage,
▷ power stage,
▷ –5 V adjustment controller.

The battery input assembly power section BEBL consists of the power transformer U1, the inductors, L1, L2 and L3, the one-way rectifier diodes G1, G3, G5

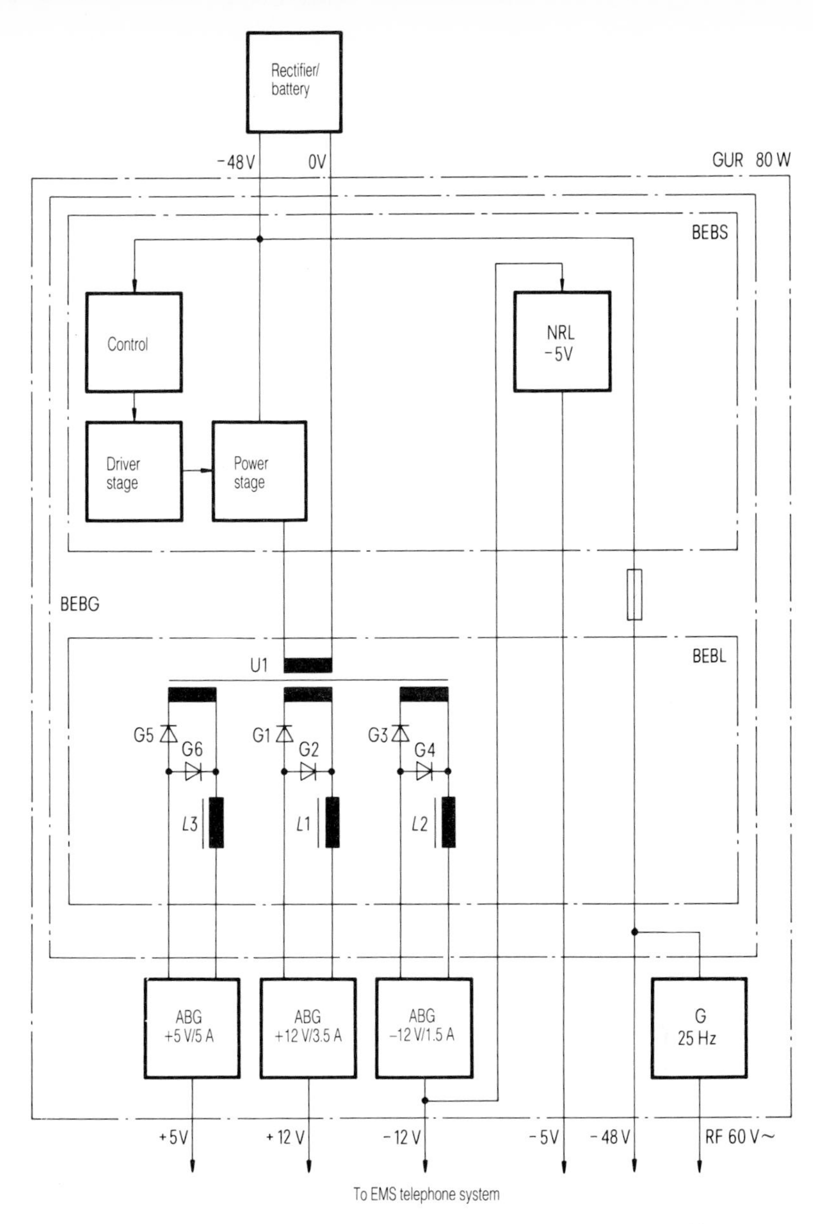

GUR	D.C./D.C. converter	ABG	Output assembly
BEBG	Battery input assembly	G	Ringing supply generator
BEBS	Battery input assembly control	RF	A.C. ringing supply
BEBL	Battery input assembly power section	NRL	Adjustment controller

Fig. 2.37 Block diagram of d.c./d.c. converter 80 W

Table 2.7 Technical data for d.c./d.c. converter type 80 W

D.C./D.C. converter		Type 80 W					
Input							
Rated direct voltage	(V)	–48 (–44 to –56)					
Degree of radio interference		Limit class B (VDE 0871)					
Outputs		Battery	ABG	ABG	ABG	Adjustment controller in BEBG	Ringing generator G
Rated direct voltage	(V)	–48	+12 ±3%	–12 ±3%	+5 ±3%	–5 ±3%	
Rated direct current	(A)	–	3.5	1.5	5	0.01	
Interference voltage	(mV)	≦0.5 (frequency-weighted with CCITT 'A' filter)	–	–	–	–	
A.C. ringing supply output							
Ringing supply voltage	(V)						60 ±8%
Frequency	(Hz)						25 ±8% sinusoidal
Rated apparent power	(VA)						8
A.C. ringing current	(A)						0.13
Dimensions (*W*)	(mm)	SIVAPAC® assembly chassis with a width of 342 (half-width or 447 (2/3-width)					
		–	40	35	40	–	BEBG 134

and the free-wheel diodes G2, G4 and G6. Further secondary windings provide alternating voltages which, after one-way rectification and smoothing, serve as auxiliary supplies for the output modules ABG (not shown).

The output assembly ABG has been described previously in Section 2.1.2.1.

The 25 Hz/8 VA generator (G 25 Hz) produces from the –48 V battery supply an isolated 60 V ringing supply for the telephone system.

Technical data

The principal technical data for the 80 W d.c./d.c. converter are listed in Table 2.7.

2.2.4 Type 125 to 250 W

The 125 to 250 W d.c./d.c. converters (SIVAPAC® built-in construction, Fig. 2.38) employ the single-ended flow-converter principle (20 kHz, with pulse-width control); they produce from an input voltage of 48 or 60 V (40 to 75 V) an output voltage of 5, 12, 24 or 34 V (of either polarity) or two outputs of ±24 V.

The converters are suitable for rack mounting, each occupying a double space, and are connected by means of terminals on the connection assemblies on the left and right at the back. A 24-way connector is mounted adjacent to each of the two light-emitting diodes on the front for connections to the indication and remote-

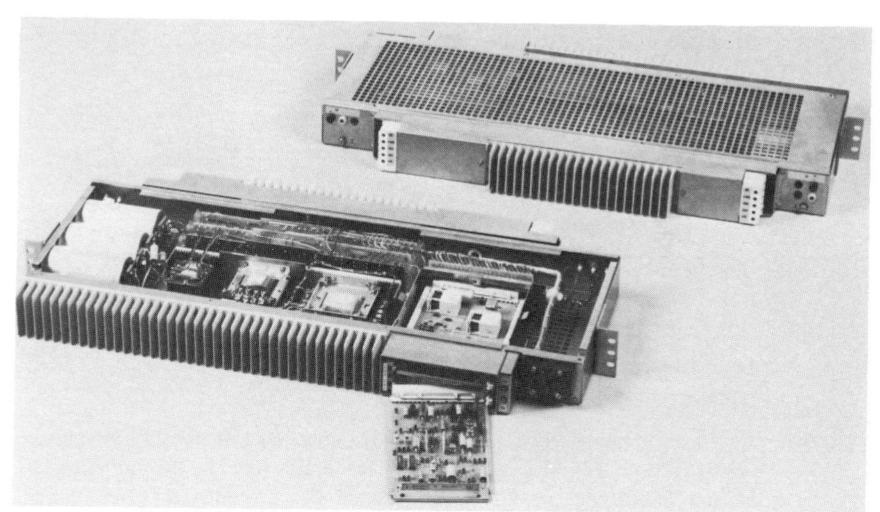

Fig. 2.38 D.C./D.C. converter (built-in equipment)

control circuits (cable plug) and for programming the operating mode (link plug). Test sockets are provided on the back for measuring the output voltage and current.

The equipments are constructed on a modular basis. Besides the two plug-in assemblies.

▷ closed-loop, open-loop control and monitoring and
▷ indication and remote control

it contains the additional assemblies:

▷ drive circuit,
▷ auxiliary converter,
▷ filtering.

All the assemblies are similar in all equipment types and in the event of a fault can be exchanged without special matching. The operating condition is indicated by light-emitting diodes. If the output voltage is within the rated range, the green LED lights. If it drops below the nominal range this LED is extinguished. If the output is switched off (due to overvoltage, undervoltage or remote switching) the red LED lights. On loss of input and/or undervoltage at the output (if the instantaneous undervoltage trip has not operated) no LEDs are illuminated.

Application

The d.c./d.c. converters are used in decentralized power supplies for loads in 48 and 60 V communications systems, such as:

▷ system EWSA,
▷ system EWSD.

Operating mode

The converters are used in the standby parallel mode (see Fig. 2.32).

Basic circuit and operation

The equipments consist of the power section, closed-loop control, open-loop control and monitor, drive circuit, auxiliary converter and remote control and indication (Fig. 2.39).

Power section

The effect on the central power supply ($+U_E$ and $-U_E$) of the 20 kHz pulses drawn by the power section is reduced to an acceptable level by the inductors L1 and L2 and the capacitors C1 to C8. Diode G1 provides reverse-polarity protection at the input and diode G2 limits the overvoltage at switch-on.

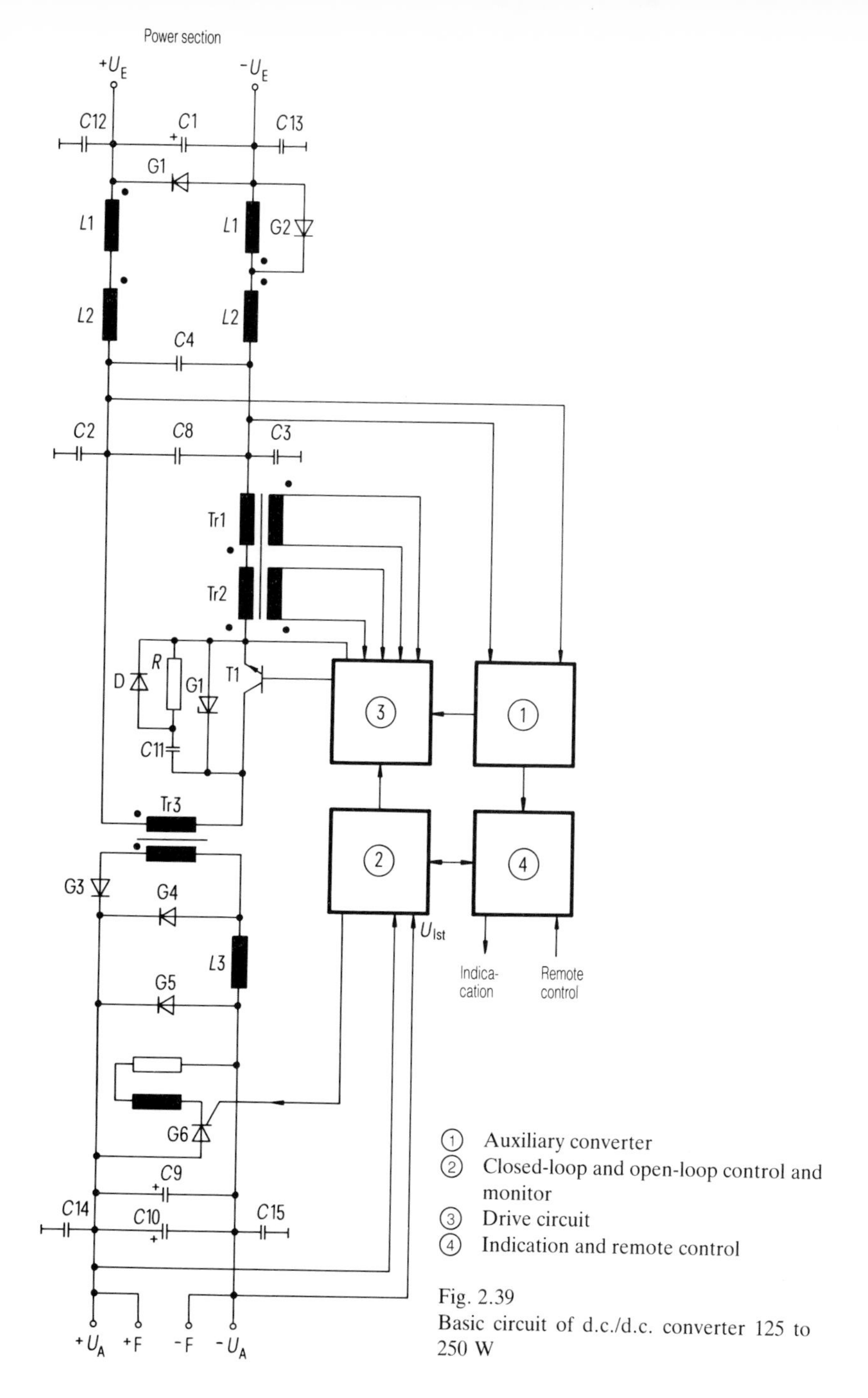

Fig. 2.39
Basic circuit of d.c./d.c. converter 125 to 250 W

The current transformer Tr1 in the drive circuit measures the primary pulse current and produces the signal required for the current limiter in the closed-loop, open-loop control section. The current transformer Tr2, also in the drive circuit, provides current feedback for the drive signals to the power transistor T1. Magnetic resetting of the power transformer Tr3 is achieved by means of a resistor–capacitor–diode network.

The switching transistor T1 (final control element) is driven by 20 kHz pulses from the drive circuit, and applies the input voltage periodically to the power transformer Tr3. The zener diode G1 limits the blocking voltage across the transistor T1 resulting from step changes in the pulse width.

After a downwards transformation, the alternating voltage produced by the periodic switching is rectified by diode G3 The output filter comprises the inductor L3 and the capacitors C9 and C10. During the non-conducting pulse periods the current in the inductor flows through the free-wheel diode G4.

To enable full use to be made of the power-handling capacity of the transformer Tr3 in unipolar operation, the transformer is reset during the non-conducting periods of the power transistor T1 by the resonant circuit consisting of the capacitor C11 and the primary inductance of Tr3. Diode G5 provides reverse-battery protection.

On overvoltage the monitoring circuit interrupts the trigger pulses and fires the limiting thyristor G6.

Capacitors C2, C3 and C12 to C15 are for EMI suppression.

Assemblies

Auxiliary supplies of –5 V, +12 V and +26 V are obtained from an *auxiliary converter* ① (blocking converter), which operates at a frequency of about 80 kHz. The 20 kHz clock frequency of the closed-loop and open-loop control is derived from the 80 kHz pulse generator. The smoothed output voltage is fed from the *closed-loop, open-loop control (regulation, control) and monitor* ② via a potential divider to the set-point-actual value comparator, where is is compared with the reference voltage and the difference applied to a control amplifier. The output signal of the regulator is superimposed on a sawtooth voltage derived from the basic 20 kHz frequency and determines the duration of the trigger pulses by reference to the input threshold of a logic circuit. The maximum conduction period is limited to 75%.

By a parallel process the pulse width of the trigger signal is also reduced if a predetermined current limit value is exceeded. For this purpose the primary current in the power section is measured by a current transformer; the signal is rectified and applied to an integrated voltage regulator. A current in excess of the set limit (1.05 to 1.3 × rated current) effects a continuous transition from constant-voltage control to current limiter. The current-limiting characteristic is designed with a degree of fold-back.

The monitor (in the closed-loop and open-loop control assembly) checks the output voltage for undervoltage and overvoltage. If the voltage drops below the rated value by 5 or 10% (selectable by links in the monitoring circuit) an appropriate signal is produced. If the acceptable output voltage is exceeded by 10 or 20% (similarly selectable) a further flow of energy is prevented by pulse inhibition. At the same time the voltage-limiting thyristor is triggered and effects a rapid suppression of the overvoltage. The closed-loop and open-loop control, the monitor and the auxiliary converter remain in operation. The switched-off state is maintained, and the power section can only be restored to operation by switching the supply off and on again or through the remote control. An instantaneous trip on the occurrence of an undervoltage alarm can be programmed in the link plug. Resumption of operation after switching off is then possible as in the case of an overvoltage trip.

The trigger pulses are amplified in the *drive circuit* ③ and applied to the base of the power transistor T1 through a transformer. A feedback arrangement matches the drive to the final control element to the required switching power. The power transistor remains in conduction until the controlling voltage pulse is terminated.

The indication and remote control ① translates the result of the monitoring into an (isolated) output signal. In addition, the output can be switched on or off remotely by a current pulse (isolated) or by an earth contact.

Technical data

The principal technical data for the 125 to 250 W d.c./d.c. converters are listed in Table 2.8.

2.2.5 Type 450 W

The 450 W d.c./d.c. converter (Fig. 2.40, module construction) employs the single-ended flow converter principle (20 kHz, with pulse-width control).

Application

The 450 W equipment is used to supply the EMS 150 and EMS 180 telephone systems. For the EMS 80 system there is a 300 W variant, which has a different complement of modules. The light-emitting diodes and test points can be seen on the front.

Operating mode

The 450 W d.c./d.c. converter is used in the standby parallel mode. In addition to the outputs shown in Fig. 2.37 it provides a further output of –60 V. The 450 W

Table 2.8 Technical data for d.c./d.c. converters 125 to 250 W (type I100 for the EWSA and EWSD systems; all others for the EWSA system)

D.C./D.C. converter	Type					
	24 V/6 A	12 V/10 A	+24 V/6 A –24 V/1 A	5 V/25 A	34 V/6 A	5 V/50 A
Abbreviated designation	B100	C100	E100	F100	A100	I100
Input supply						
Rated direct voltage (V)	–48 or –60 (–40 to –75)					
Rated direct current (A)	2.9	2.55	3.4	3.05	4.05	6.1
Degree of radio interference	Limit class B (VDE 0871)					
Output						
Rated direct voltage (V)	24 ±3%	12 ±3%	+24 ±3%/ –24 ±3%	5 ±3%	34 ±3%	5 ±3%
Rated direct current (A)	6	10	6/1	25	6	50
Polarity	Either	Either	Prescribed (rel. to 0 V)	Either	Either	Either
Dimensions ($H \times W \times D$) (mm)	SIVAPAC® construction (built-in equipment): 58 × 656 × 330					

Fig. 2.40 D.C/D.C. converter 450 W

equipment is used in conjunction with, for example, the rectifiers described in Section 2.1.3.2.

The 450 W d.c./d.c. converter is similar in construction to the 650 W rectifier (see Section 2.1.3.1), but contains a battery input module in place of the mains input module and battery converter modules in place of the mains converter modules.

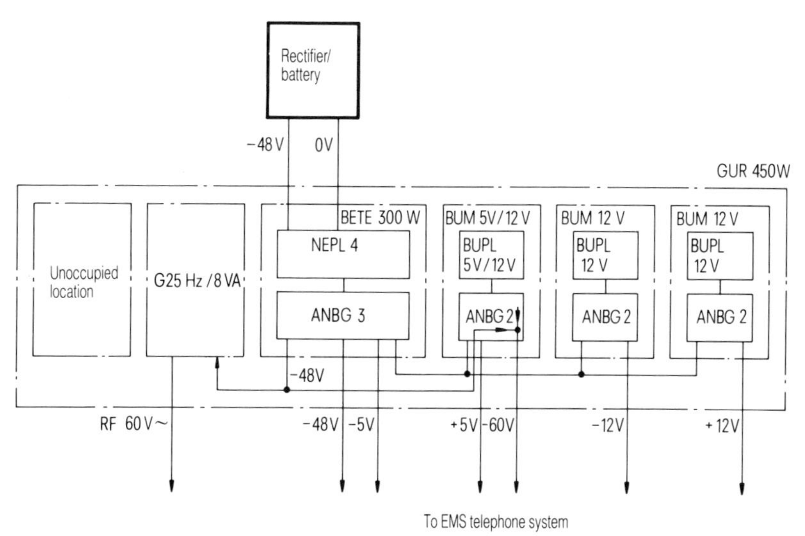

G	Ringing supply generator	NEPL	Mains circuit board
ANBG	Connection assembly	BUM	Battery converter module
RF	A.C. ringing supply	BUPL	Battery converter board
BETE	Battery input module	GUR	D.C./D.C. converter

Fig. 2.41
Arrangement of d.c./d.c. converter 450 W and component supply voltages

Block diagrams, basic circuit and operation

Figure 2.41 shows the equipment of the 450 W d.c./d.c. converter with the corresponding modules and the subassemblies contained in them. These include a battery input module BETE (300 W), a ringing generator G 25 Hz, a 'dual-output' battery converter module BUM and two 'single-output' battery converter modules BUM. The battery input module BETE passes the battery voltage of –48 V directly as a partial power supply to the telephone system and also to the ringing generator G 25 Hz and the battery converter module BUM 5 V/12 V. The G 25 Hz generator converts the –48 V to the 60 V alternating ringing voltage, and the battery converter module BUM adds the –12 V supply to the –48 V supply. This produces the required –60 V supply for the telephone system.

As shown in the block diagram of the 450 W d.c./d.c. converter (Fig. 2.42), this comprises an assembly of one battery input module BETE (Fig. 2.43) and a number of battery converter modules BUM (Fig. 2.44). The battery input module BETE is the central unit necessary for the operation of the battery converter modules BUM. These operate with pulse-width-controlled single-ended flow converters (20 kHz).

Battery input module

The battery input module BETE consists essentially of a power section with protective devices and EMI suppression filters and a mains circuit board NEPL 4.

The module is connected to the battery voltage of –48 V (permissible tolerance 44 to 56 V). The direct loads are supplied from separate –48 V outputs via fuses in the power section. The –48 V supply is passed through an EMI suppression filter and constitutes an input voltage U_E for the battery converter modules BUM and for the internal +10 V power supply.

The operation of the battery input module BETE (300 W) is explained more fully below with reference to the basic circuit diagram of Fig. 2.45.

The main functions of the battery input module BETE are:

- ▷ distribution of the 48 V load power to a number of separately protected branches,
- ▷ supply of power for the operation of individual battery converter modules BUM,
- ▷ production of a –5 V stabilized d.c. supply for small loads,
- ▷ indication of operating condition,
- ▷ generation of auxiliary supplies and overvoltage monitoring with detection of permanent faults for the operation of battery converter modules BUM.

The battery input module BETE (300 W) incorporated into the 450 W d.c./d.c. converter contains the mains circuit board NEPL 4, the connection assembly ANBG 3, peripheral components and test sockets. The battery input module

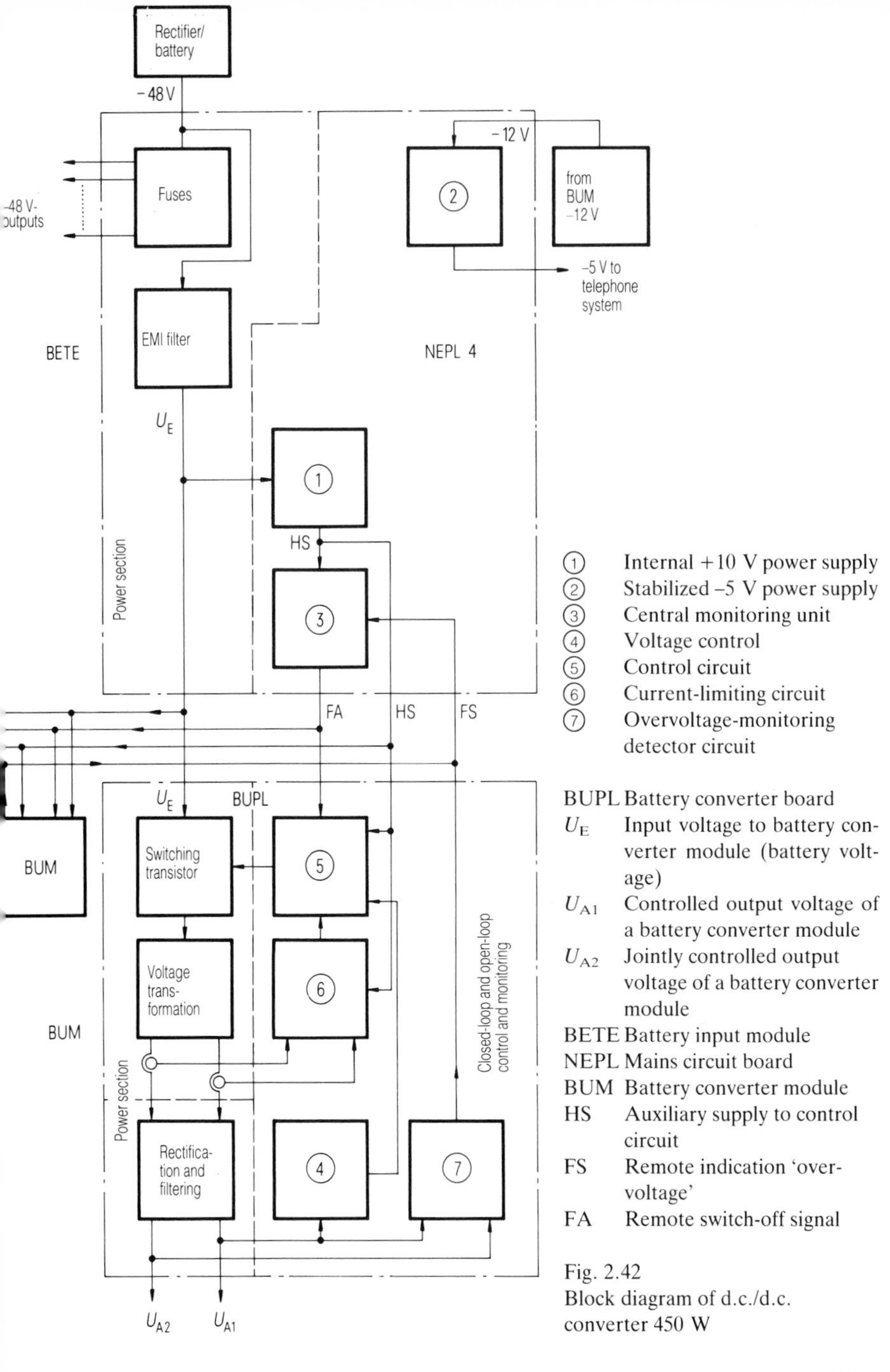

① Internal +10 V power supply
② Stabilized –5 V power supply
③ Central monitoring unit
④ Voltage control
⑤ Control circuit
⑥ Current-limiting circuit
⑦ Overvoltage-monitoring detector circuit

BUPL Battery converter board
U_E Input voltage to battery converter module (battery voltage)
U_{A1} Controlled output voltage of a battery converter module
U_{A2} Jointly controlled output voltage of a battery converter module
BETE Battery input module
NEPL Mains circuit board
BUM Battery converter module
HS Auxiliary supply to control circuit
FS Remote indication ‘overvoltage’
FA Remote switch-off signal

Fig. 2.42
Block diagram of d.c./d.c. converter 450 W

Fig. 2.43 Battery input module BETE (300 W)

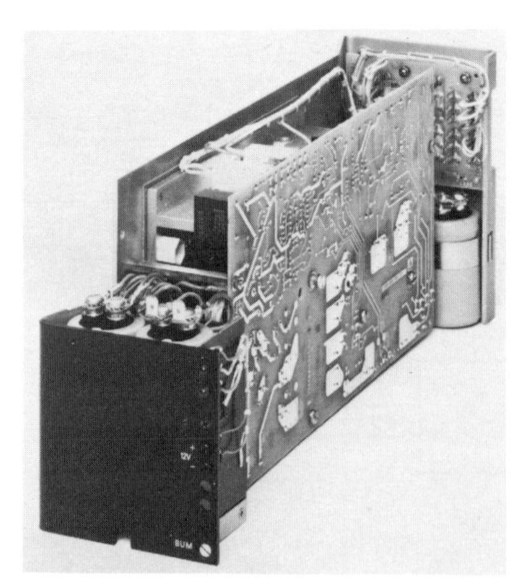

Fig. 2.44 Battery converter module BUM 12 V

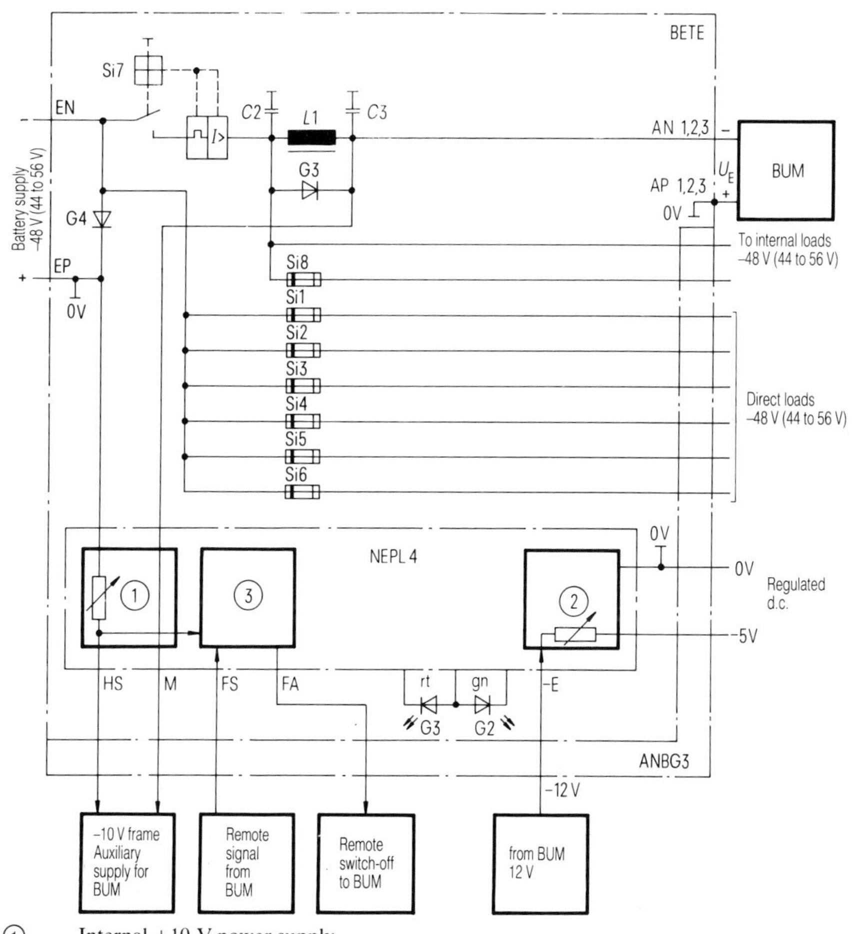

①	Internal +10 V power supply
②	Stabilized –5 V power supply
③	Central monitoring unit
U_E	Battery converter module terminal voltage –48 V (44 to 56 V)
BETE	Battery input module
NEPL	Mains circuit board
ANBG	Connection assembly
BUM	Battery converter module
G2	Light-emitting diode (green), auxiliary +10 V supply available
G3	Light-emitting diode (red), overvoltage on one of the outputs (battery converter module) and/or permanent fault

Fig. 2.45
Basic circuit of battery input module BETE (300 W)

Table 2.9 Technical data for d.c./d.c. converter type 450 W

D.C./D.C. converter	Type 450 W						
Input							
Rated direct voltage (V) Degree of radio interference	−48 (−44 to −56) Limit class B (VDE 0871)						
Outputs	Battery	BUM		BUM	BUM	Stabilized power supply in BETE 300 W	Ringing generator G
Rated direct voltage (V)	−48	+5 ±4%	−60 (−55 to −70), produced by −12 ±4% and battery voltage	−12 ±4%	+12 ±4%	−5 ±4%	
Rated direct current (A)	–	20	2.5	12.5	12.5	0.03	
Interference voltage (mV)	≦0.5 (frequency-weighted with CCITT 'A' filter)	–	≦0.5 (frequency-weighted with CCITT 'A' filter)	–	–	–	
A.C. ringing supply output							
Ringing supply voltage (V)							60 ±8%
Frequency (Hz)							25 ±8%, sinusoidal
Rated apparent power (VA)							8
Ringing supply current (A)							0.13
Dimensions ($H \times W \times D$) (mm)	The power supply chassis (single-width) matches the system cubicles. The mounting dimensions correspond to those of the single-height SIVAPAC® assembly chassis except for the somewhat greater depth: 148 × 658 × 358						

receives its supply from the battery through terminals EP and EN. The positive pole EP is connected to the frame of the module (functional earth); the negative pole is connected via the communication protection circuit breakers Si1 to Si6 to the connection assembly ANBG 3.

The input supply is taken to further outputs for connections within the converter through the automatic circuit breaker Si7 and the inductor L1 (EMI suppression filter C2, L1, C3). The power supplies to the battery converter modules BUM are conveyed by plug cables from the front of the battery input module to the fronts of the battery converter modules. The connections to the battery input module are numbered AP1 to 3 and AN1 to 3.

Battery converter module

The battery converter modules BUM are similar to the mains converter modules NUM (see Section 2.1.3.1). Working on the same principle, they produce from the battery supply obtained via the battery input module BETE either one or two constant direct voltages for the power supplies to the installation (see Fig. 2.26).

Like the mains converter module NUM, the battery converter module BUM provides 'potential-free' outputs, isolated from the input, and is short-circuit and and overload-proof, being protected against overvoltages by the central monitoring unit with detection of permanent faults (battery input module). Fault-free operation and overload are indicated by light-emitting diodes. Both outputs have reverse-battery protection by shunt diodes. The outputs may be of either polarity.

Technical data

The principal technical data for the 450 W d.c./d.c. converter are listed in Table 2.9.

2.3 Inverters with Transistor Power Sections

2.3.1 Clock-modulated equipments

2.3.1.1 Type 1000 VA

The 1000 VA inverter (Fig. 2.46) employs the flow converter principle (20 kHz, with pulse-width control). It produces a single-phase a.c. output of 220 V, 50 Hz from the d.c. input of 48 or 60 V.

Application

The inverter is used in decentralized power systems to supply 220 V loads, such

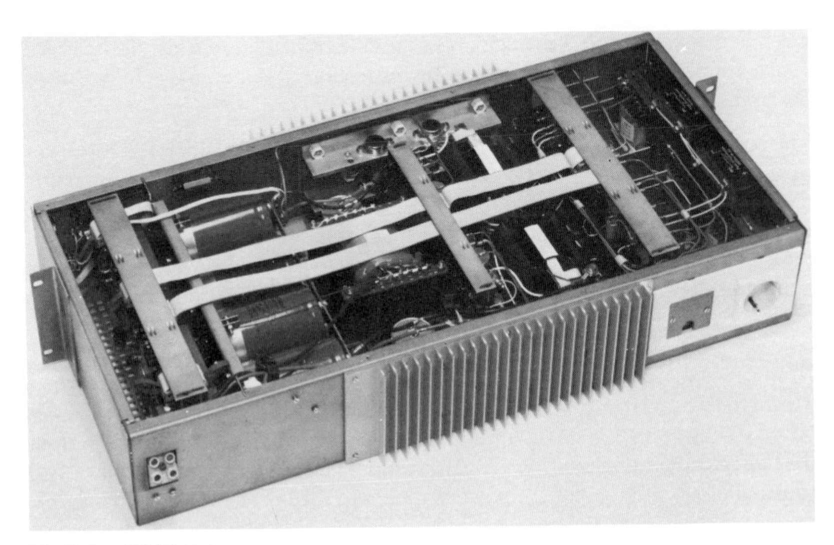

Fig. 2.46 1000 VA inverter

as:

- ▷ storage devices,
- ▷ teletype machines,
- ▷ printers,

in 48 and 60 V communications systems (e.g. system EWSD, text and data-transmission systems and loads supplied from solar generators).

Besides the 1000 VA inverter there are also 2.5 kVA and 100 VA equipments working on the same principle. In each case several inverters can be connected in parallel.

Operating mode

The inverters are normally used in the standby parallel operating mode. In this arrangement they are connected in the same way as the d.c./d.c. converters (see, for example, Fig. 2.32).

Basic circuit and operation

As shown in Fig. 2.47, the d.c. input voltage U_E (–40 to –75 V) is applied to the converter through a filter (low frequency), which is provided to attenuate the interference voltage. The converter is driven at a constant frequency (20 kHz) by regulator 1; the reference sine-wave is rectified and applied to the regulator 1 as a set-point voltage value U_{RM}. The modulated converter voltage U_M closely resembles a rectified a.c. mains voltage. The alternating output voltage U_A is

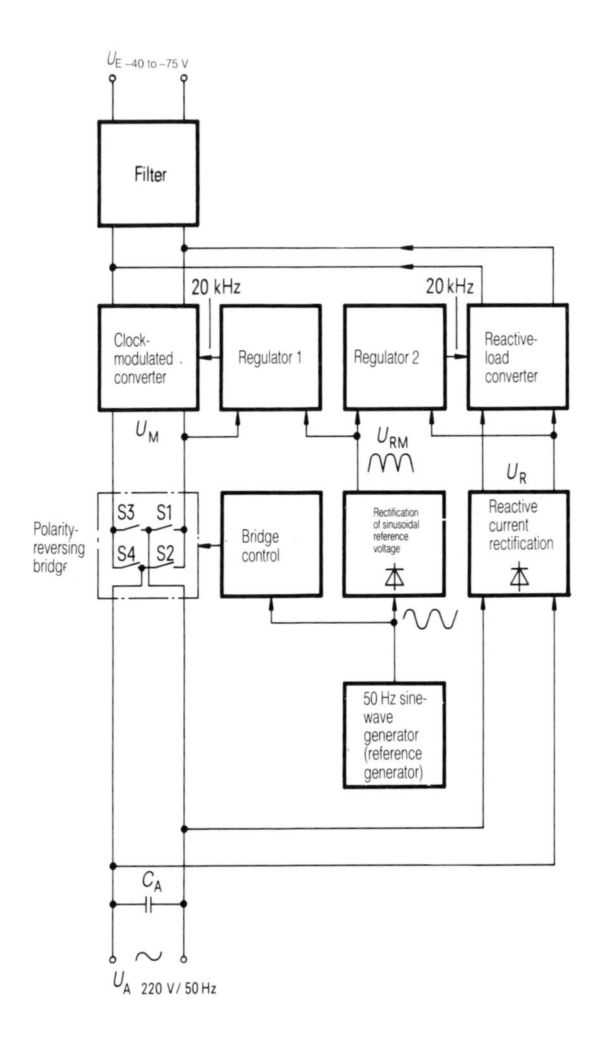

C_A Output capacitance
S1 to S4 Controlled bridge switch
U_A A.C. output voltage
U_E D.C. input voltage
U_M Modulated converter voltage
U_R Rectified output voltage
U_{RM} Reference voltage for the modulated converter

Fig. 2.47
Basic circuit of clock-modulated 1000 VA inverter

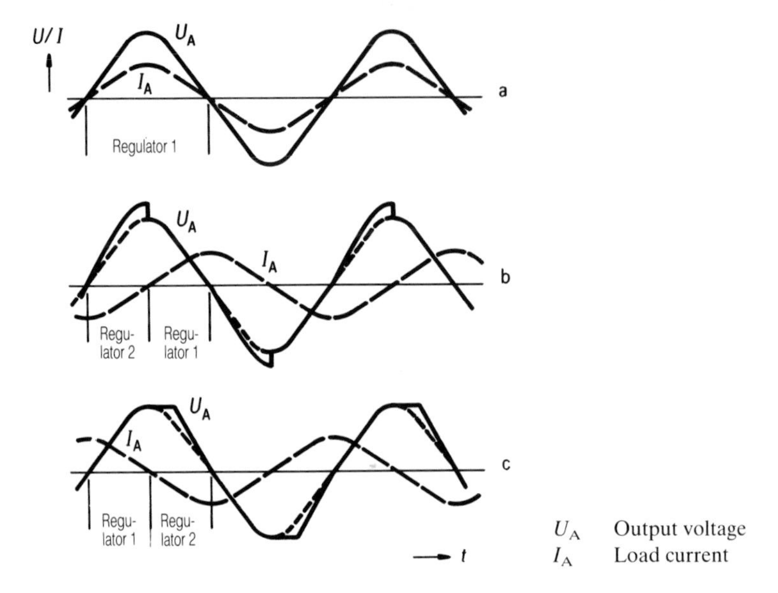

Fig. 2.48
Operating ranges of inverter regulators 1 and 2 with different kinds of load

subsequently produced by means of the polarity-reversing bridge, in which the bridge control circuit closes the arms S1/S4 and S2/S3 alternately at each zero-crossing of the sine-wave generator.

In general, allowance has to be made for complex loads. The energy stored in the reactive parts of the load distorts the output voltage waveform and deteriorates the distortion factor. A reactive compensation circuit is therefore provided; this takes the form of a reactive-load converter, which feeds the surplus energy back to the input supply. Figure 2.48 illustrates the basic voltage waveforms, according to the type of load:

a resistive load ($\phi = 0°$),
b inductive load ($\phi = -90°$),
c capacitive load ($\phi = +90°$).

The oscillogram in Fig. 2.49 shows the voltage waveform distortion at the output of the inverter with cos ϕ equal to about 0.7 (inductive). The distortion factor under this condition is less than 8%.

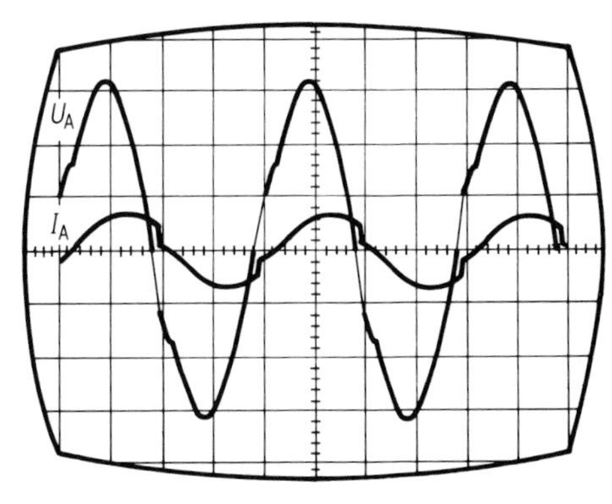

U_A Output voltage
I_A Load current

Fig. 2.49
Inverter output voltage and load current with cos ϕ of approximately 0.7 (inductive); the voltage distortion factor is less than 8% (each division: U_A = 100 V, I_A = 10 A, t = 12.5 ms)

Technical data

The principal technical data for the inverter type 1000 VA are listed in Table 2.10.

Table 2.10 Technical data for the 1000 VA inverter

Input		
Rated direct voltage	(V)	–48 or –60 (–40 to –75)
Maximum input current (with rated load and rated direct voltage)	(A)	28
Degree of radio interference		Limit class B (VDE 0871)
Output		
Rated alternating voltage	(V)	220 ±10%
Frequency	(Hz)	50 ±1%
Rated apparent power	(VA)	1000
Peak power	(VA)	1750
cos ϕ inductive		0.7
capacitive		0.86
Distortion factor at rated load		<10%
Interference voltage	(mV)	≦0.775 (frequency-weighted with CCITT 'A' filter)
Dimensions ($H \times W \times D$)	(mm)	118 × 660 × 330 (suitable for SIVAPAC® chassis

3 Energy Stores

Supplementing the discussion of basic principles in Part 1, the following chapter is concerned with the construction and application of energy storage devices.

3.1 Stationary Batteries

3.1.1 Requirements for lead-acid batteries

Stationary batteries are expected to meet the following requirements:

▷ long shelf life before commissioning,
▷ ease of installation,
▷ low commissioning cost,
▷ simple charging technique,
▷ high efficiency,
▷ reliability (high availability of capacity at any time),
▷ long life,
▷ low maintenance cost,
▷ safety against explosion,
▷ mechanical durability,
▷ ability to withstand short-circuits,
▷ environmental acceptability.

3.1.2 Lead–acid batteries types OPzS, OGi-block, OGiE and GroE

Any lead-acid battery consists of at least one cell containing dilute sulphuric acid and a set each of positive and negative plates; these are welded in alternating sequence to connector bars. The outer plates are always negative; the negative plate assembly therefore has one more plate than the positive.

The construction of the negative and positive electrodes depends upon the particular type of design. The designation of a lead–acid battery generally derives from the type of positive plate used.

Three types of construction are used in telecommunications power supply systems:

▷ OPzS batteries with tubular plates,[1])
▷ OGi-block or OGiE batteries with rod plates[2]) and
▷ GroE batteries with large-surface-area plates.[3])

In all cases the negative plates are of grid construction.

These batteries are available up to the largest capacities in *closed* containers; open cells are not considered nowadays.

Batteries of the OPzS design, and also the OGi-block and OGiE batteries, are now made with positive plates of lead containing less than 2% of antimony (previously 9 to 12%). The strength of the plates is ensured by means of special alloying additions (e.g. selenium). As a result of the reduction in the proportion of antimony, maintenance intervals of up to three years are possible with the new batteries. Within the stipulated interval, maintenance-free operation is guaranteed, provided that the battery is used in accordance with the relevant recommendations; e.g. the battery must not be overcharged. In particular, operation in the parallel mode (standby parallel mode) or changeover mode with continuous float charging of the battery (at 2.23 V/cell) is a prerequisite.

It is also possible to use an antimony-free alloy. To obtain the necessary strength, a proportion of calcium – less than 0.1% – is added to the lead for the grid.

Batteries incorporating these kinds of alloy have the same low water consumption as batteries with low-antimony alloys, but it is possible for them to get into a state of apparent passivity, in which the charge–discharge reaction is inhibited; to restore the normal charge–discharge behaviour, special treatment (an equalizing charge) is necessary. In addition, the number of possible charge–discharge cycles is reduced, but this is not too significant in the case of batteries which are used in the standby parallel mode or changeover mode, and only occasionally discharged (in the event of a mains failure or for the purposes of a functional test).

3.1.2.1 Data

A summary of the principal data appertaining to stationary batteries used in telecommunications power supplies is given in Table 3.1.

3.1.2.2 OPzS-type batteries

Lead–acid batteries type OPzS are the most frequently used in telecommunications power supplies.

---

[1]) OPzS: stationary (fixed) armoured cells (ortsfeste Panzerzellen) of special design; it is more usual nowadays to speak of 'tubular' than of 'armoured' cells or plates.

[2]) OGi: stationary (fixed) grid-plate (rod-plate) battery (ortsfeste Gitterplattenbatterie).

OGiE: stationary grid-plate (rod-plate) battery – individual-cell construction (Einzelzellenbauweise).

[3]) GroE: large-surface-area plate battery in fixed narrow construction (Großoberflächenplattenbatterie in Engbauweise).

Table 3.1 Data for stationary lead–acid batteries for telecommunications power supplies

	Blocks	Individual cells		
	OGi-block battery	OGiE battery	OPzS battery	GroE battery
Nominal voltage (V/cell)	2		2	2
Nominal capacity (Ah)	C_{10} 12,5 to 200	C_{10} 250 to 2000	C_{10} 250 to 12 000	C_{10} 200 to 2800
Nominal capacity refers to a nominal discharge current to a final discharge voltage of … V/cell at a temperature of 20 °C	(for C_{10} and I_{10}) 1.87		(for C_{10} and I_{10}) Depending on type, 1.78 to 1.82	(for C_{10} and I_{10}) 1.85
Trickle-charging voltage (V/cell)	2.23		2.23	2.23
Charging voltage (V/cell)	2.33		2.33	2.33
Initial charging voltage: end-of-charging voltage with I_a characteristic (V/cell)	2.6 to 2.8[1])		2.6 to 2.8[1])	2.6 to 2.8[1])
Charging factor	1.2		1.2	1.2
Internal resistance: approximate value after discharge of 10% of capacity (mΩ/100Ah)	1.2		3	1.4
Nominal acid density in charged condition (kg/l ±0.01 kg/l at 20 °C)	1.24		1.24	1.22
Fall in acid density from nominal value with 100% discharge (kg/l at 20° C)	Discharge current … /hours		Discharge current … /hours	Discharge current … /hours
	I_{10} I_5 I_3 I_1 $I_{1/6}$		I_{10} I_5 I_3 I_2 I_1	I_{10} I_5 I_3 I_2 I_1
	1.050 1.095 1.130 1.195 1.225		1.09 1.12 1.135 1.15 1.17	1.08 1.095 1.11 1.12 1.15
Application area/standby time (h): short-term loading	≦1		–	≦1
and/or long-term loading	1 to 10		1 to 10	–
Maintenance-free for … years (treated in accordance with the relevant directions)	3		3	2
Life (years)	12 to 15		12 to 15	15 to 20
Cell container	Translucent plastic		Transparent or translucent plastic Hard rubber	Transparent or translucent plastic

[1]) Applicable only to an unformed battery.

Example of type designation

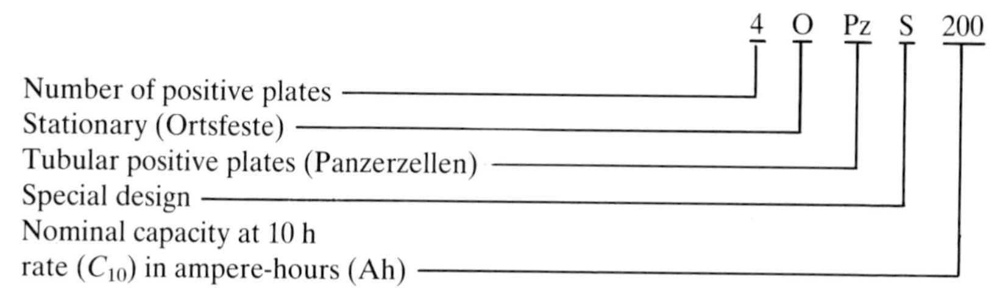

Depending on their length, a pair of plates of an OPzS battery has a capacity of 50, 70, 100, 125 or 200 Ah.

In the example given above, four 50 Ah plates would be needed to give the required nominal capacity of 200 Ah.

Construction

An OPzS battery is illustrated in Fig. 3.1(a). Figure 3.1(b) shows the internal construction of the OPzS cell. The OPzS cell embodies tubular (armoured) positive plates in conjuction with negative grid plates. The positive tubular plates consist of a side-by-side arrangement of thin lead rods – again, of a special low-

(a)

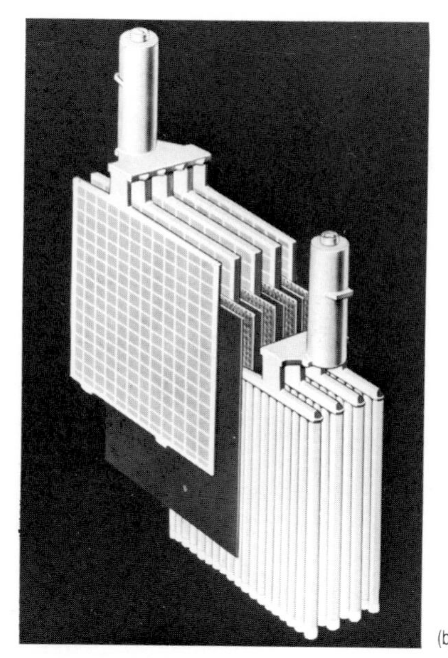

(b)

Fig. 3.1 OPzS-type lead–acid battery. (Photo by courtesy of VARTA AG)

antimony alloy – provided with centering pieces, over which are drawn woven pockets of a highly acid-permeable insulating material. The active material (lead dioxide) is contained in the space between the lead rod and the pocket. The pockets prevent shedding of the active material, so that no sludge is formed.

In the negative plates the active material (initially lead dioxide) is pressed into a grid of hard lead. Through the forming process the lead dioxide is converted electrochemically into finely divided spongy lead (Pb).

A microporous separator is used to insulate the plates, in conjunction with a spacer of corrugated PVC. The electrolyte is dilute sulphuric acid with a density of 1.24 kg/l at 20 °C.

Design

Individual OPzS cells are manufactured in transparent acid-proof plastic containers for capacities from 250 to 1500 Ah, in translucent acid-proof plastic containers for capacities of 2000, 2500 and 3000 Ah and in hard rubber containers for capacities from 3500 to 12 000 Ah.

Cells with capacities of from 250 to 3000 Ah are designed for bolted connections (with insulated copper links).

Cells of from 3500 to 12 000 Ah are currently available only with welded connections. The links are of hard lead, welded onto the cell terminals.

Application

OPzS batteries are very suitable for long discharge periods. They are therefore installed when the battery is required to cater for a mains supply interruption of from 1 to 10 h. Other application areas, apart from telecommunications, are in, for example:

- alarm systems,
- signalling systems,
- fire alarm systems,
- equivalent power supply systems conforming to VDE 0107 and VDE 0108.

Installation

OPzS batteries are installed on stillages, or in stepped or tiered racks (see Section 3.5).

Life

The life of OPzS batteries can be quoted as 12 to 15 years (for a reduction in capacity to 80%).

3.1.2.3 Type OGi-block and OGiE batteries

The block batteries type OGi and OGiE batteries are rapidly gaining favour for telecommunications power supplies, particularly for small-to-medium-sized telephone systems. Worthy of note are their low internal resistance, with correspondingly good voltage characteristics (even under rapid-discharge conditions with high loading), and the fully insulated construction.

Construction

Block batteries of the OGi type and OGiE batteries are illustrated in Fig. 3.2(a). The internal construction is shown in Fig. 3.2(b).

Both the OGi-block battery and the OGiE battery incorporate positive rod plates in association with negative grid plates. The rod plates are made of a special low-antimony lead alloy and are a development of the tubular positive plate. The active material (initially lead dioxide) is pasted round the rods and lead dioxide (PbO_2) is formed from it by an electrochemical process. To prevent the shedding of active material and the consequent formation of sludge, the positive rod plate is enclosed in a highly acid-permeable fabric pocket. To locate the vertical bars and to ensure mechanical stability the positive plates are provided with horizontal braces.

In the negative plates the active material (initially lead dioxide) is pressed into a grid of hard lead. Through the forming process the lead dioxide is converted

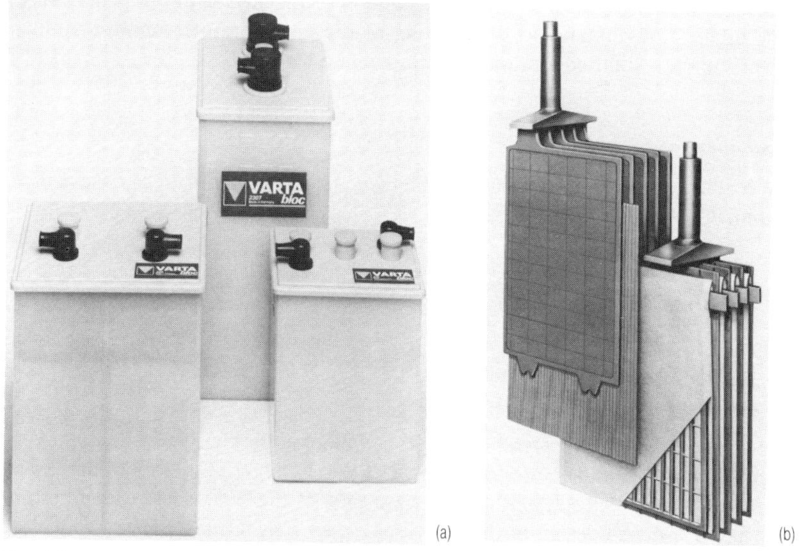

Fig. 3.2
OGi-block-type and OGiE-type lead–acid batteries. (Photo by courtesy of VARTA AG)

electrochemically into finely divided spongy lead (Pb). Positive and negative plates are assembled into plate stacks. Insulation is provided by a microporous separator and the electrolyte is dilute sulphuric acid with a density of 1.24 kg/l at 20 °C.

Design

The OGi-block batteries and the OGiE batteries are supplied in translucent plastic containers; they are in the form of three-cell blocks constituting 6 V units with capacities from 12.5 to 150 Ah, and two-cell blocks, as 4 V units, with capacities of 175 and 200 Ah (OGi-block). They are also available as individual cells with capacities from 250 to 2000 Ah (OGiE).

The interconnection of the block batteries or individual cells is effected by means of plastic-insulated copper links bolted to the terminals. External short-circuiting of the installed battery can be prevented by the use of additional rubber terminal caps.

Application

By virtue, particularly, of their previously mentioned low internal resistance the OGi-block batteries and the OGiE batteries can be used universally both for rapid-discharge operation in the range of minutes and for long discharge periods of from 1 to 10 h. As well as those indicated in connection with OPzS batteries, their applications extend to the starting of diesel generators, uninterruptible power supplies, solar generators and wind generators.

Installation

The block batteries are installed on stillages, or in stepped or tiered racks (see Section 3.5).

Life

The life of the OGi-block and OGiE batteries may be quoted as 12 to 15 years.

3.1.2.4 Lead–acid batteries type GroE

The GroE design represents a further development of the well-known and proven large-surface-area plate batteries (Gro). The Gro battery of the traditional type is no longer used, because, among other reasons, of the high water consumption associated with the design of the individual open cells. Other respects in which they fall short of the present state of battery technology are, for example, short maintenance intervals, large volume and high weight.

Example of type designation

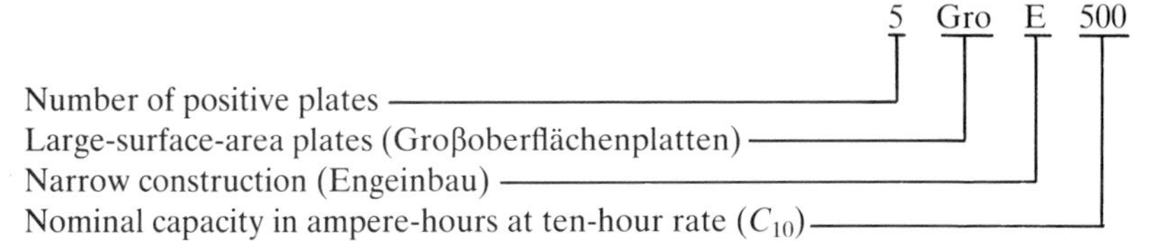

Depending upon the plate dimensions, a pair of plates in a GroE battery has a capacity of 25 or 100 Ah. In the example shown above, five 100 Ah plates would be required for a nominal capacity of 500 Ah.

Construction

A GroE battery is illustrated in Fig. 3.3. The GroE cell contains positive large-surface-area plates in association with negative grid plates.

The positive plates consist of soft cast lead. The surface area of the plate is increased by means of a special lamellar structure, affording a large area of contact with the acid. The active material is produced by an electrochemical process.

A microporous separator is introduced to insulate the plates. The electrolyte is dilute sulphuric acid with a density of 1.22 kg/l at 20 °C.

Fig. 3.3
GroE-type lead–acid battery. (Photo by courtesy of HOPPECKE, Carl Zoellner u. Sohn GmbH & Co KG)

Design

GroE batteries are constructed in transparent or translucent plastic containers for capacities of from 200 to 2800 Ah.

Application

The principal areas of application (mainly in the short-time range $\leqq 1$ h) are, for example:

▷ equivalent power-supply systems in power stations,
▷ UPS (uninterruptible (a.c.) power supply) systems.

Installation

GroE batteries are installed on stillages, or in tiered or stepped racks (see Section 3.5).

Life

The life is between 15 and 20 years.

3.1.3 Comparison of lead–acid batteries

Internal resistance

Lead–acid batteries differ significantly in their internal resistance R_i according to the various kinds of plate construction (Fig. 3.4).

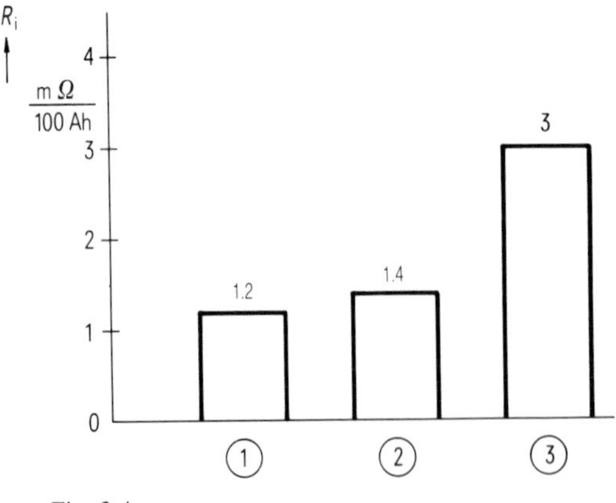

Fig. 3.4
Internal resistances R_i of stationary lead–acid batteries used in telecommunication power supplies

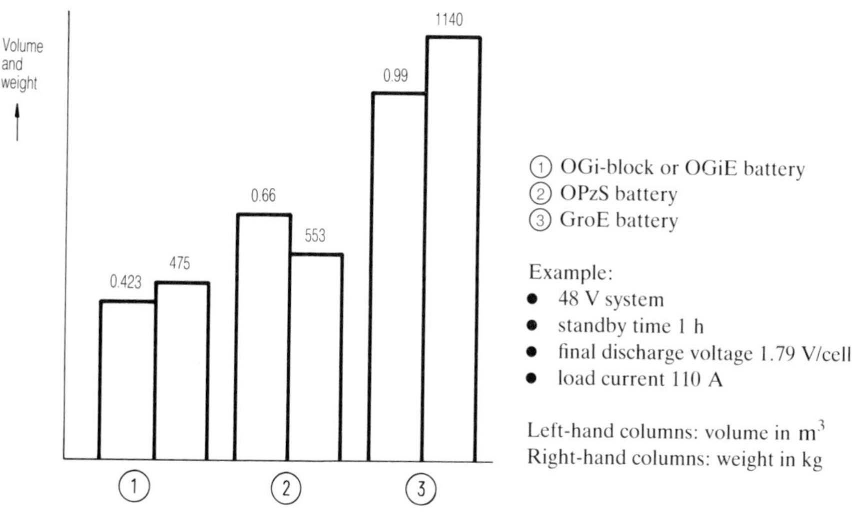

Fig. 3.5
Comparison of volume and weight between OGi or OGiE, OPzS and GroE batteries

Comparison of volume and weight

Figure 3.5 illustrates the variations in volume and weight, based on a comparison between the OGi or OGiE batteries and the OPzS and GroE batteries.

3.1.4 Vent plugs

There are essentially three types of vent plug in stationary batteries today:

▷ gas-drying plug
In this plug a filling of granules prevents the egress of acid droplets (acid mist), which can be produced by gas evolution during charging. The gas mixture itself can escape freely. These plugs are standard fittings on the OGi, OPzS and GroE cells.

▷ ceramic plug
This 'flame-trap' or 'explosion-proof' plug prevents ignition of gases present in the cell due to external causes. By 'ceramic funnel plug' is understood a variant of the flame-trap plug. Topping-up water can be added through a tube, in the form of a funnel, which reaches directly into the electrolyte, without removing the plug.

▷ recombination plug
The recombination plug ('Rekombinator') converts the explosive gas produced by charging back to water by a catalytic method, thus reducing the loss of water. This results in longer maintenance intervals for stationary batteries.

Due consideration must be given, in accordance with the manufacturer's recommendations, to the relationship between the quantity of gas given off, in the trickle-charging condition as well as in charging, and the capacity of the recombination plug. Under normal conditions the use of recombination plugs in stationary batteries embodying low-antimony alloys is unnecessary, since maintenance intervals of up to three years are in any case achievable.

3.2 Stationary Alkaline Batteries

Alkaline batteries ('steel accumulators') are only employed for telecommunications power supplies where they offer advantages over lead–acid batteries. Particularly noteworthy are their lower susceptibility to the effects of low temperatures and relative tolerance of deep discharging. As against these advantages, they are commonly more expensive, have less satisfactory voltage characteristics and are only available with capacities of up to 1250 Ah. Only a brief review of the alkaline battery will, therefore, be presented here. The description relates to nickel–cadmium batteries with pocket plate cell types TP and T.

3.2.1 Nickel–cadmium batteries types TP and T

Construction

The types TP and T nickel–cadmium batteries are illustrated in Fig. 3.6(a). Figure 3.6(b) shows their internal construction, with pocket plates. Both the positive and

(a)

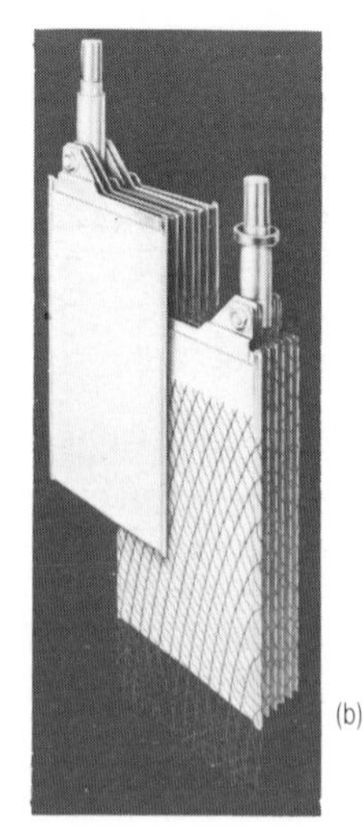
(b)

Fig. 3.6 Nickel–Cadmium batteries type TP and T. (Photo by courtesy of VARTA AG)

the negative electrodes are pocket plates, with active material consisting of nickel and cadmium hydroxides. The supporting structure is in the form of rectangular perforated metal pockets arranged in a grid pattern. The plates are insulated by a plastic mesh or a perforated corrugated separator. The electrolyte is dilute potassium hydroxide with a density of 1.20 kg/l.

Design

Individual cells type TP are supplied in impact-resisting, temperature-stable, translucent plastic containers with capacities of from 10 to 315 Ah, and those of type T in steel containers with capacities of from 380 to 1250 Ah.

Application

Ni–Cd pocket-plate cells types TP and T are very suitable for long discharge periods; they are preferably applied, therefore, in cases where mains supply interruptions of 1 to 10 h are to be catered for.

Charging

Charging of TP and T-type batteries is typically carried out according to an *IU* (constant current/constant voltage) characteristic.

The charging factor amounts to 1.4: that is to say, a given discharge has to be replaced by 140% of the equivalent charge.

In general, an initial charging current of $1.5 \times I_5$ up to a voltage of 1.55 V/cell is recommended. The maximum continuous voltage permitted in the EMS telephone system is 58 V. A battery of 38 cells can therefore be charged only up to 1.52 V/cell.

TP cells should be floated at 1.4 V/cell to maintain their charge. Under this condition they draw a current of 30 to 60 mA/100 Ah (C_5). The permissible operating temperature range is from –20 to +45 °C. An equalizing charge is necessary, in accordance with the manufacturer's instructions, in order to maintain the capacity.

Characteristic data

- ▷ nominal voltage: 1.2 V/cell,
- ▷ nominal capacity (C_5), related to a nominal charging current I_5 of $0.2 \times C_5$ (A) up to a final discharge voltage of 1.0 V/cell at 20 °C,
- ▷ charging voltage: 1.55 V/cell,
- ▷ float charging voltage: 1.4 V/cell,
- ▷ charging factor: 1.4,
- ▷ electrolyte: potassium hydroxide with a density of 1.20 ±0.02 kg/l,
- ▷ internal resistance: approximately 1.4 mΩ/100 Ah.

Installation

Blocks of multi-cell arrangements of TP batteries and individual T-type cells are installed on stillages, or in tiered or stepped racks.

Life

The life of TP and T-type batteries can be taken as about 10 to 15 years.

3.3 Design Pointers

3.3.1 Standby time

The following indications apply exclusively to lead–acid batteries. In d.c. telecommunications power supply systems the standby time specified may range from minutes to hours. A period of one to four hours is typical. A standby time of six hours is generally provided for the German Federal Republic Postal Administration. In installations which include a stationary standby generator, a period of four hours is allowed. In unmanned local exchanges a battery reserve of up to 72 hours is provided.

3.3.2 Number of cells

3.3.2.1 D.C. telecommunications power supply systems

The number of battery cells is determined by consideration of the upper and lower voltage limits of the communications system.

In 48 V systems, lead–acid batteries of 24, 25 or 26 cells are used. In 60 V systems, it is customary to use either 30 or 31 cells.

In preference to 25 cells, 26 may be chosen, or 31 in preference to 30. This enables the battery to be more deeply discharged, and thus increases its usable capacity. It must be borne in mind, however, that at the beginning of the discharge the voltage applied to the load is higher by the voltage of one cell.

It is alternatively possible to choose, for example, 24 rather than 25 cells. A lead–acid battery of 24 cells can be used in the parallel mode without voltage-reducing diodes if the permissible upper limit of continuous voltage is not exceeded in either the trickle-charging condition ($2.23 \text{ V} \times 24 = 53.52 \text{ V}$) or the charging condition ($2.33 \text{ V} \times 24 = 55.92 \text{ V}$).

This is the situation in the EMS telephone system (voltage range 42 to 58 V).

The somewhat lower discharge voltage must be taken into account in determining the permissible voltage drop in the supply cables. In the EMS telephone system the total voltage drop amounts to 2 V.

In the example quoted, the system-conditioned final voltage U_{Smin} is 44 V/24 = 1.83 V.

3.3.2.2 UPS systems

In UPS systems the nominal voltage may be, for example, 220 or 440 V, and the corresponding number of cells 110 or 220. The capacity is normally based upon standby times of from 10 to 60 minutes.

3.3.3 Capacity

The capacity depends upon the required standby time and the permitted voltage limits for the communications system; it can be arrived at approximately as follows:

Nominal capacity = standby time × I_{HV} × mean correction factor

Standby time — Period for which the battery is required to supply the communications system when necessary.

I_{HV} — Mean current drawn by the communications system during the busy hour.

Mean correction factor — Depends upon the specified standby time, the system-conditioned final voltage and the battery construction; it is given in Table 3.2 for OPzS batteries with commonly adopted standby times and system-conditioned final voltages.

The mean current I_{HV} in a communications system during the busy hour is determined mainly by the amount of traffic and the number of connected lines (subscribers' or junction lines). This value is obtained by multiplying the current consumption of the individual circuit elements by the traffic intensity (erlang) with which these circuit elements are loaded. The values can usually be obtained from the trunking diagram of the communications system.

Example of the determination of battery capacity for a digital electronic switching system (EWSD)

Given:

- ▷ mean current during the busy hour I_{HV} = 320 A (typical for 5000 subscribers),
- ▷ required standby time: 2h,

Table 3.2 Mean correction factor for the determination of OPzS battery capacity (accurate for 100 Ah plates)

Communications system	ESK 10 000 E and EWSA	EMS or EWSD and KN system	EMD Export 48 V	EMD Export 60 V	Direct loads
System-conditioned final voltage U_{Smin} (V)	46.6	44 or 45.8	46	58	57.4
System-conditioned final voltage U_{Smin}/cell (V)	1.79	1.83	1.84	1.87	1.92
Number of cells	26	24 or 25	25	31	30
Standby time (h)	Mean correction factor				
1	2.41	2.63	2.87	3.10	4.31
2	1.70	1.84	1.94	2.11	2.89
3	1.44	1.54	1.61	1.73	2.25
4	1.30	1.38	1.42	1.52	1.92
5	1.19	1.28	1.31	1.41	1.75
6	1.13	1.22	1.23	1.35	1.67

▷ nominal system voltage: 48 V,

▷ permissible continuous tolerance on operating voltage, measured at the assemblies:

(a) for apparatus connected directly to the operating voltage (direct loads): 48 to 58 V,

(b) for d.c./d.c. converters and inverters: 40 to 75 V,

▷ maximum permitted voltage drop in distribution system: 1.8 V.

Required:

▷ battery capacity.

Since both the direct loads and the d.c./d.c. converters and inverters are fed from a *single* power supply system, the significant voltage limits for the dimensioning of the battery are the narrower ones for the direct loads. Taking the voltage drops into account, the system-conditioned final voltage is

$$U_{Smin} = 44\text{ V} + 1.8\text{ V} = 45.8\text{ V}.$$

In terms of one battery cell, this becomes 45.8 V/25 $\approx$ 1.83 V/cell.

Having regard to the required standby time and I_{HV}, the capacity in ampere-hours can now be read from the battery manufacturer's data.

Table 3.3 shows, as an example, the range of types and the selection guide for the OPzS individual-cell batteries.

Table 3.3 Types of OPzS individual-cell batteries with electrical data (approximate values)

Container	Capacity per pair of plates (Ah)	Type OPzS Number of positive plates	Nominal capacity (Ah)	Capacity (Ah)			Discharge current (A)				Final discharge voltage U_s (V/cell) Discharge time (h)			
			C_{10}	C_5	C_3	C_1	I_{10}	I_5	I_3	I_1	10	5	3	1
Individual cells in transparent plastic containers	50	5	250	215	187	135	25	43	62	135	1,8	1,78	1,76	1,71
		6	300	258	225	162	30	51	75	162				
	70	5	350	303	267	193	35	61	89	193	1,79	1,76	1,74	1,67
		6	420	363	329	231	42	73	107	231				
		7	490	424	374	270	49	85	125	270				
	100	6	600	520	450	324	60	104	150	324	1.79	1.77	1.74	1.67
		8	800	690	600	432	80	138	200	432				
		10	1000	865	750	540	100	173	250	540				
		12	1200	1040	900	648	120	208	300	648				
	125	12	1500	1302	1152	828	150	260	384	828	1.8	1.77	1.73	1.61
Individual cells in translucent plastic containers		16	2000	1736	1536	1104	200	347	512	1104				
		20	2500	2170	1920	1380	250	434	640	1380				
		24	3000	2604	2304	1656	300	521	768	1656				
Individual cells in hard rubber containers	250	14	3500	2940	2520	1820	350	588	840	1820	1.78	1.75	1.72	1.6
		16	4000	3360	2880	2080	400	672	960	2080				
		18	4500	3780	3240	2340	450	756	1080	2340				
		20	5000	4200	3600	2600	500	840	1200	2600				
		22	5500	4620	3960	2860	550	924	1320	2860				
		24	6000	5040	4320	3120	600	1008	1440	3120				
		28	7000	5880	5040	3640	700	1176	1680	3640				
		32	8000	6270	5760	4160	800	1254	1920	4160				
		36	9000	7560	6480	4680	900	1512	2160	4680				
		40	10000	8400	7200	5200	1000	1680	2400	5200				
		44	11000	9240	7920	5720	1100	1848	2640	5720				
		48	12000	10080	8640	6240	1200	2016	2880	6240				

Table 3.4 Discharge currents for a system-conditioned final discharge voltage $U_{S_{min}}$ of 1.83 V/cell (approximate values)

Container	Capacity per pair of plates (Ah)	Type OPzS Number of positive plates	Nominal capacity (Ah)	Discharge time (h)								
			C_{10}	0.5	1	2	3	4	5	6	8	10
Individual cells in transparent plastic containers	50	5	250	135	105	72	55	46	40	35	28	24
		6	300	162	126	86	66	55	48	43	34	28
	70	5	350	170	134	96	75	63	54	47	39	33
		6	420	204	160	115	90	76	65	57	47	40
		7	490	238	187	135	105	88	76	66	55	46
	100	6	600	271	217	160	127	107	93	82	67	57
		8	800	362	290	213	170	143	124	109	89	76
		10	1000	453	362	267	213	179	155	137	111	95
		12	1200	543	443	320	255	214	186	164	133	114
	125	12	1500	558	504	388	315	265	234	208	169	139
Individual cells in translucent plastic containers		16	2000	880	721	546	434	365	320	281	225	185
		20	2500	1100	901	683	543	456	400	351	281	231
		24	3000	1320	1082	820	652	548	480	422	338	278
Individual cells in hard rubber containers	250	14	3500	1075	999	798	665	569	502	446	369	323
		16	4000	1229	1142	912	760	651	574	510	422	369
		18	4500	1382	1285	1026	855	732	646	574	475	415
		20	5000	1536	1428	1140	950	814	718	638	528	462
		22	5500	1690	1570	1254	1045	895	789	701	580	508
		24	6000	1843	1713	1368	1140	976	861	765	633	554
		28	7000	2150	1999	1596	1330	1139	1005	893	739	646
		32	8000	2458	2284	1824	1520	1302	1148	1020	844	739
		36	9000	2765	2570	2052	1710	1465	1292	1148	950	831
		40	10000	3072	2856	2280	1900	1628	1436	1276	1056	924
		44	11000	3379	3141	2508	2090	1790	1579	1403	1161	1016
		48	12000	3686	3427	2736	2280	1953	1723	1531	1267	1108

In addition, the variation in capacity with discharge current and final discharge voltage U_S (relative to periods of one, three, five and ten hours) is indicated for each battery type. It will also be observed that in the OPzS individual-cell battery range capacities from 250 to 12 000 Ah can be obtained with five plate sizes (50 to 250 Ah).

The discharge currents given in Table 3.4 correspond to the system-conditioned final voltage U_{Smin} of 1.83 V/cell (for a 25-cell lead–acid battery) derived above for the example of a communications system, based upon standby periods of from 0.5 to 10 h. In the example, an I_{HV} of 320 A was assumed. Here a discharge current of 320 A is found in association with a standby time of 2 h. If the corresponding value of nominal capacity C_{10} is traced to the left of the table, it is found to be 1200 Ah (at the ten-hour rate).

In power supply systems with battery capacities greater than 240 Ah, e.g. for public communications systems, the battery will be divided into two groups for safety reasons. Instead of one battery of the calculated capacity, two batteries are selected, each of half the capacity.

If the battery is divided into two 600 Ah groups of 25 cells each, the battery type, with its specification, is as follows:

▷ OPzS individual-cell battery in transparent plastic containers,
▷ 100 Ah plates (per pair of plates),
▷ 6 OPzS 600,
▷ nominal capacity C_{10} 600 Ah (at the ten-hour rate),
▷ discharge current 160 A for 2 h standby time and a system-conditioned final voltage U_{Smin} of 1.83 V/cell.

3.4 Maintenance

In the use and maintenance of lead–acid batteries, the following points should be particularly observed:

▷ batteries should not be left discharged for long periods,
▷ the battery room, cell containers, racks and connections should be kept clean and dry,
▷ the terminals should be smeared with non-acid grease,
▷ all metal parts in the battery room, such as steel racks, insulator studs and conductors, should be protected by acid-resisting paint, which should be reapplied from time to time,
▷ the electrolyte level should be checked and the cell voltages measured at intervals of one to three months,
▷ the density of the electrolyte in every cell should be measured with a hydrometer at regular intervals; a density of 1.30 kg/l should not be exceeded,

▷ the battery should be kept at the trickle-charging voltage (2.23 V/cell) whenever possible, avoiding undercharging,
▷ the battery should not be too deeply discharged (not beyond the final discharge voltage U_S), since this can lead to sulphation of the plates or loosening of the active material through excessive swelling, and hence to a reduced efficiency,
▷ the battery should not be overcharged,
▷ the battery should as far as possible be protected against cold; its utilization is impaired by low temperature.

3.5 Installation of Batteries

There is an advantage in mounting the batteries in one plane. In stepped or tiered racks sufficient separation must be allowed between the rows of cells. If the

Fig. 3.7
Battery room with OPzS lead–acid batteries on floor stillages. (Photo by courtesy of VARTA AG)

batteries are mounted in two or more vertical tiers, a space of at least 50 mm must be left free in front of and behind the racks for air circulation. The rows of batteries must be accessible from suitable service gangways. The width of the gangways should be appropriate to the size of the batteries, but not less than 0.5 m. Electrolyte-resistant insulation must be provided between the individual cells and earth or frame.

Recently plastic-insulated racks have been used for mounting (also stillages).

Three different types of battery mounting are in current use:

▷ floor stillages (Fig. 3.7),
▷ tiered racks (Fig. 3.8),
▷ stepped racks (Fig. 3.9).

Fig. 3.8
GroE-type lead–acid batteries on tiered racks. (Photo by courtesy of VARTA AG)

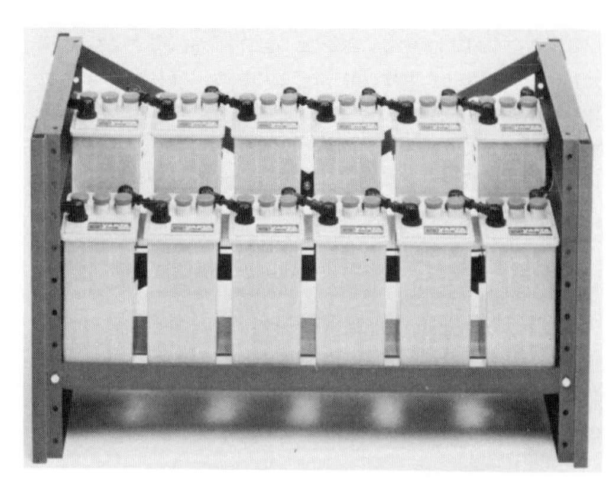

Fig. 3.9
OGi-block-type lead–acid batteries on a stepped rack. (Photo by courtesy of VARTA AG)

4 Earthing and Potential Equalization

Earthing embraces all the means and methods whereby conducting parts are connected to earth through an earthing system.

By *potential equalization* is understood 'bonding' to eliminate potential differences between conducting parts. In telecommunications practice, potential equalization is mostly at earth potential.

The system of earthing and potential equalization must:

- ▷ in the event of faults in electrical equipment and installations, prevent the occurrence of dangerous contact potentials and effect the disconnection of the faulty circuit,[1])
- ▷ ensure the freedom from disturbance of all central and decentralized functions of the telecommunications system,
- ▷ provide a sufficiently low earth resistance for systems that use earth as a return conductor,
- ▷ be suitable for screening electronic equipment,
- ▷ assist in reducing the effects of lightning strokes, particularly in tall telecommunications buildings.

A distinction is made between earthing for protective purposes and earthing for functional reasons.

4.1 Protective Earthing

In telecommunications practice the term 'protective earthing' denotes the measures which in power engineering may be referred to as 'cut-off protection' against indirect contact (see Chapter 5).

As the term 'protective earthing' conveys, the earthing of the conductive housings of power equipment is intended to prevent the appearance of dangerous contact potentials. This is achieved, on the one hand, by potential equalization through the earth conductor and, on the other hand, by disconnecting the faulty equipment. In the event of an insulation failure, a short-circuit current flows in the connection to the earth conductor, which causes either a fuse or an automatic circuit breaker to operate.

Equipment with a protective earth connection is classified in VDE 0106 as

[1]) DIN 57 100/VDE 0100 and DIN 57 800/VDE 0800 are especially applicable to the protection of persons and equipment.

protective class I equipment. Such equipment may be used with supplies in the following categories:[1])

- ▷ TT supply system (formerly protective earthing according to VDE 0100 §9),
- ▷ TN supply system (formerly neutral earthing according to VDE 0100 §10),
- ▷ IT supply system (formerly protective line system according to VDE 0100 §11).

In the TT supply system, in most cases, the neutral point of the three-phase supply is earthed directly. Equipment in protective class I is earthed directly or through an earth conductor. Fault current flows back to the power source via earth. Disconnection is effected either by a fuse or, preferably, by a fault current-operated protective switch (circuit breaker).

In the TN supply system also, the neutral point is usually earthed directly. Equipment in protective class I is earthed directly through a connection with a cross-sectional area of at least 10 mm^2 (TN C supply system) or through a protective conductor to the PEN conductor (formerly neutral conductor). Fault currents flow back to the power source mainly through the PEN conductor. To prevent the PEN conductor from assuming a dangerous contact potential in the event of its being broken, it should be earthed at as many points as possible, particularly at the entrance to the building. For the overall resistance of the supply system to earth, a value of 2Ω is adequate. Disconnection depends upon a fuse. Fault current-operated protective switches are acceptable in a TN supply system as a means of improving the disconnection characteristics.

The IT supply system is operated without an earth connection. Protection depends entirely upon potential equalization. IT supplies are not widely adopted, and are only used in special circumstances, where disconnection is not acceptable.

The voltage drop to earth permitted by the protection arrangements in a TT supply system should not exceed 50 V a.c.

These protective measures cannot be applied beyond a certain load power level, because the necessary earth resistances cannot generally be achieved economically. This limitation does not apply when fault current-operated protective switches are used. The earth resistance R_E can in this case be significantly higher, depending upon the trip current (≈10 mA to 1 A):

$$R_E = \frac{\text{permissible contact potential (V)}}{\text{fault current (A)}} = \frac{U_B}{I_F}$$

(see Table 4.1).

Higher powers in the medium-voltage range are in any case supplied through a transformer. A local TN supply system is then provided on the low-voltage side.

[1]) These symbols are used by international agreement to describe the earthing of supplies and loads (see Chaper 5).

Table 4.1 Maximum values of earth resistance in telecommunications systems

Protective earthing	Functional earthing		Functional and protective earthing[2]			Lightning-protection earthing
Medium-voltage supply through transformer with TN system on low-voltage side $R_E \leqq 2\ \Omega$	Single-wire ground signalling		Single-wire ground signalling	With lightning protection	Without lightning protection	$\leqq 5\ \Omega$
	LX and PABX up to 500 LU	$\leqq 10\ \Omega$	LX and PABX up to 500 LU	$\leqq 5\ \Omega$[3]	$\leqq 10\ \Omega$[3]	
	LX and PABX from 500 to 1000 LU	$\leqq 5\ \Omega$	LX and PABX from 500 to 1000 LU	$\leqq 5\ \Omega$[3]	$\leqq 5\ \Omega$[3]	
	LX from 1000 to 2000 LU and PABX with more than 1000 LU	$\leqq 2\ \Omega$	LX from 1000 to 2000 LU and PABX with more than 1000 LU	$\leqq 2\ \Omega$	$\leqq 2\ \Omega$	
	LX with more than 2000 LU	$\leqq 0.5\ \Omega$[1]	LX with more than 2000 LU	$\leqq 0.5\ \Omega$[1]	$\leqq 0.5\ \Omega$[1]	
	RS, RT	$\leqq 2\ \Omega$	RS, RT	$\leqq 2\ \Omega$	$\leqq 2\ \Omega$	
	TX, DCS	$\leqq 0.5\ \Omega$	TX, DCS	$\leqq 0.5\ \Omega$	$\leqq 0.5\ \Omega$	
Current-operated protection circuit,	Loop signalling		Loop signalling	With lightning protection	Without lightning protection	
e.g. $R_E = 50/0.3 \approx 166\ \Omega$ with $I_F = 0.3$ A	Up to 1000 LU	$\leqq 10\ \Omega$	Up to 1000 LU	$\leqq 5\ \Omega$[3]	$\leqq 10\ \Omega$[3]	
	Up to 2000 LU	$\leqq 5\ \Omega$	Up to 2000 LU	$\leqq 5\ \Omega$[3]	$\leqq 5\ \Omega$[3]	
	More than 2000 LU	$\leqq 2\ \Omega$	More than 2000 LU	$\leqq 2\ \Omega$	$\leqq 2\ \Omega$	

LU Line units
RT Radio transmission station
LX Local exchange
RS Repeater station
TX Trunk exchange
DCS Data conversion station
PABX Private automatic branch exchange

[1]) Provided that there is *no* three- or two-wire junction traffic with other systems through d.c. repeaters, the resistance in systems with more than 2000 subscribers may be as high as 2 Ω.

[2]) Measured at the earthing ring main or busbar, allowing for all connected earths.

[3]) 2 Ω in the case of connection to a medium-voltage supply.

Potential equalization represents an effective extension of 'protective disconnection arrangements'; it is prescribed at every house lead-in as the 'main potential equalization', and connects the PEN or earth conductor, via the potential equalization busbar, to the foundation earth, the metal structures and the lightning earth.

A 'supplementary potential equalization' is prescribed, in addition to the main potential equalization, if the conditions laid down for automatic disconnection as a protection against indirect contact cannot be met. Supplementary potential equalization requires the bonding of metal structures, reinforcing bars in reinforced-concrete structures, etc., to all conducting parts of fixed equipment that can be touched simultaneously. This equalization corresponds to the functional and protective earthing that is normally provided in any case in telecommunications systems (see Section 4.3).

Table 4.1 indicates the maximum earth resistance for protective earthing.

4.2 Functional Earthing

As is well known, signal circuits in telecommunications engineering are often arranged as single-wire circuits with a common earth return. This offers technical, but primarily economic, advantages. This common signal earth return uses, within the telecommunications system, the mesh network of reinforcing mesh, rack suites and the earthed 48 or 60 V distribution system. Because of the interconnected areas of all the earthed parts (hence the alternative term, area earth, as opposed to the radial earth), the result is a mesh network with low impedance, which gives rise to very low potential differences with both d.c. loading and a.c. loading up to the high-frequency region. It thus meets the requirements for use as a potential reference plane. The earth is necessary in order to stabilize the reference potential in relation to the surroundings, to protect against corrosion and to make use of the earth as a return conductor between geographically separated telecommunications installations. Common return paths, however, introduce the possibility of interference, especially from power circuits. Nevertheless, the use of earth returns in conjunction with single-wire circuits and signal circuits is widespread – e.g. between exchanges in local telephone networks. With the introduction of digital exchange systems (e.g. the EWSD system) the use of earth return signal paths is avoided by using PCM links between exchanges.

A further reason for earthing telecommunications systems is the reduction of interference.

External interference can, for example, be reduced by the equalization of potentials or by diverting voltages to earth. This kind of interference reduction embraces also the suppression of overvoltages resulting from atmospheric effects or from the effects of power circuits.

To enable the requirements of the functional earth to be met, the earth resistance must not exceed a certain maximum value (Table 4.1). If the necessary low values are not achieved, a loss of quality may result. This relates especially to noise pick-up and message-switching performance.

With increasing earth resistance, the introduction of noise deteriorates the signal-to-noise ratio.

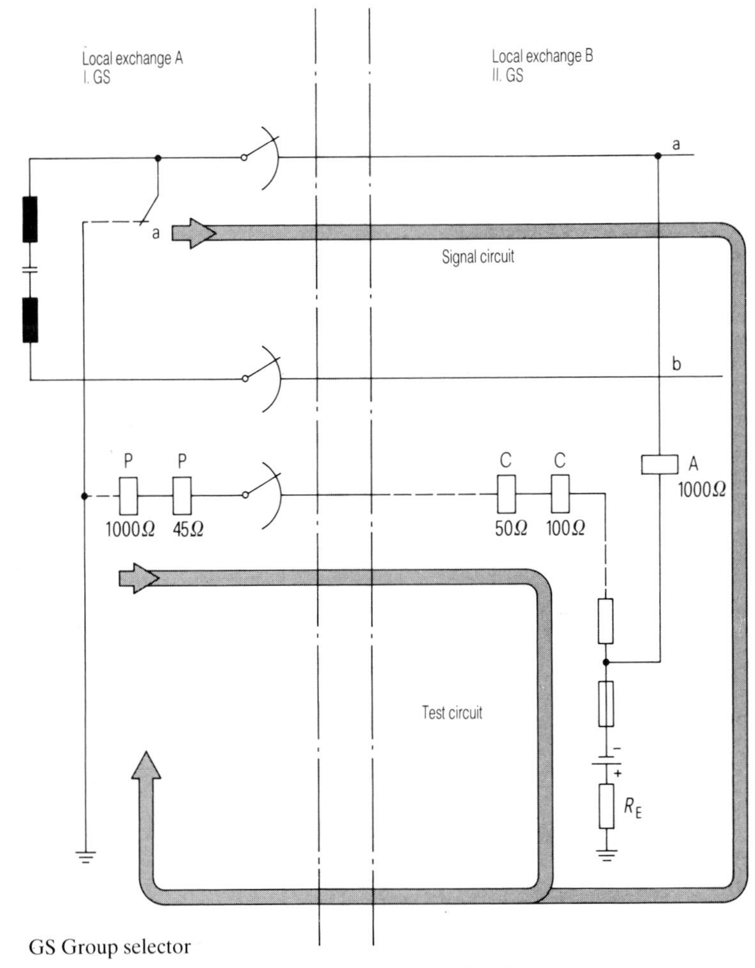

GS Group selector

Fig. 4.1 Signal and test circuits between two local exchanges

The message-switching characteristics are adversely affected in systems with three-wire communications traffic, where the earth serves as a return conductor because of signal asymmetry (Fig. 4.1). In this case current in the earth connection produces a voltage drop which is in opposition to the exchange voltage. This can lead to malfunctions, such as double testing, faulty subscriber number, line interruption, metering errors and line blocking.

An increase in the earth resistance by one step (Table 4.1, functional earthing) is acceptable if truly symmetrical signal processes are employed, with or without unbalanced earth relationships in the junction or exchange lines.

4.3 Functional and Protective Earthing

As explained in Section 4.2, the earth resistance, from the point of view of the operation of a telecommunications system, must not exceed a certain value (see Table 4.1, functional earthing). This value is usually significantly lower than that required for protective earthing. The telecommunications equipment is earthed either through the power supply conductors or by means of a separate functional earth conductor FE. Because of the low permissible voltage drop, these conductors usually have a large cross-sectional area and can therefore be extended to a combined functional and protective earth (Table 4.1). The functional and protective earth[1]) is used in German Federal Republic Postal Administration practice for all exchange and transmission installations.

This method of earthing has the following characteristic features:

▷ the functional and protective earthing are combined in one conductor or conductor system,

▷ all conducting parts that require to be earthed for protective or functional reasons are connected to the functional and protective earth conductor FPE (area earthing),

▷ through the connection of the earthed positive conductor to the racks, the rack structure in telecommunications equipment becomes part of the earthing system and assumes the protective earthing function for power system loads in protective class I. The earth conductor PE of the power system is then not connected,

▷ the housings of the power supply equipment – e.g. rectifiers, mains switch panel and battery switching panel – are similarly bonded to the earthed positive conductor. The earthing connection is provided by the FPE conductor between the earthed conductor in the power supply system and the earthing main conductor. Also effected at the earthing main conductor is the

[1]) The functional and protective earth of a telecommunications system is identical with that previously described as an operational and protective earth (BSE).

matching to the protective devices of the supply system (see Sections 5.1.1 to 5.1.3). Thus, for example, the PE earth conductor of the house wiring and the potential equalization conductor of the building are similarly connected at this point (see Figs 4.2 and 4.3).

4.4 Selection and Design of Earths and Earthing Systems

The choice and design of the earthing system is determined by the requirements placed upon it and the local conditions. The components of an earthing system are:

▷ the *actual earth* (an earth electrode or earthy metalwork, pipes, etc.) which conducts current into the ground,
▷ the *earthing main conductor*, to which the earth and the parts of the installation that are to be earthed are connected,
▷ the *earthing conductors* (or potential equalization conductors), which connect the parts of the installation to the earthing main conductor.

4.4.1 Types of earth

In the case of a newly-erected building, the electricity supply undertaking will provide a foundation earth. This should be connected to the main potential equalization system and can be used for protective and functional earthing purposes. To the extent that the resistance of the foundation earth does not meet the requirements, additional earths have to be provided.

These earths may take the form of surface earths (earth bands), vertical (strip) earths or a combination of both.

Besides these types of earth, pipe and cable networks, provided that they consist of conducting material and are laid in contact with the earth, function effectively as earths because of the large area which they cover; it is then a question of so-called 'natural earth'. Included in this category are the metal sheathing of telecommunications cable and water pipelines, but not gas pipelines; these may be used only within buildings, and only for potential equalization. For small telecommunications systems the water main (or the heating pipes connected to it) is frequently an adequate functional earth.

In recent times the use of metal pipeline systems and cable sheathing has become progressively less common. Water pipes are often of plastic material, and lead-sheathed cable has mostly been replaced by communications cable with plastic-covered aluminium sheathing. For these reasons, the German Postal Administration, for example, specifies a foundation earth for new telecommunications installations.

By *foundation earth* is meant a conductor which is embedded in the foundation level of a building, and therefore makes a large-area contact with the ground; it is incorporated as a closed ring in the external foundations and consists of galvanized steel strip. In general, foundation earths provide sufficiently low earth resistances.

Surface earths consist of galvanized steel strip or rod, and are generally installed at a depth of 0.5 to 1 m. The length of these earths depends upon the necessary earth resistance; they can be arranged as radial, ring or mesh earths, as well as straight. Radial earths should have equal divisions; more than three branches (with an angle not less than 60°) are not recommended because of mutual interference.

Vertical earths can be installed at a depth of up to 30 m, depending on the subsoil; they consist of rod, tube or profiled bars and are driven into the earth as nearly as possible vertically; their length depends upon the necessary earth resistance. If several vertical earths are necessary (in order to obtain a sufficiently low earth resistance) a separation of at least twice the length of an individual electrode is desirable.

Plate earths were formerly often employed. They entail a greater expense than other types of earth and will not be described here.

4.4.2 Earthing main conductor

In the simplest case, the earthing main conductor consists of an *earth terminal* on the equipment. In small or medium-sized systems an *earth busbar* of copper, brass or galvanized steel is used, whose length and cross-sectional area are chosen in accordance with the number of earths and conductors to be connected to it.

The most effective form of earthing main conductor is represented by the earthing ring main conductor, which is installed in the basement or the ground floor of buildings containing extensive telecommunications systems. The metal sheaths of cables, conduits, water and heating pipes and the like are connected to the earthing ring main conductor by the shortest paths.

Because of its cross-sectional copper area of a least 50 mm^2 (usually 95 mm^2), the earthing ring main conductor has a very low resistance and effects a *potential equalization* in respect of all potentials acting upon it, external as well as internal.

The earthing main conductor may also be installed close to the telecommunications system in a storey of a high office building, in which case the connection to the earth line of the power system – assuming that one is provided for functional and protective earthing purposes – is made at the floor distribution board of the low-voltage supply.

4.4.3 Earthing and potential-equalization conductors

Certain stipulations should be noted in regard to the installation of earth conductors, particularly their cross-sectional area (Table 4.2).

In most cases earth conductors serve at the same time the purpose of potential equalization for functions associated with communications operations, in that it provides a practically uniform reference potential within the telecommunications system. This is particularly the case when one pole of the power supply line is also used as an earth conductor over long distances.

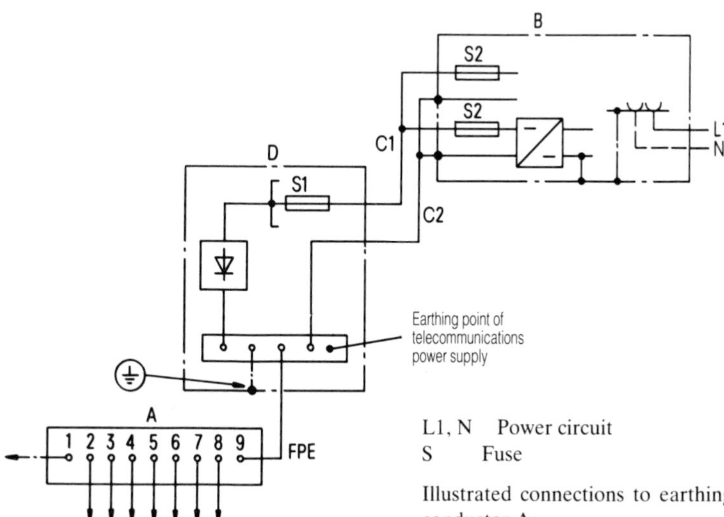

L1, N Power circuit
S Fuse

Illustrated connections to earthing main conductor A:

1 Protective conductor (PE) of interior installation or potential equalization conductor of building
2 Telecommunications earth
3 Foundation earth
4 Conducting sheath of telecommunications cable
5 Reinforcement of building
6 Conducting water pipes in building
7 Heating system
8 Lightning-protection earth
9 Functional and protective earth conductor (FPE)

A Earthing main conductor
B Telecommunications equipment
C1, C2 Telecommunications operating circuit: earthed conductor C2 connected to FPE and the reference conductors of the telecommunications equipment
D Telecommunications power supply: the nominal direct and alternating voltages may exceed 120 V and 50 V respectively

Fig. 4.2
Functional and protective earthing of a telecommunications system (in accordance with VDE 0800 Part 2)

This kind of earth is illustrated in Fig. 4.2.

In transmission practice it is customary to earth the equipments by means of a separate conductor, which is kept free of power supply currents (Fig. 4.3).

By virtue of the exclusive use – in this case – of d.c./d.c. converters, 'galvanic separation' is achieved between the power supply and the communications equipment, which is especially advantageous in high buildings susceptible to lightning.

In dimensioning the earth conductor C3, however, it is necessary to check whether, in the event of an earth short-circuit, the resistance of the fault path is such that the fuse can be relied upon to rupture in an acceptable time. In systems such as that shown in Fig. 4.2 this check is not necessary, because the earth conductor C2 carries the power supply current, and therefore has the same cross-sectional area as the conductor C1.

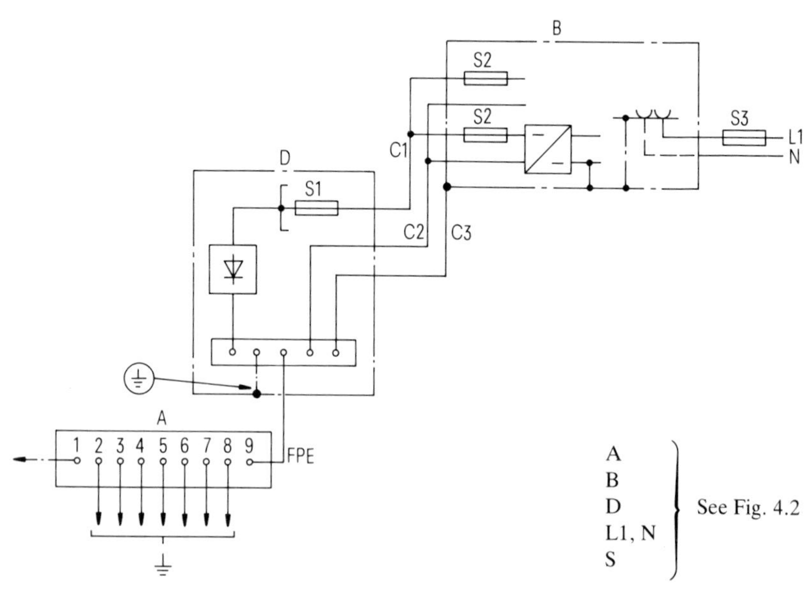

A, B, D, L1, N, S — See Fig. 4.2

C1, C2 Telecommunications operating circuit, earthed conductor C2 insulated and *not* connected to the reference conductors of the telecommunications equipment

C3 Additional earth conductor, not carrying supply current, connected to FPE and the reference conductors of the telecommunications equipment

Fig. 4.3
Functional and protective earthing, with an additional earth conductor, of a telecommunications system (in accordance with VDE 0800 Part 2)

Table 4.2 Minimum cross-sectional area of functional earth conductor FE (according to VDE 0800 Part 2)

Current rating of overcurrent protection device (A)[1]	Minimum cross-sectional area of copper conductor (mm^2)	Current rating of overcurrent protection device (A)[1]	Minimum cross-sectional area of copper conductor (mm^2)
Up to 25	2.5	Up to 224	35
Up to 35	4	Up to 250	50
Up to 50	6	Up to 630	70
Up to 63	10	Up to 800	95
Up to 125	16	Up to 1000	120
Up to 160	25		

[1]) The overcurrent protection device is not in the earth conductor path.

From Table 4.2 the cross-sectional areas can be chosen according to the current rating of the operative overcurrent protection device. This will be explained with reference to Figs 4.2 and 4.3.

In the event of an earth short-circuit fault on conductor C1 in Fig. 4.2, the fault current would flow through the fuse S1, through conductor C1 to the fault location in the communications equipment B and back to the power supply through conductor C2. The conductor C2 must therefore be specified in conformity with the fuse S1. In the case of an earth short-circuit fault after fuse S2, the earth conductor concerned has, of course, only to be dimensioned to correspond to the fuse S2.

If an earth short-circuit fault occurs on the conductor C1, not to the communications equipment B but to a part of the building (not shown) – an earthed rack, for example – the conductor shown as FPE between the earthing main conductor and the power supply equipment carries the earth fault current, and so must be dimensioned in conformity with fuse S1.

The cross-section of conductor C3 in Fig. 4.3 is specified in accordance with the larger of the two fuses S1 and S3.

4.5 Design Pointers for Earthing Systems

In designing an earthing system it is necessary to determine whether the earth that is available for the connection of the main earthing conductor, such as a foundation earth, lightning earth or cable sheath, has a sufficiently low resistance. This entails measurement. Inasmuch as these considerations arise before the building construction begins, an estimate has to be made, for which purpose the specific earth resistance must be known.

4.5.1 Specific earth resistance

The specific earth resistance can vary greatly according to the depth at which the earthing electrode is buried; it also depends upon the composition of the soil, its degree of dampness and the temperature. Figure 4.4 shows values of specific earth resistance ρ_E that can be used to estimate the resistance of an earth electrode. In each case the lower values relate to damp ground conditions.

4.5.2 Measurement of specific earth resistance

The specific earth resistance ρ_E is quoted in ohm-metres. Special earth-measuring equipment is used to determine it; this usually employs the voltage-balance method. A measuring current is caused to flow in the ground by means of an earth electrode and an auxiliary earth electrode. The voltage drop in the earth resistance is compared with that across an adjustable resistor. The measurement is made with a.c. (e.g. from a hand generator) to avoid errors due to polarization which could occur with d.c. Probes are required to measure the voltage drops (Fig. 4.5).

The value of the resistance between the probes S_1 and S_2 can be measured on the balancing resistor 3.

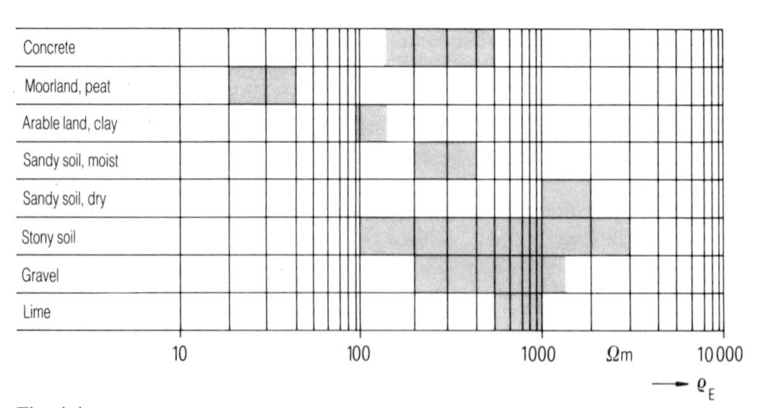

Fig. 4.4
Specific earth resistances ρ_E in various ground conditions

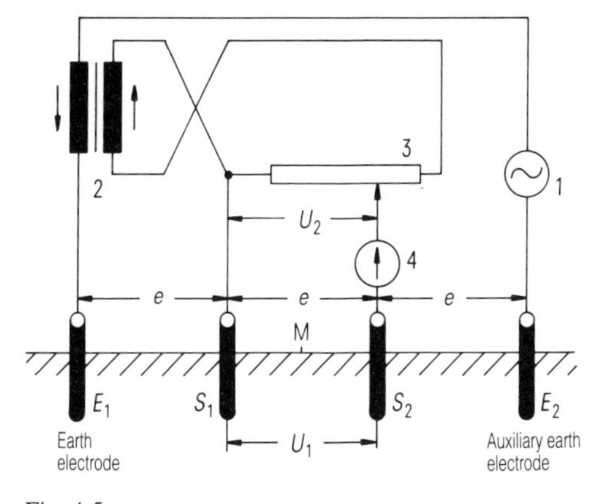

1 A.C. source (generator)
2 Transformer
3 Balancing resistor
4 Null detector
E_1, E_2 Current electrodes
S_1, S_2 Voltage probes

Fig. 4.5
Arrangement for the measurement of soil specific resistivity

The *mean specific earth resistance* ρ_E is given by:

$$\rho_E = 2\pi eR$$

where

R = measured resistance (Ω)
e = probe distance (m)
ρ_E = mean specific earth resistance (Ωm) up to a depth equal to the probe distance e.

Four earth spikes are driven into the ground at equal distances in a straight line with its mid-point at M. Measurements are made initially with a distance e of 1 m. For further measurements the distance of the spikes is increased in steps of 1 m with the same mid-point M. If the value of earth resistance ρ_E so measured is constant with varying probe distance, the soil is homogeneous. If ρ_E increases, the resistance of the deep soil strata is higher, and vice versa.

A rapidly falling value of resistance indicates that the water table has been reached.

In large towns lower values of specific earth resistance than are accounted for by the constitution of the soil are often obtained. The reason for this is the large number of conductors, such as water pipes, cables, foundation earths, etc., laid in the ground.

4.5.3 Measurement of earthing resistance

The earthing equipment described is also suitable for measuring earthing resistances. This measurement – of practical importance – is carried out with the aid of an auxiliary earth electrode and a probe.

As in the measurement of specific earth resistance (see Fig. 4.5) a measuring current is passed between the earth electrode and the auxiliary electrode, and the voltage drop is determined by means of the probe. To obtain reliable results, the probe must be applied outside the areas of high potential gradient surrounding the earth and auxiliary earth electrodes, in which large variations of potential occur as a result of the proximity of the electrodes or non-uniformity of the soil. It is therefore necessary first to establish the size and shape of these areas by exploratory measurements.

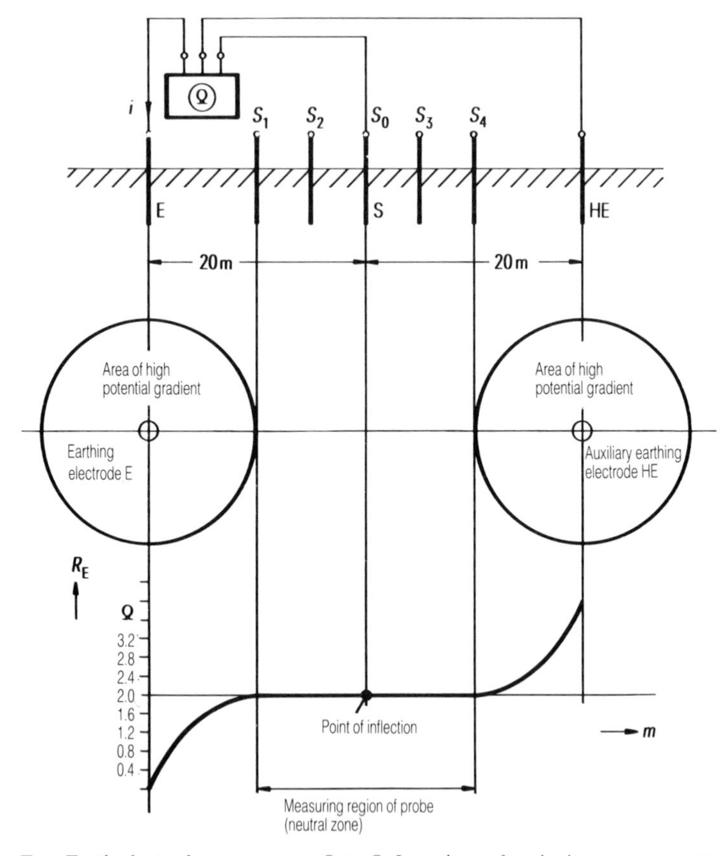

E Earth electrode

S Probe

HE Auxiliary earth electrode

R_E Earth resistance at particular location

S_1 to S_4 Locations of probe in measurement sequence

Fig. 4.6 Variation of resistance obtained from earth resistance measurements

Beyond the surrounding areas of high potential gradient the potential is approximately constant (neutral zone, reference potential; see Fig. 4.6). The probe is then moved in steps of about 5 m, beginning at the earth electrode, towards the auxiliary electrode, and the measurements are plotted on a curve. The earth resistance is obtained by drawing a straight line parallel to the horizontal axis through the point of inflection. The lower part of the curve then gives the resistance of the earth electrode.

4.5.4 Calculation of earth resistance

Rule-of-thumb formulae for calculating earth resistances are given in Table 4.3.

The formulae are valid so long as the specific earth resistance is constant along the length of the earth electrode. In the case of vertical electrodes this is often not so. For the purpose of a preliminary calculation, with a known specific earth resistance ρ_E, the vertical earth electrode may in this case be considered as divided into sections 1 m long. The resistance of one section is then equal to the specific earth resistance ρ_E at the relative position.

The total resistance of a vertical earth is obtained by adding the conductances of the individual sections.

Thus the earth resistance is given by:

$$\frac{1}{R_E} = \frac{1}{\rho_{E1}} + \frac{1}{\rho_{E2}} + \cdots + \frac{1}{\rho_{En}}.$$

The relationship holds for d.c. and low-frequency a.c.

With impulse (lightning) currents, the impedance of an extended earth electrode (>30 m) can be considerably increased. From this point of view a number of short electrodes connected together are more satisfactory.

Table 4.3 Earth resistances obtained with various kinds of earth electrode

Type of earth electrode	Formula	Type of earth electrode	Formula
Vertical electrode	$R_E \approx \frac{\rho_E}{l}$	Ring electrode	$R_E \approx \frac{2\rho_E}{3D}$
Surface electrode	$R_E \approx \frac{2\rho_E}{l}$	Mesh electrode	$R_E \approx \frac{\rho_E}{2D}$

R_E earth resistance (Ω)
l length of earth electrode (m)
D diameter of ring electrode or equivalent circular area (m) (equivalent diameter) = $1.13\sqrt{A}$
A area enclosed by ring or mesh electrode (m^2)

The resistance of a *combined earth* can be calculated sufficiently accurately for practical purposes by first determining, in the case of a vertical electrode and a surface electrode coupled together, their separate resistances, but taking into consideration only a part of the conductance of the surface electrode. To this end the length of the surface electrode is reduced to half the length of the vertical electrode. The total resistance R_G is then calculated as:

$$R_G = \frac{R_1 R_2}{R_1 + R_2}$$

The resistance of a *foundation earth* is given approximately by the formula for the ring or mesh electrode (see Table 4.3), according to the nature of the foundations, if the value of A is taken as the area enclosed by the foundation earth. If the resistance of the foundation earth is not sufficiently low, it must be supplemented by a vertical or surface earth electrode.

The question of which kind of earth electrode to use can only be decided from a knowledge of the subsoil. Given a relatively homogeneous soil, the cost of earthing with vertical or surface earths is about the same.

In most cases, however, vertical earths are more economical, because the deeper soil strata are generally of higher conductivity. For the same earth resistance, a vertical earth needs to be only about half the length of a surface earth.

Surface earths are useful when the subsoil is stony or rocky, when the specific earth resistance increases with increasing depth, or when it is necessary to extend the area of the earthing system because of an expansion of technical equipment and buildings.

In mountainous country, surface earths frequently offer the only possibility of installing an earthing system.

4.6 Protection against Overvoltage and Interference

An aspect of earthing previously mentioned in passing is its role as an important part of the measures adopted by way of protection against overvoltage and interference.

The introduction of electronic devices and systems increased the demands made upon overvoltage protection.

Since that time it has been necessary to adopt a comprehensive, coordinated protection system against overvoltage; it embraces the overall and specific protection of the conductor system, constructional features in the design of apparatus and the incorporation of earthed parts of the building, such as reinforcing steel, structural steelwork, lightning protection systems, etc., in the total earthing scheme.

If this system of earthed parts is sufficiently fine-meshed, it can also be used as a screen to enhance EMC (electromagnetic compatibility). By the term EMC is meant here the limitation of emission of electromagnetic interference and immunity to external fields (e.g. radio and broadcasting).

Through the interference path, the *source of interference* is coupled to the *affected apparatus* either 'galvanically', inductively, capacitively or by radiation.

There are three possible ways of reducing interference when necessary:

▷ reducing the interference at source – e.g. by suitable suppression of relays or contactor coils,
▷ reducing the degree of coupling; this can be done, for example, by screening, modification of wiring layout, etc.,
▷ increasing the immunity to interference of the system; this may, for example, be achieved by circuit techniques, such as the incorporation of voltage-limiting components.

At this point the earthing and potential equalization is being considered primarily as part of the means of reducing interference. It is, however, worth considering briefly the sources of interference and the methods of increasing immunity to interference.

4.6.1 Overvoltage sources

External conductor networks

Overhead lines are by their nature the most frequently affected by atmospheric discharges.

More numerous than overvoltages due to close discharges (a direct stroke on the conductor or flashover in nearby apparatus) are those caused by remote strokes. In the case of mains cables these are strokes in the medium-voltage or high-voltage system, or they can be produced in any conductors by lightning discharges between two clouds. The overvoltage spreads from its place of origin in both directions with the speed of light. The magnitude of the transmitted voltage is determined by the insulation level of the conductor system. In the event of a lightning stroke on a low-voltage conductor, a flashover to earth will occur on the nearby insulators. In low-voltage networks and overhead telecommunications cables, overvoltages higher than 10 kV occur but rarely.

Telecommunications cables laid in the ground are affected by power systems as well as lightning strokes.

A lightning stroke raises the potential of the ground at the striking point relative to a remote earth. If a cable route passes through the area of high potential gradient at the location of the stroke and the cable sheath, in the course of the cable run, is in contact with the remote earth, a transient current flows in the sheath. Through the inductive effect of the lightning current and the capacitance of the conductors, voltages are also induced in the cores of the cable. Depending upon the magnitude of the potential difference between the cable sheath and the cores, the result may be breakdowns or even destruction of the cable.

Particularly frequent is interference with telecommunications cables from the indirect effect of lightning – i.e. through the effect of lightning discharges parallel to the cable route.

In relation to interference from power systems, electric railways should be particularly mentioned; in these, a portion of the operating current returns to the transformer or rectifier substation through the ground. As a result, the sheaths of telecommunications cables laid in the vicinity of the railway installation can assume a proportion of these currents, which give rise to interference and overvoltage of longer duration.

Short-duration interference (of 0.1 to 1 s duration) with high-voltage transients is produced mainly by short-circuits in power systems.

The overvoltages in the cable cores are also dependent upon the construction of the cable.

Conductor networks within buildings

Because of the smaller area covered by conductor networks within buildings, and the screening effect of modern concrete structures, consideration is principally directed at this point, leaving aside direct lightning effects, to interference caused by the switching of currents.

Every conductor possesses an inductance, which depends upon the length of the conductor, its geometrical arrangement and the magnetic properties of the surrounding materials. Every change in current produces a voltage in the inductance of the conductor, which opposes the applied voltage. The magnitude of the voltage depends upon the rate-of-change of current and the inductance. When a short-circuit occurs, large current variations are caused by the rupture of the fuse, and high overvoltages are consequently produced. Overvoltages of up to 230 V have been measured in 60 V telecommunications systems where the arrangement of the conductors is not satisfactory. Even in well-laid-out systems with low inductance, considerable overvoltages still occur; e.g. the following typical levels were measured in a 60 V system (before the fuse):

▷ rupture of a 10 A rack fuse: 150 V, duration 0.4 ms,
▷ rupture of a 63 A rack row fuse: 130 V, duration 0.7 ms.

Significantly higher overvoltages occur as a result of short-circuits in 220/380 V systems. The following average overvoltages were measured in tests on low-voltage systems, according to the fuse rating and the length of cable between the distribution transformer and the short circuit:

- ▷ 10 A fuse: 2 to 7 × nominal voltage,
- ▷ 35 or 100 A fuses: 1.5 to 4 × nominal voltage.

The overvoltages measured with the 10 A fuses were surprisingly high: the reason lies in the rapid current interruption in these fuses.

Overvoltages do not occur only on d.c. or a.c. power supply conductors; they are also produced on signal conductors and earth conductors through inductive effects or 'galvanic' coupling.

Through the earthing of one pole of the d.c. power supply (e.g. +60 V) for functional reasons, this is connected at many points to structural parts of the telecommunications installation, and hence to earthed structural parts of the building. If the building is struck by lightning, components of the lightning current may flow in the earthed power supply conductors and, as a result of the voltage drop across the impedance of these conductors, give rise to potential differences between parts of the telecommunications system.

At the same time, the lightning current may induce voltages in unearthed signal conductors. The magnitude of these voltages depends upon the extent of the conductor loop. Widely ramified conductor systems are therefore especially affected.

4.6.2 Interference in telecommunications systems

Telecommunications installations must be protected against interference voltages deriving from the conductor network. Limiting values for acceptable interference levels were given in VDE recommendations at a very early stage (from 1920). The permissible limits in regard to risk from power installations are contained in VDE 0228 Part 1.

If the interference levels exceed the limits, or if atmospheric overvoltages have to be allowed for, protective measures are necessary (in accordance with VDE 0845); these may be:

- ▷ in the case of conductor networks, a suitable choice of cable routes to avoid proximity,
- ▷ the use of cables or conductors of lightning-protected construction,
- ▷ the use of isolating transformers,
- ▷ the use of overvoltage suppressors (surge diverters).

With the change to switching systems with central electronic control and electronic digital systems, the question of EMC has become serious.

In the peripheral circuits such systems operate with signal levels of ±12 V, and in the processors with ±5 V, at speeds in the high-frequency range. The clock frequency in the processor of the SSP103D, for example, is 5 MHz. Transient overvoltages on earth and signal lines thus lie, in terms of voltage and frequency, in the operating range of the signal sequences of the system, and can therefore lead to errors in signal processing.

Measures must accordingly be adopted to make the system immune to interference.

These are principally design features, such as the disposition of conductors, potential equalization, screening and the incorporation of overvoltage protection devices.

The fundamental problem area in digital systems is not so much in the control of the interference on the live power supply conductors and unearthed signal leads – this can be achieved by means of suppression circuits and screening – as in the effect on the earthing system. It must be ensured, for example, that transient currents due to the discharge of overvoltage suppressors (surge diverters) do not induce voltages in the earthing system of such a magnitude as to cause signal-processing errors. Precautions against interference are described below.

Limitation of overvoltages by overvoltage protection devices

To protect against externally derived overvoltages which could affect the telecommunications system, voltage-limiting components must be provided. The classical overvoltage protection device is the *spark gap*, originally an air gap – in modern practice a *gas-discharge surge diverter* (arrester, overvoltage suppressor). This is a suppressor in which two electrodes are mounted with a small separation in a capsule filled with an inert gas. When an overvoltage occurs, the voltage across the suppressor first rises to the triggering level, without a significant flow of current. Ionization of the gas in the discharge path leads to a breakdown between the electrodes, and consequently a highly conducting connection ($<0.1\Omega$), such that the voltage across the suppressor is brought down to the arc burning voltage of 10 to 15 V. At the conclusion of the discharge process, the suppressor reverts to the non-conducting condition. A disadvantage of gas-discharge surge diverters is that the triggering voltage depends upon the rate-of-rise of the voltage.

When gas-discharge surge diverters are applied to power conductors, triggering may lead to a follow-up current through the surge diverter to earth. Such diverters may not therefore be used on a supply with a short-circuit current greater than 0.5 A or an operating voltage greater than 20 V without an appropriate means of current interruption; this purpose may be served either by fuses or by (series-connected) varistors.

To suppress overvoltages below the triggering level of the gas-discharge surge diverter, it may be combined with other voltage-limiting devices. For this purpose a number of components are available – e.g.:

▷ metal oxide varistors,
▷ surge-suppressor diodes (transient-absorption zener diodes),
▷ zener diodes.

Metal oxide varistors are voltage-dependent resistors with symmetrical characteristics.

A block of sintered zinc oxide, with an admixture of other metal oxides, is mounted between two contact plates. The voltage dependency is due to the variable contact resistance between the sintered oxide crystals. The resistance (>1 MΩ) collapses very rapidly (<25 ns) to values in the region of 1Ω. Metal oxide varistors are used in combination with gas-discharge surge diverters for primary protection, while specific protection is afforded in the circuits by means of suppressor diodes or zener diodes (see Part 1 of this book).

The characteristics of overvoltage protection components are compared in Table 4.4.

As an example of 'graded protection' with various components, Fig. 4.7 shows the protection scheme against lightning interference for an exposed telemetry cable. The gas-discharge surge diverter, as primary protection, absorbs the main energy associated with the overvoltage, while the following combination, consisting of a varistor and a suppressor diode, clips the voltage wavefront up to the point where the gas-discharge surge diverter triggers.

In this connection, the *diverter valve* should be mentioned. This consists of a combination of a spark gap and a varistor and is used, for example, to protect mains supplies against the effects of lightning strokes.

4.6.3 Design of the distribution network for EWSD systems

In conventional switching systems one common line (main load line) serves to distribute current to all the equipments in a room. From this line, lines of smaller cross-section branch off to the rack rows (branch load lines). Because of the reduction in cross-section, fuses are inserted into the branches. In this distribution system, if the installation has to be extended, work has to be carried out on the live main load or branch load lines. Undervoltages or overvoltages caused by fuse blowing as a result of short-circuits or earth faults, which cannot be avoided in this kind of work, affect the whole distribution system. Under unfavourable conditions an operational breakdown of the electronic system is then possible. To limit the incidence of operational failures of the system due to short-circuits in the distribution network, therefore, the power distribution for the EWSD system has to be arranged in a fundamentally different way (Fig. 4.8).

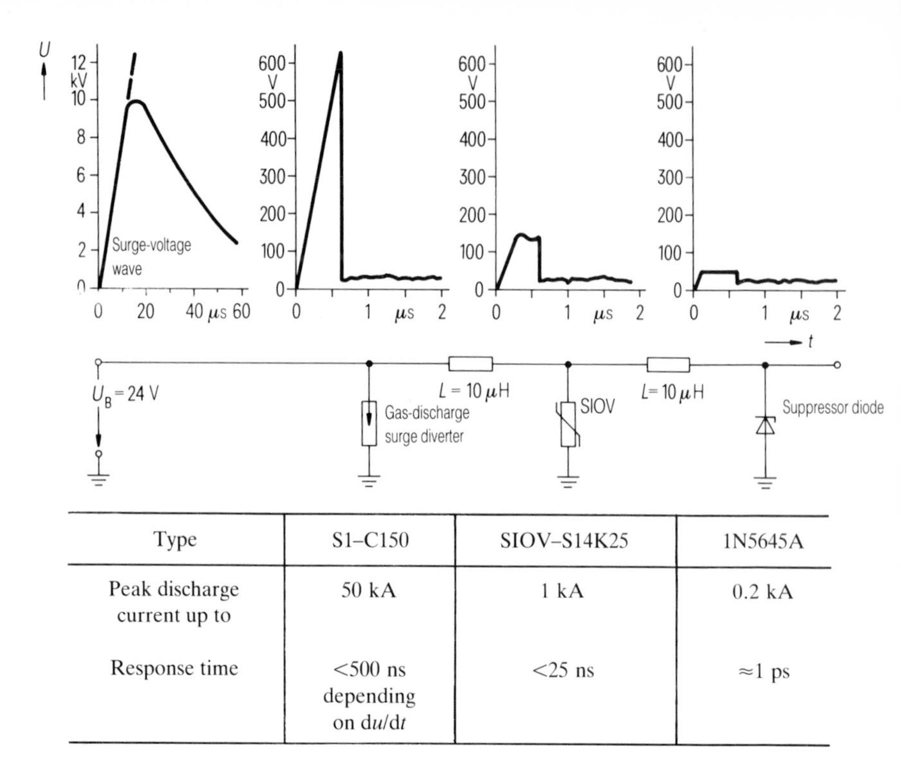

Type	S1–C150	SIOV–S14K25	1N5645A
Peak discharge current up to	50 kA	1 kA	0.2 kA
Response time	<500 ns depending on du/dt	<25 ns	≈1 ps

Fig. 4.7
Example of graded protection with voltage-limiting components

Each cubicle row has its own feeder. A choice of two standard cross-sections (95 and 150 mm^2) is available. To facilitate the laying of cables for the purpose of extensions, feeder cables with fine-strand cores are used. To keep the d.c. voltage drop within acceptable limits, up to four cables can be connected in parallel.

To minimize the overvoltages which can be produced by the blowing of a 10 A fuse, the distribution network must be so arranged as to introduce the lowest possible impedance. This is achieved in the case of the positive line by means of large-surface-area return conductors, and in general through close proximity of the positive and negative conductors. In addition, the total combination of the 2.2 mF capacitors incorporated in each of the system cubicles effects a useful degree of suppression. The separate power circuit for each cubicle row also affords good decoupling. The voltage dip caused by the blowing of a fuse affects principally the associated current-carrying negative line.

Table 4.4 Comparison of the principal characteristics of overvoltage protection components

		Inert-gas-filled discharge surge diverter	Metal oxide varistor	Suppressor diode	Zener diode
Range of protection level	(V)	65 to 12 000	20 to 2000	6 to 400	2.4 to 200
Surge current	(A)	60 000[1])	25 000[1])	1000	200
Energy adsorption capacity	(J)	60	1800	1	0.05
Range of continuous loading	(W)	800	<1.5	>1	0.5 to 50
Response time	(ns, ps)	du/dt-dependent, approx 500 ns	<25 ns	<10 ps[2])	10 to 1000 ps
Capacitance	(pF)	1 to 7	40 to 15 000	300 to 12 000	8 to 1500
Leakage current	(nA)	<15	<200 000	<5000	<100
Application		Primary protection for following close-protection devices	Primary and close protection in mains circuits	Close protection	Close protection
Advantages		High surge current, low capacitance	High surge current, very good extinction characteristics, no follow-on current	Very short response time, very good extinction characteristics, no follow-on current	Short response time, relatively high loading capability
Disadvantages		Low-impedance supplies require assisted extinction, moderate response time falling with increasing du/dt	High capacitance, affected by aging	High capacitance, relatively low surge current capability, dependent on voltage	Relatively low current loading capability, low protection level

[1]) With 8/20 μs overvoltage impulse.
[2]) With low-inductance connections.

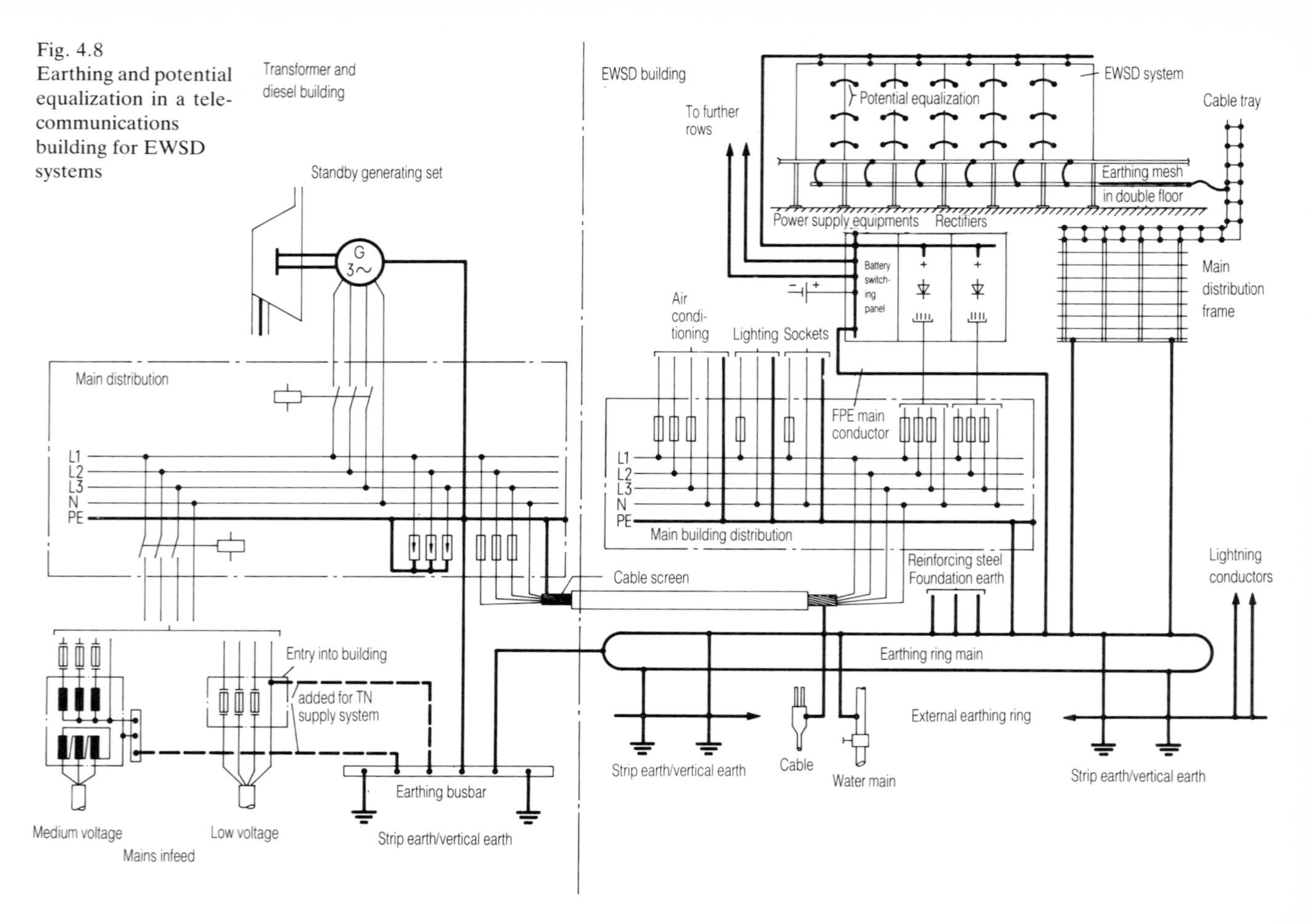

Fig. 4.8
Earthing and potential equalization in a tele-communications building for EWSD systems

4.6.4 Design of earthing, potential equalization and lightning protection for EWSD systems

Internal system provisions

To ensure undisturbed operation of all the central and decentralized functions of the system, and for the purposes of screening, a potential reference plane is necessary, which also permits no appreciable potential differences within the plane as a result of pulse currents (fuse-blowing, atmospheric overvoltages). In former systems the design of the potential reference plane was characterized by cross-connection of the earthed positive lines with the rack structure and the floor reinforcing mesh. In the case of digital systems, because of the compact construction that is possible and the consequent high heat loss per unit area, a false floor is sometimes provided in the exchange room for air conditioning; since this is consequently available also for the cabling, the floor mesh in its structural form would not be effective in respect of the cabling. The electrical function of the mesh as a significant part of the potential reference plane cannot, however, be dispensed with.

To fulfil the electrical function of the floor mesh, a large-area conducting network of aluminium strips is laid in the double floor.

Each system cubicle is connected to this earthing screen. An earthing network of typical dimensions can limit the potential difference produced by a peak current of 1 kA (straight across the area of the exchange room) to less than 10 V. This is the highest value, according to experience, that can result from interference in earth cables from atmospheric discharges.

Figure 4.9 shows the relationships that apply on the occurrence of an overvoltage due to an atmospheric discharge. The current that flows as a result of the triggering of a gas-discharge surge diverter is divided according to the impedances of the conducting paths represented by the main distribution frame (HVT)–system–power supply equipment–earth and the main distribution frame earth. The voltage drop due to the peak current in Z_2 (cable tray between the main distribution frame and the system) is applied to the subscribers' or junction lines, while the impedance Z_1 (the earthing line to the main distribution frame) determines the overall magnitude of the overvoltage. That means that these connections must be made with the minimum impedance. The main distribution frame must therefore be provided with multiple earths. The cable trays between the main distribution frame and the system or between parts of the system must be bonded with low-resistance area joints. If the cables have to pass through walls, and it is not possible to carry the longitudinal tray members through, wide copper strips must be provided for electrical connection. Where they enter the exchange room, the trays should be connected to the earthing network over a broad area.

With the interconnections described a substantial uniformity of potential in the

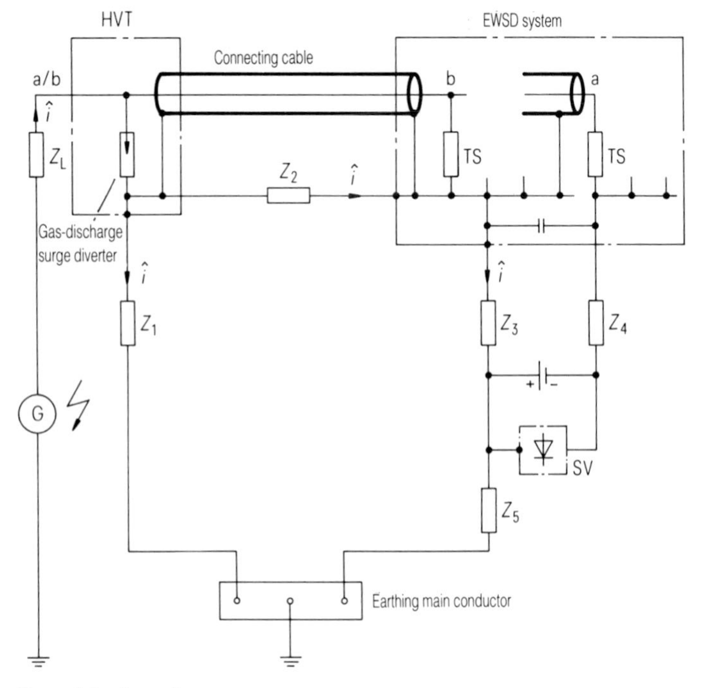

Z_L Line impedance
Z_1 Impedance of earth conductor between main distribution frame and earthing main conductor
Z_2 Impedance of cable tray between main distribution frame and system
Z_3 Impedance of positive conductor between system and power supply installation
Z_4 Impedance of negative conductor between system and power supply installation
Z_5 Impedance of FPE conductor between power supply installation and earthing main conductor
HVT Main distribution frame
TS Subscriber circuit
SV Power supply installation

Fig. 4.9 Essential representation of conditions on overvoltage

d.c. area will be achieved. This is a prerequisite for the two-way connection of the screens of distribution cables in order to utilize their attenuating properties.

To make further use of the attenuation of the earthing network, the connecting cables must be laid directly on the metal sheets; in addition, the cable runs must lie within the area of the network. To the extent that in large installations individual cables are laid above the cubicles – e.g. bus connections – cable with a braided screen must be used for this purpose.

Provisions in the building

The erection of lightning-protection systems is covered by VDE regulations DIN 57185/VDE 0185 Parts 1 and 2; these have replaced the previous general lightning-protection regulations (published by the lightning conductor construction committee, ABB).

Lightning-protection systems installed in accordance with the former ABB recommendations ensured protection only for persons and against fire.

The design of lightning protection described below takes into consideration both the 'normal' building protection and the additional measures necessary for the protection of electronic systems.

Lightning conductor installation

On flat roofs, or on the ground under ridged roofs, lightning conductors should be erected at a regular spacing of about 0.6 m. A conducting roof curb, in conjunction with the closely interconnected network of lightning conductors, affords the best distribution of the lightning current to the diverter discharge cage.

Diverters

In steel-framed concrete buildings the individual stanchions of the outer walls are included in the earthing system and used as diverters. To this end it is necessary to weld the vertical reinforcing bars or girders. Where this is not possible from considerations of strength, 10 mm galvanized steel rod should be laid with the reinforcement of the reinforced-concrete pillars and bonded. The distance between the diverters must be less than the distance between floors.

In framed buildings with wider spacing, additional diverters should be introduced into the brick bays of the walls. For horizontal potential equalization the reinforcing bars of the floors and ceilings must be bonded to the vertical diverters. The same applies to the reinforcing bars in the foundations; these should be carefully welded or tied with binding wire to the individual mats.

Foundation earth

A foundation earth should be laid in the footings of the outer walls. The reinforcing bars of the reinforced-concrete stanchions must be bonded to the foundation earth. For connection to the external lightning conductors terminal lugs should be brought out at the spacing of the diverters. Terminal lugs should also be provided for the internal earthing ring main conductor.

Earthing ring main conductor

For the lowest floor of the building an earthing ring main conductor through all the rooms is recommended. The ring main takes the form of an insulated copper conductor of 95 mm^2 cross-section.

To this conductor, for the purposes of potential equalization, are connected metal piping systems (except gas pipes), ventilating ducts, cable trays and the conducting sheaths of the external cables.

To avoid introducing potential differences in these external conductor systems into the building, where the risk from lightning is high they should, as far as possible, be brought into the building at one point, where the sheaths can be connected directly to one another and to the earthing ring main.

Screening

Insofar as precast reinforced-concrete parts or conducting cladding panels are used for outer walls and included in the earthing system, a screening attenuation of about 20 dB is to be expected in the broadcast radio-frequency range. This attenuation is sufficient, on the one hand, to meet the requirements of interference class N and, on the other hand, to render the EWSD apparatus, for example, immune to an external field of more than 3 V/m.

Brick-built outer walls or bays between concrete stanchions, at least in the vicinity of telecommunications apparatus, necessitate additional screening. This can be achieved with suitable metal cladding, or else the separation between the diverters must be reduced to about 1.2 m.

The diverters should be joined at roof level through the lightning rods and at ground level through a ring main.

In special cases an equipment room can be lined with copper foil as a retrospective measure.

In larger installations, accommodated in separate parts of the building, the connecting cable must be installed with adequate protection. The cables should be laid either in steel conduit, in ducts with conducting linings or in reinforced-concrete cable ducts. In each case, at the point of entry into the building, the screen (conduit, reinforcement) must be connected with a large-area contact to the earthing system of the building.

The earth lines of the exchange and transmission apparatus and the power supply equipment should be connected to the earthing ring main in the basement by the shortest route. Since with the desired dense network of diverters the close proximity between the diverters and the system cannot be avoided, the potential reference plane of a system larger than 20 m in height should be connected to the diverters.

With the measures described, substantial potential differences within a building are avoided. Cables entering the building from outside, however, are connected to a remote earth. If the building is struck by lightning, its potential can be raised by the voltage drop in the earthing system.

For telecommunications buildings, which are particularly susceptible to direct lightning strokes because of their position and height, an especially low earth resistance is desirable; in addition, the cable cores should be protected by diverters.

Since mains supply cables also represent a remote earth in the event of a lightning stroke, the unearthed conductors must be protected by diverter valves.

4.6.5 Special requirements for lightning protection in telecommunications towers

Usually the power supply installation for the transmission system is in the operating building beside the base of the tower, and the very long power supply conductors introduce the risk of induced overvoltages.

A lightning stroke on a telecommunications tower produces a large voltage drop in the steel reinforcement. The overvoltage thus produced in the cable may damage the input circuit of the d.c/d.c. converter.

To avoid this, the following protective measures must be adopted:

- ▷ the d.c. cables between the operating building and the base of the tower are run in a metal tube. This tube, which is 'galvanically' connected to the reinforcement of the tower and the operations building and also, via a specially formed cable duct in the operating building, to the battery switching panel, ensures that the tower and the building, including the power supply system, are always at approximately the same potential with respect to earth (Fig. 4.10);
- ▷ the d.c. power supply conductors between the battery switching panel in the operating building and the operating floor in the telecommunications tower are in the form of screened cables. The screen of the cable is connected by the shortest route to the structural metalwork of the switching panel, to the steel tube in the cable duct and the base of the tower, and in the operating floor to the floor mesh.

By these expedients the earth resistance is reduced considerably.

Overvoltages which exceed the permissible level in spite of the measures described above are limited in the operating floor by means of a lightning-protection assembly.

In the case of d.c./d.c. converters with 'galvanic' separation, adequate protection is obtained if the two lines are balanced with respect to earth; interference

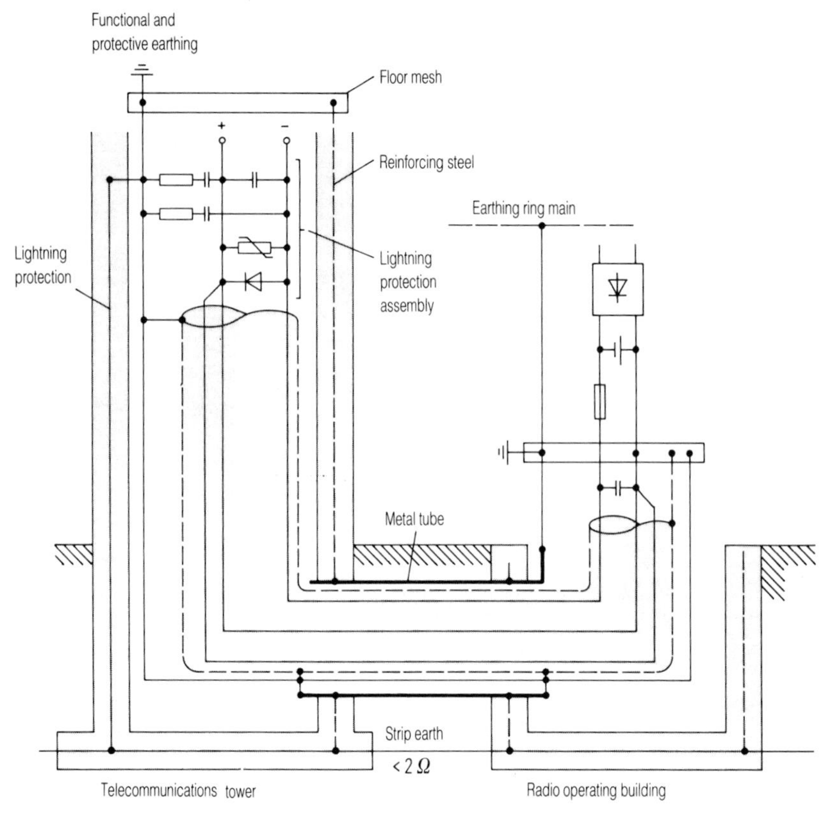

Fig. 4.10
Diagrammatic representation of earthing system in a telecommunications tower and an operating building

voltages between the positive and negative poles are then sufficiently small. This balancing is achieved by the capacitors C013 and C014 (Fig. 4.11).

For the protection of converters without 'galvanic' separation, the positive pole should be connected to the functional and protective earth within the exchange wiring. Additional protective measures therefore have to be applied to prevent an unacceptable rise of voltage between the negative pole and the functional and protective earth; these take the form of a capacitor (C012) and a number of varistors (*R*010) connected between the positive and negative lines, In this arrangement the capacitor limits the high-frequency component of the interference voltage while the varistors limit the low-frequency component which represents greater energy.

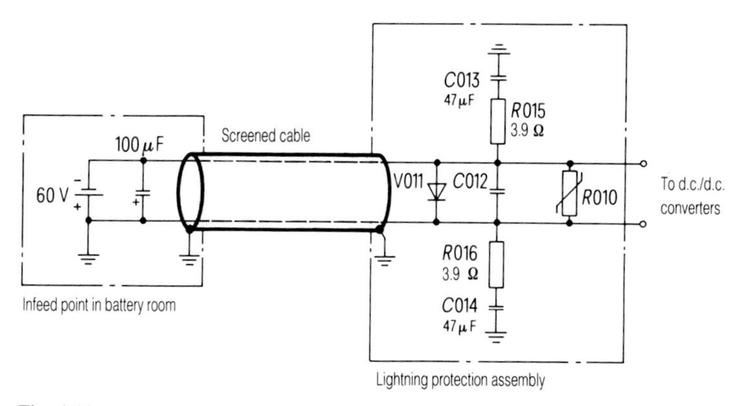

Fig. 4.11
Lightning-protection assembly for d.c./d.c. converters with and without 'galvanical' separation

It is possible, as a result of the polarity of the interference voltage, for the polarity at the input to the converter to be reversed. The possibility that the input transistors may be destroyed before the varistors respond cannot be ruled out; a diode (V011) is therefore provided as additional reverse-polarity protection.

5 Protective Measures

To avoid accidents from electrical causes, the *rules for the prevention of accidents* of the professional association, the *VDE regulations* (e.g. DIN 57 100/ VDE 0100) and all other *safety rules* should be observed.

The VDE regulations are divided into:

▷ equipment regulations, for the manufacture of equipment,
▷ erection regulations, for installation, etc., on site,
▷ operating regulations, for operation and maintenance of equipment or installations.

Decisive factors affecting the magnitude of the consequences of an electrical fault are the current level, the duration of the effects of the current, the current path and the nature of the current. Mains alternating currents with a magnitude of more than 50 mA are particularly dangerous if they persist for more than 0.1 s and the current path includes the heart.

The regulations pertinent to protection against dangerous body currents should be strictly observed.

The corresponding measures are divided into:

▷ protection against direct contact. These measures should prevent contact with live parts,
▷ protection against indirect contact. These measures should protect against danger arising from contact with conducting parts, referred to as 'bodies', which may become live under fault conditions,
▷ protective low voltage or functional low voltage. This method rests upon the operation of circuits at voltages of not more than 50 V a.c. or 120 V d.c.

Working on live parts or systems represents an especial hazard. Such work is therefore not permitted. Exceptions are possible if the installation cannot be made dead and only qualified people are working on it (VDE 0105).

Protection against direct contact

These regulations should prevent the inadvertent touching of live parts, specifically through the 'insulation' and 'covering' of such parts.

The requirements for insulation (test voltage, leakage currents, etc.) are laid down in the relative equipment specifications. Covers must comply with creepage distance and air-gap requirements and must not anywhere permit the ingress of a hand.

In rooms through which unqualified people may pass, covers must be such that they can only be removed with the aid of tools.

Less onerous conditions are permitted in electrical operating locations. Thus, for example, a cover is adequate which protects against inadvertent contact and can be removed without tools.

In locked electrical operating locations, protection against contact may be completely dispensed with. However, such relaxations are only permissible when these special operating locations comply with all the requirements stipulated by the VDE regulations.

Protection against indirect contact

In electrical installations, parts which in normal operation carry no voltage with respect to earth (e.g. housings) may assume dangerous contact potentials in the event of faults. A means of protection for indirect contact is therefore necessary (formerly known as 'protection against excessive contact potential').

Protective measures against indirect contact are required in installations and equipment with voltages in excess of 50 V a.c. or 120 V d.c. with respect to earth.

On the other hand, no protective measures are required in relation to indirect contact in installations and equipment:

▷ with voltages below 50 V a.c. or 120 V d.c. with respect to earth (protective low voltage or functional low voltage),
▷ with voltages up to 250 V with respect to earth in public supply mains equipment for electrical energy and power measurement,
▷ in domestic installations with insulating floors, where no coincidental contact with earthed water, gas or heating equipment is possible,
▷ with alternating voltages below 1000 V and direct voltages below 1500 V for steel or reinforced-concrete poles, metal conduits and metal cases with insultating coatings, metal conduits for the protection of multiple conductors or multi-core cables, and sheathing or armouring of conductors and cables, so long as the cables are not laid in the ground.

Protective measures for indirect contact may be classified as follows:

Protection by disconnection or indication

▷ Disconnection by overcurrent protection device:[1])
TN supply system (neutral earthing),
TT supply system (protective earthing),
IT supply system (protective-line system),

[1]) Supply systems in which protective measures involving protective conductors are used are designated by two letters. The first letter indicates the relationship to earth of the power source: T, direct earth connection; I, insulation. The second letter indicates the relationship to earth of the loads: T, direct earth connection; N, direct connection to star or neutral point.

▷ disconnection by protective switch:
fault voltage-operated (FU) protective circuit (not described),
fault current-operated (FI) protective circuit.

One of these protective measures is a prerequisite for the use of equipments in protective class I.[1]) All such equipments are provided with an earth connection.

Protection without disconnection or indication

▷ Protective insulation (protective class II[1])),
▷ equipments with protectively insulated parts,
▷ protective isolation.

Equipment in protective class II can be used on any supply system.

Protective low voltage and functional low voltage

It is stipulated that it must not be possible for the voltage on the low-voltage system to exceed 50 V a.c. or 120 V d.c. This necessitates the use of specified power sources and the observance of regulations covering the arrangement of circuits (e.g. isolation from higher-voltage circuits; see Section 5.2.3).

In general, protection against direct contact can be dispensed with if the nominal voltage does not exceed 25 V a.c. or 60 V d.c.

5.1 Protection by Disconnection or Indication

Bodies[2]) which do not form part of the functional circuit may, as a result of insulation failure, assume a potential relative to earth or to other conducting parts, referred to as the contact potential U_B (Fig. 5.1); this can drive a dangerous current through the human body.

Common to all protection methods relying on disconnection is a *protective conductor* PE[3]) (protective class I). The protective conductor connects the body to earth (see, for example, Fig. 5.4) or with the PEN conductor.[4])

In the event of a fault, the fault current is carried from the body by the protective

[1]) Protective classes I and II are defined in several VDE specifications – among others, VDE 0106 and VDE 0804.

[2]) Bodies, in the terminology of the VDE specifications, are touchable conducting parts of an installation which are not live in normal operation (e.g. equipment housings).

[3]) The designation SL was formerly used for this.

[4]) The PEN conductor combines the functions of the neutral conductor N (previously designated Mp) and the protective conductor PE. The description 'Nulleiter' ('zero or neutral conductor') (SL/Mp) was previously applied to the PEN conductor.

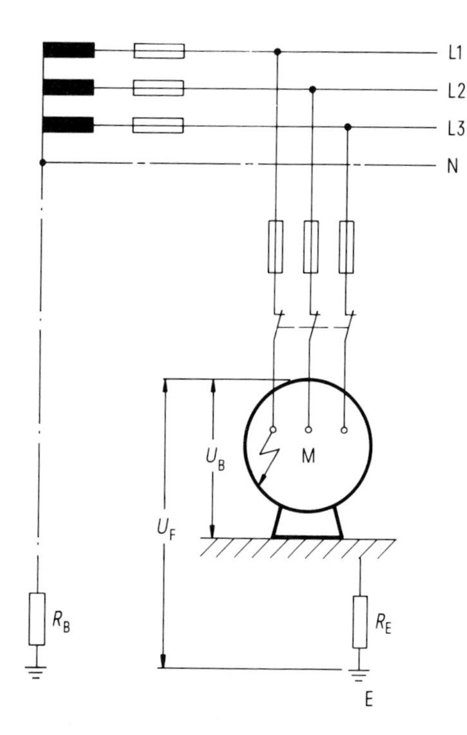

E Reference earth
R_B Operating earth resistance
R_E Sum of earth resistances

Fig. 5.1
Fault voltage U_F and contact potential U_B with a non-insulated floor[1])

conductor. The disconnection must be so affected that the supply voltage is removed as quickly as possible.

All bodies must be connected to the protective conductor. The terminal for the connection of the protective conductor is distinguished by the symbol ⏚ (DIN 40 011). To prevent rupture of the protective conductor, it must have a certain minimum cross-sectional area (depending upon its length and the rating of the fuse which precedes it). In the case of flexible lines it should be contained within the common sheath and coded green/yellow. Structural parts may be used as protective conductors if they fulfil the requirements as to cross-sectional area, provided that the protective connection is not broken by the removal of structural parts and the connection remains highly conductive with the passage of time.

In telecommunications equipment, the insulation between power circuits and accessible conducting parts must be specified for a test voltage of 1.5 kV (in accordance with VDE 0804 for equipment in protective class I). For functional reasons one pole of the d.c. supply line – usually the positive pole – is directly

[1]) The fault voltage U_F is the voltage that appears between a body and the reference earth, or between bodies, in the event of a fault. By contact potential U_B is meant that part of the fault voltage that can be embraced by the human body.

earthed. From the point of view of the functional efficiency of the system, this line should not introduce more than a certain voltage drop. Depending on the current level and the length of the line in question, this can lead to large cross-sections.

Since the conductors thus earthed are connected together in each rack, through the rack structure, for example, the result is a very low-resistance (earthed) conductor network. Hence this 'area earth', apart from its operational function, is also applied in general to protective purposes, in that the enclosures of equipments in protective class I are bonded to the rack structure. The whole system is designated as functional and protective earthing (see Chapter 4). This represents a generic term for all protection measures based on disconnection, and should not be confused with the special protective measures associated with the TT supply system (protective earthing) in accordance with DIN 57 100/VDE 0100 (see Section 5.1.2).

By virtue of the functional and protective earthing, telecommunications systems can be designed *in standard form*, irrespective of the protective measure of the public mains supply. The coordination with the various protective systems is effected centrally at the earthing main conductor.

In the following sections, 5.1.1 to 5.1.4, the power system protection measures according to DIN 57 100/VDE 0100 are compared with the corresponding implementation of the functional and protective earthing scheme in accordance with VDE 0800 Part 2.

5.1.1 Protective measures in the TN supply system

The TN system corresponds to the type of supply in which neutral earthing was formerly applied; it refers to the most widely used protection method. In this scheme all bodies are connected, either directly or through a special protective conductor PE, to the PEN conductor (formerly the neutral conductor). The connection to the protective line ensures that all accessible parts are at the same potential in the event of a fault to frame in an equipment. The short-circuit current to which it gives rise causes the preceding fuse to operate and thereby to disconnect the circuit from the supply.

The following time-delays are appropriate for this purpose:

▷ 0.2 s for socket-outlet circuits of up to 35 A current rating,
▷ 5 s for all other circuits.

To ensure that the potential of the PEN or protective conductor deviates as little as possible from earth potential, this conductor should be earthed at numerous points to the public distribution network, particularly at the point of entry into the building.

Two examples of telecommunications systems connected to low-voltage power systems, with protection arrangements in the TN supply system, may now be considered.

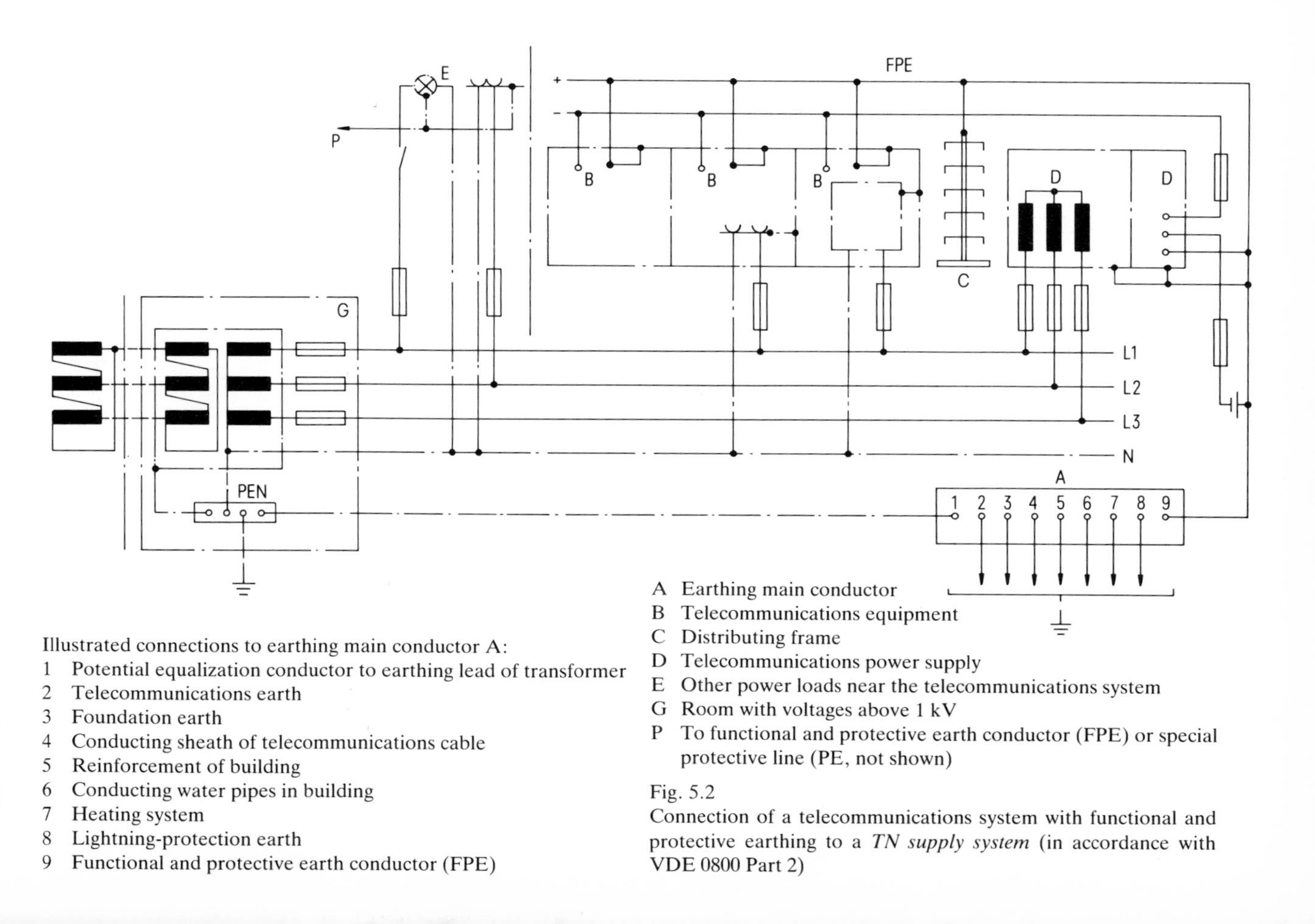

Illustrated connections to earthing main conductor A:

1 Potential equalization conductor to earthing lead of transformer
2 Telecommunications earth
3 Foundation earth
4 Conducting sheath of telecommunications cable
5 Reinforcement of building
6 Conducting water pipes in building
7 Heating system
8 Lightning-protection earth
9 Functional and protective earth conductor (FPE)

A Earthing main conductor
B Telecommunications equipment
C Distributing frame
D Telecommunications power supply
E Other power loads near the telecommunications system
G Room with voltages above 1 kV
P To functional and protective earth conductor (FPE) or special protective line (PE, not shown)

Fig. 5.2
Connection of a telecommunications system with functional and protective earthing to a *TN supply system* (in accordance with VDE 0800 Part 2)

In the case of connection to a medium-voltage supply through a transformer substation in the region of the telecommunications system, the low-voltage earthing system of the substation (installed in accordance with DIN 57 100/VDE 0100) should be connected to the earthing main conductor of the telecommunications system (Fig. 5.2).

Only a *single connection* is permissible between the directly earthed star point of the low-voltage side of the transformer substation and the telecommunications earthing system. The cross-sectional area of this connection must be such that the overcurrent protection device connected in the feeder to the telecommunications power supply or to the power system loads in the telecommunications racks operates in the event of a fault to frame or to earth in the feeder. To this end the cross-section should be specified in accordance with VDE 0800, but it must in any case be at least 16 mm^2 (copper).

When a protective measure is applied to a TN supply system *with* a separate earth conductor PE in accordance with DIN 57 100/VDE 0100, a connection should be made between the PEN conductor terminal at the connection point of the incoming low-voltage mains supply and the earthing main conductor of the telecommunications system (Fig. 5.3). The cross-sectional area of this connection must be the same as that of the PEN conductor of the incoming feeder, but in any case at least 16 mm^2 (copper). Further connections between the PEN conductor and the telecommunications earthing system (the functional earth) at other points in the building are not permitted, because they would introduce the risk of interference in the telecommunications system. The separate earth conductor PE is connected to the PEN conductor only at the point of entry into the building; it is thus connected also to the earthing main conductor of the telecommunications system through the required connection.

VDE 0800 Part 2 also describes the protective arrangement of the TN supply system *without* the separate earth conductor PE (the 'classical neutral earthing') in accordance with DIN 57 100/VDE 0100 in telecommunications systems connected to a low-voltage mains supply. Since this variant is hardly ever employed nowadays, it is not illustrated here. It corresponds substantially, however, to Fig. 5.3, but with the separate earth conductor omitted.

5.1.2 Protective measures in the TT supply system

The protective measures in the TT supply system (protective earthing) entail the direct connection of bodies to earth, or earthed parts, so that, as with the TN supply system (see Section 5.1.1), the supply is disconnected by overcurrent protective devices in the event of an insulation failure (a direct breakdown to frame). In the example shown in Fig. 5.4 the faulty circuit must be isolated by the fuse F1. In a TT supply system one line of the power supply is directly earthed, which means that the fault current flows through the ground. The contact

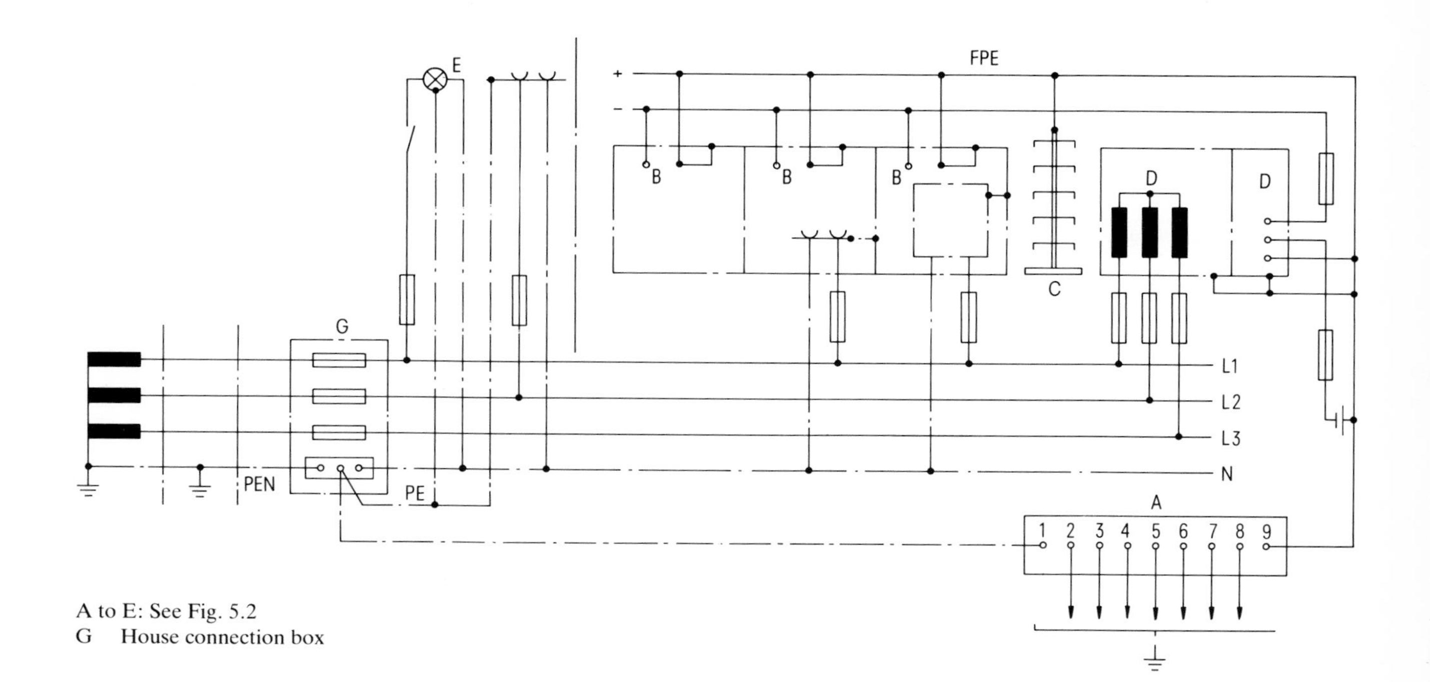

A to E: See Fig. 5.2
G House connection box

Fig. 5.3
Connection of a telecommunications system with functional and protective earthing to a *TN supply system with special protective line* PE (in accordance with VDE 0800 Part 2)

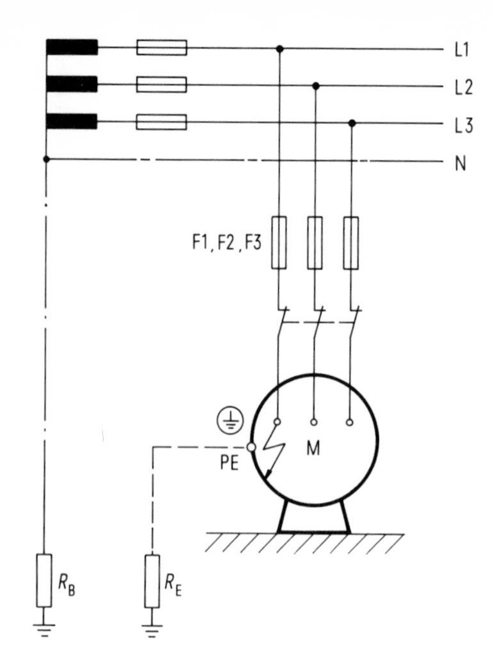

R_B Operating earth resistance
PE Protective line
R_E Earth resistance

Fig. 5.4 Principle of protective measures in the TT supply system

potential U_B may not exceed a level of 50 V a.c. To this end the earth resistance R_E (in ohms) must be calculated according to the following relationship:

$$R_E \leqq \frac{U_B}{I_A} \ .$$

Nowadays the protective device is generally the FI protective circuit (see Section 5.1.4). If overcurrent protective devices are used, disconnection is required additionally in the neutral line. If this is not possible, a 'supplementary potential equalization' is used; its requirements are fulfilled in telecommunications systems by the functional and protective earth (see Section 4.3).

An example of the protective measures in a *TT supply system* in telecommunications installations with functional and protective earthing (see Section 4.3) and connected to a low-voltage mains supply is shown in Fig. 5.5. Telecommunications installations incorporating functional and protective earthing can be connected without any special precautions to TT supply systems that comply with DIN 57 100/VDE 0100.

If there are earth conductors (PE) for other power loads in the vicinity of the telecommunications installation (in the same room), they also may be connected directly to the earthing main conductor instead of the functional and protective earth. The cross-sectional area for this purpose should be in accordance with VDE 0800 Part 1.

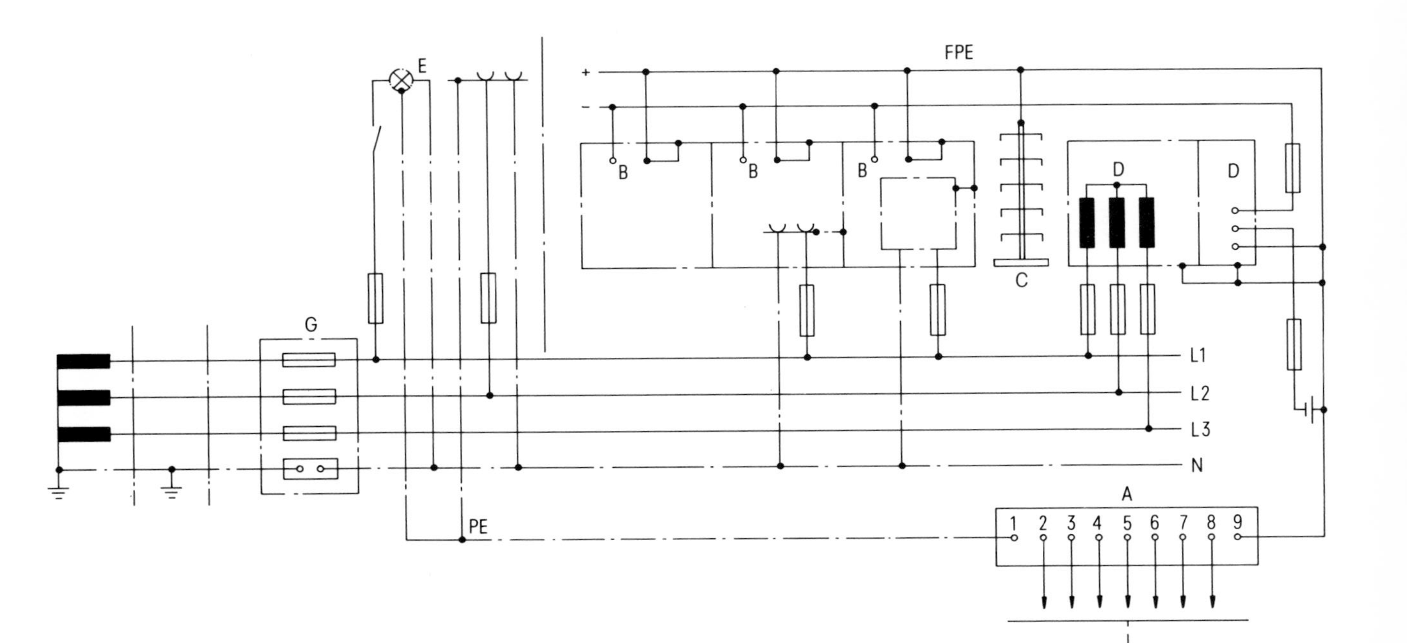

A to E: See Fig. 5.2
G House connection box

Fig. 5.5
Connection of a telecommunications system with functional and protective earthing to a *TT supply system* (in accordance with VDE 0800 Part 2)

The neutral conductor N associated with a TT supply system must *not* be connected to the functional earth of the telecommunications system (see DIN 57 100/VDE 0100).

5.1.3 Protective measures in the IT supply system

In the IT supply (protective-line) system, to prevent the occurrence of excessive contact potentials, all bodies are connected together and also to the accessible conducting parts of the building, pipe systems, etc., and to earthing electrodes in the case of unearthed supplies (no diagram).

A detailed explanation of the protective measures in the IT supply system is omitted here, since the arrangement is only used in special supply systems – in hospitals, for example, in which it is not acceptable to interrupt the low-voltage supply immediately on the occurrence of a fault.

5.1.4 Fault current-operated protective circuit

In the fault current-operated (FI) protective circuit, a fault current-operated protective circuit breaker (switch) opens automatically on the occurrence of a fault current in excess of its rated current (the circuit being interrupted within a period of not more than 0.2 s).

The basic mode of operation of the fault current-operated protective circuit breaker SS may be explained with reference to Fig. 5.6.

All the a.c. lines to the equipment to be protected are passed through a summing current transformer. In the absence of a fault (not illustrated) current flows from the main conductor L1 through the fuse, fault current-operated protective circuit breaker and the load, and returns completely to the neutral conductor.

The current-transformer core is not magnetized, since the magnetizing effects of the currents cancel. No voltage, therefore, is induced in the secondary winding of the transformer and the protective circuit breaker is not tripped.

If an insulation failure occurs in the protected equipment, the fault current I_F flows through the earth to the star point of the supply transformer (Fig. 5.6). The current that flows to the load is larger than the return current by the amount of the fault current. This difference produces a magnetic field in the core of the current transformer, which energizes the relay of the protective circuit breaker. If the fault current reaches the rated value, the relay operates and the protective circuit breaker trips. The circuit is then broken by the opening of the contacts.

Modern fault current-operated protective circuit breakers trip very rapidly ($\leqq$0.03 s) in the event of a fault. With high-sensitivity types, operating on, for

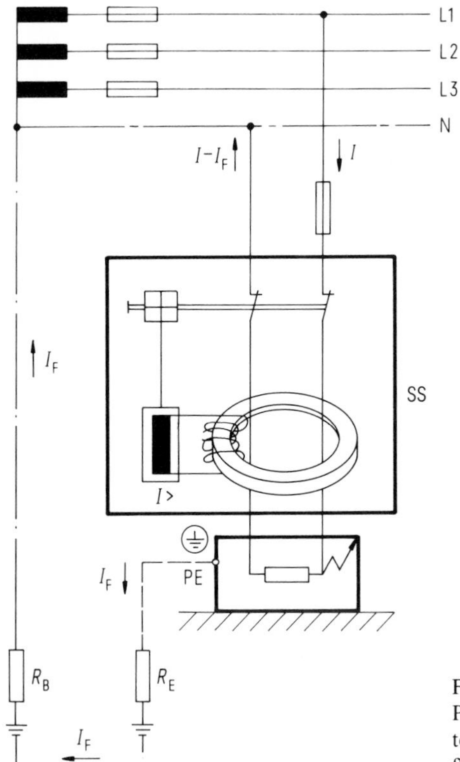

Fig. 5.6
Principle of fault current-operated protective circuit (shown with fault current flowing)

example, 30 mA fault current, it only requires an 'earthed' person to touch a live part to cause the protective circuit breaker to trip.

Figure 5.7 shows an example of a fault current-operated protective circuit breaker in a telecommunications system connected to a low-voltage mains supply.

When a fault current-operated protective device is used, the neutral conductor N of the low-voltage mains supply must not be connected to the earthing system of the telecommunications installation behind the protective circuit breaker, since it would in that case by-pass the circuit breaker and render it ineffective. The earth conductor PE for other power system loads in the vicinity of the telecommunications system should be connected to its earthing main conductor. The cross-sectional area must be in accordance with VDE 0800 Part 1.

The fault current-operated protective circuit can also be used to obtain improved tripping in TN supply systems.

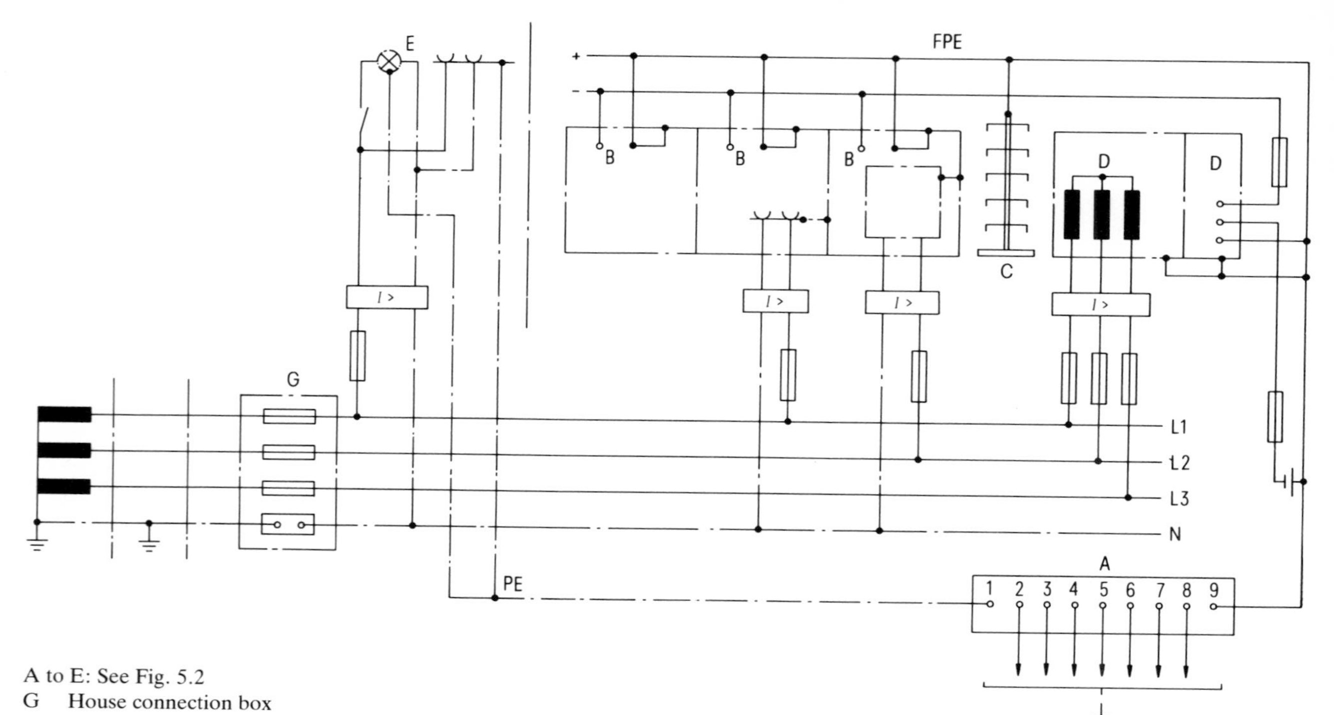

A to E: See Fig. 5.2
G House connection box

Fig. 5.7
Connection of a telecommunications system with functional and protective earthing to a low-voltage supply with *fault current-operated protective circuit-breakers*

5.2 Protection without Disconnection or Indication

Protective measures without any disconnection device relate to equipments which have no protective conductor connection; they are 'intrinsically safe'. Protection in this case depends neither upon an effective potential equalization nor upon disconnection of the deranged circuit.

5.2.1 Protective insulation

The regulations covering 'protective insulation' (protective class II) indicate the measures to be adopted in equipments and in the construction of electrical installations in order to ensure protection against excessive contact potential through insulation.

In protectively insulated equipment, protection depends not only upon primary (basic) insulation; additional, or at least increased, insulation is required (Fig. 5.8). No connection is provided for a protective conductor.

Such an equipment must have a housing of insulating material which encloses all metal parts, apart from those such as nameplates, screws, rivets, etc., which must, however, be isolated by extra insulation.

If for functional reasons a metal housing or accessible metal parts are unavoidable (e.g. a chuck on a hand-drill), these parts must, without exception, have 'double insulation' (basic plus additional insulation) from live parts. The insulation must not become ineffective in the event of the breaking and springing apart of wiring.

The leakage current must not exceed 0.5 mA in telecommunications equipment (VDE 0804) or 0.25 mA in domestic equipment (VDE 0700 Part 1).

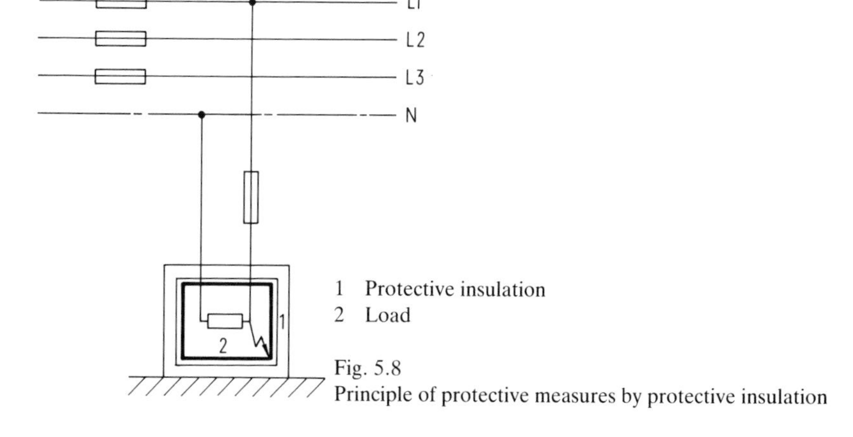

1 Protective insulation
2 Load

Fig. 5.8
Principle of protective measures by protective insulation

Protectively insulated equipment must be marked with the appropriate symbol ⧈ (in accordance with DIN 40 014).

The statements above refer to telecommunications and domestic equipment. Protective insulation can also be applied to the construction of systems with equipments which are accommodated within an installation from the outset. In this case several further requirements apply:

▷ parts that are conducting, but not live, within the protectively insulated housing must not be connected to a protective conductor inserted or looped into it,
▷ the protectively insulated housing must not at any point be pierced by conducting parts in such a way that a voltage can be introduced,
▷ unused access openings should be closed in such a way that a tool is necessary to open them.

The requirements listed above do not, however, preclude the possibility of looping a protective conductor into protectively insulated equipments so long as the associated live conductors are carried with it (e.g. in protectively insulated distribution boxes including protective conductor bars).

5.2.2 Equipments with protectively insulated parts

Equipments with protectively insulated parts, in accordance with VDE 0804, are also to be found in communications engineering (Fig. 5.9).

These equipments are similar in construction to equipments with protective insulation (protective class II) – see Section 5.2.1. Unlike protectively insulated equipments, they have a protective conductor connection or *protective conductor*, and must *not*, therefore, be classified as protectively insulated equipments.

The protective conductor in this case is for the purpose of EMI suppression or screening, or is required for following equipment in protective class I.

In this equipment the following conditions must be met:

▷ telecommunications circuits and any accessible conducting parts (frame) connected to them must be insulated from the protective conductor and accessible conducting parts connected to it for a test voltage of 500 V,
▷ the protective conductor must be coded green/yellow and its connecting terminal marked with the symbol ⏚ in accordance with DIN 40 011,
▷ equipments with protectively insulated parts must *not* be marked with the symbol ⧈ of DIN 40 014.

The same requirements apply to protectively insulated parts of equipment as to protectively insulated equipments (see Section 5.2.1).

Equipments with protectively insulated parts may have accessible conducting

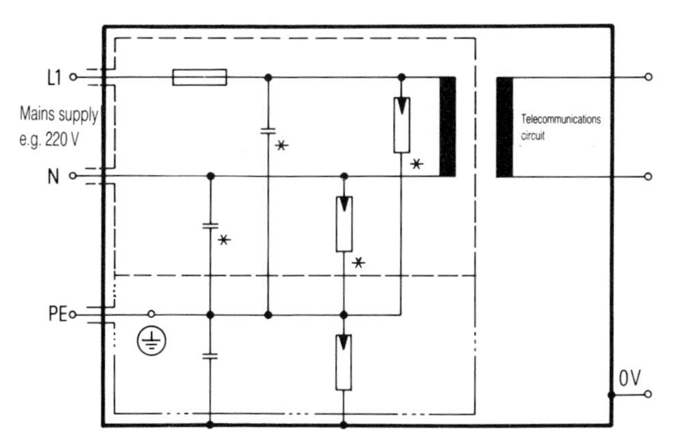

Insulation test voltage:

- - - - At least 2.5 kV between mains and telecommunications circuits or between mains and frame (double insulation)

* 1.5 kV between mains and PE

– ··· – ··· – 0.5 kV between PE and frame

———— Accessible metal parts (frame)

Fig. 5.9 Power supply equipment with protectively insulated parts

parts and an accessible conducting enclosure, which are not part of the protection arrangement against excessive contact potential (by indirect contact). The following should be observed for such parts:

▷ if the parts referred to possess a connection terminal for functional earthing or potential equalization, it must be marked with the frame symbol ⊥ in accordance with DIN 40 016,

▷ instead of the functional earth, a functional and protective earth (in accordance with VDE 0800 Part 2) may be connected to this terminal,

▷ conductors used to interconnect accessible conducting parts of the frame must *not* be coded green/yellow.

5.2.3 Protective low voltage and functional low voltage

Protective low voltage or functional low voltage refers to protective measures in which the circuit operates with a rated voltage of up to 50 V a.c. or 120 V d.c.; hence an excessive contact potential cannot arise as a result of an insulation failure. The protective low voltage may not be obtained by means of series resistors, autotransformers or voltage dividers, but only through safety transformers (protective isolating transformers) complying with VDE 0551 (Fig. 5.10)

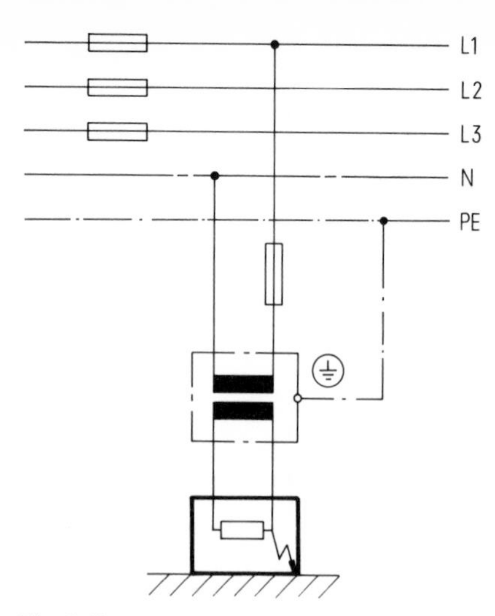

Fig. 5.10
Principle of protective measures by protective low voltage and protective isolation

or from motor generators with separate windings, diesel generators and electrochemical power sources (e.g. secondary batteries). With these power sources are included certain electronic equipments in which the voltage at the output terminals does not exceed the permissible level, even under fault conditions. The disposition of the circuits must be such that the low-voltage circuits (e.g. 24 V) are reliably separated from higher-voltage circuits. If a point in the circuit is earthed for functional reasons, the system is classified as *functional low voltage* (e.g. the supply to a communications system at rated voltages of 48 or 60 V).

5.2.4 Protective isolation

By protective isolation is understood the 'galvanic' isolation of an operating equipment, also as shown in Fig. 5.10, but with a rated voltage of up to 1000 V, by means of an isolating transformer.

In the event of a fault to the frame, protective isolation prevents the occurrence of a contact potential derived from the supply mains between earth and the faulty operating equipment.

Protective isolation is only effective so long as no earth fault occurs on the secondary side – e.g. through damage to cables; it is applied in telecommunications engineering mainly in mobile systems, such as military equipment or broadcasting vans. The advantage lies in the fact that the equipment can be operated without dependence upon the kind of protective measures embodied in the supply mains, making it unnecessary to provide an earthing system with defined characteristics at the point of installation.

Bibliography

(Note: In most instances the cited articles, standards etc. are available only in German.)

Books

Team of authors: *Electronics*, 2nd Part: *Industrial Electronics*, Verlag Europa-Lehrmittel, Wuppertal-Barmen (1970).

Team of authors: *Converter Engineering*, AEG-Telefunken, Elitera Verlag, Berlin (1977).

Benz, W., Heinks, P., and Starke, L.: *Tables for Electronics and Communications Engineering*, Kohl + Noltemeyer Verlag GmbH, Dossenheim, and Frankfurter Verlag, Frankfurt on Main (1980).

Bergtold, F.: *Dealing with Operational Amplifiers*, Verlag R. Oldenbourg, Munich and Vienna (1975).

Beuth, K.: *Components in Electronics; Electronics* 2, Vogel-Verlag, Würzburg (1980).

Ernst, D., and Ströle, D.: *Industrial Electronics; Principles – Method – Applications*, Springer-Verlag, Berlin, Heidelberg, New York (1973).

Fleischer, D.: *Digital Circuit Elements; Switching Elements, Storage Elements*, 3rd ed., Siemens AG, Berlin, Munich (1979).

Fröhr, F.: *Electronic Controllers for Drive and Power Engineering*, 2nd ed., Siemens AG, Berlin, Munich (1977).

Fröhr, F., and Orttenburger, F.: *Introduction to Electronic Control Engineering*, 5th ed., Siemens AG, Berlin, Munich (1981).

Garten, W.: *Lead Storage Batteries*, Varta Series of Technical Books, Vol. 1, Varta Batterie AG, Hannover, VDI Verlag GmbH, Düsseldorf (1974).

Gerlach, W.: *Thyristors*, Springer-Verlag, Berlin, Heidelberg, New York (1981).

Hartel, W.: *Power Converter Circuits; Introduction to the Circuits of Mains-Operated Power Converters*, Springer-Verlag, Berlin, Heidelberg, New York (1977).

Herhahn, A.: *First Reader on Safety for Electric Installations; to VDE 0100 with Safety Glossary*, Vogel-Verlag, Würzburg (1978).

Heumann, K.: *Principles of Power Electronics*, Verlag B. G. Teubner, Stuttgart (1975).

Heumann, K., and Stumpe, A.: *Thyristors; Properties and Application*, Verlag B.G. Teubner, Stuttgart (1974).

Hille, W., and Schneider, O.: *Technical Book on Electrical Professions*, Verlag B.G. Teubner, Stuttgart (1972).

Hirschmann, W.: *Electronic Circuits*, Siemens AG, Berlin, Munich (1982).

Hoffmann, A., and Stocker, K.: *Thyristor Manual*, 4th ed., Siemens AG, Munich (1976).

Integrated Phase-Angle Control Systems TCA 780 for Power Electronics: Technical Reports, Components Division, Siemens AG, Berlin, Munich (1978).

Integrated Switching Power Pack Control Circuits TDA 4700/TDA 4718: Technical Reports, Components Division, Siemens AG, Berlin, Munich (1980).

Kinzelbach, R.: *Alkaline Batteries*, Varta Series of Technical Books, Vol. 3, Varta Batterie AG, Hannover, VDI Verlag GmbH, Düsseldorf (1974).

Kloss, A.: *Power Electronics without Padding; A Practical Introduction to Circuit Engineering in Power Electronics*, Franzis-Verlag, Munich (1980).

Kolb, O.: *Introduction to Power Converter Engineering*, Vol. 1: *Principles*, AT-Fachverlag GmbH, Stuttgart (1976).

Krakowski, H.: *Telecommunications Power Supply*, Telecommunications Engineering Textbook, Verlag Schiele & Schön GmbH, Berlin (1978).

Kurscheidt, P.: *Power Electronics*, Berliner Union Verlag, Stuttgart, and Verlag W. Kohlhammer, Stuttgart, Berlin, Cologne, Mainz (1979).

Lappe, R., *et al.*: *Thyristor Power Converters for Drive Control Systems*, VEB-Verlag TECHNIK, Berlin (1975).

Möltgen, G.: *Mains-Operated Power Converters with Thyristors; Introduction to Mode of Operation and Theory*, 3rd ed., Siemens AG, Berlin, Munich (1974).

Müller-Schwarz, W.: *Principles of Electronics*, 3rd ed., Siemens AG, Berlin, Munich (1974).

MWM Diesel: *Generating Sets Manual; Guide for the Planning and Installation of Fixed Diesel-Electric Systems*, Motoren-Werke AG, Mannheim (1975).

Pabst, D.: *Operational Amplifiers; Principles and Application Examples*, Dr Alfred Hüthig Verlag, Heidelberg (1976).

Patzschke, U.: *Applied Thyristor Engineering; Principles and Application of Power Electronics*, Telekosmos-Verlag, Franckh'sche Verlagshandlung, Stuttgart (1970).

Paul, R.: *Semiconductor Diodes; Principles and Application*, Dr Alfred Hüthig Verlag, Heidelberg (1976).

Paul, R.: *Transistors and Thyristors; Principles and Application*, Dr Alfred Hüthig Verlag, Heidelberg (1977).

Schilling, W.: *Thyristor Engineering; An Introduction to the Application of Semiconductors in Power Engineering*, Verlag R. Oldenbourg, Munich, Vienna (1968).

Schlotheim, G.: *Power Electronics; Tasks and Solutions*, Vogel-Verlag, Würzburg (1980).

Tholl, H.: *Components in Semiconductor Electronics*; Part 2: *Field-Effect Transistors, Thyristors and Optoelectronics*, Verlag B.G. Teubner, Stuttgart (1978).

Watzinger, H.: *Application of Mains-Operated Power Converters for Direct Current Drives*, Siemens AG, Berlin, Munich (1976).

Watzinger, H.: *Mains-Operated Power Converters with Direct Current Output; Circuits and Mode of Operation*, Siemens AG, Berlin, Munich (1972).

Watzinger, H.: *Power Converters – Direct Current Drives; Measuring, Recording, Fault Finding in Start-up and in Operation*, Dr Alfred Hüthig Verlag, Heidelberg (1980).

Westphal, H.: *Selenium Today*, ITT Components, Verlag Gruppe Europa (1973).

Witte, E.: *Lead–Acid and Alkaline Accumulators*, Varta Series of Technical Books, Vol. 4, Varta Batterie AG, Hannover, VDI Verlag GmbH, Düsseldorf (1977).

Wolf, G.: *Digital Electronics; The Mode of Operation of Integrated Logic and Storage Elements*, Franzis-Verlag, Munich (1977).

Wüstehube, J., *et al.*: *Switching Power Packs; Principles, Design, Examples of Circuits*, expert-Verlag, Grafenau/Württ, and VDE-Verlag, Berlin (1979).

Zach, F.: *Power Electronics; Components, Power Circuits, Control Circuits, Influences*, Springer-Verlag, Vienna, New York (1979).

Journals

Baer, G., and Wollschläger, M.: Power supply equipment for teleprint and data switching, *telcom report*, **3**, 203–206 (1980).

Becker, H., and Schmalzl, F.: Small power supply systems independent of the mains, *telcom report*, **3**, 212–217 (1980).

Engert, S., and Weller, K.: Construction and fault finding of power supply equipment for long-distance communication systems, *telcom report*, **3**, 48–51 (1980).

Forstbauer, W., Kublick, C., Schultze, W., Schwarz, R., and Vau, G.: Alternative power supply for all applications, Extended Reprint from *Siemens Journal*, **52**, July 1978, No. 7, 393–444 (1978).

Ganzer, E., Knesewitsch, J., and Ziegler, A.: Power supply equipment for long-distance communication systems, *telcom report*, **3**, 207–212 (1980).

Herfurth, M.: Selection circuits for SIPMOS transistors in switching mode, *Siemens components*, **18**, No. 5, 218–224 (1980).

Hermanspann, F., and Probst, H.: Power supply equipment for public and private telephone switching technology, *telcom report*, **3** 197–202 (1980).

Hertneck, K., Kübler, K., and Ponzer, F.: Central power supply for telecommunications systems, *telcom report*, **3**, 190–196 (1980).

Krakowski, H.: Telecommunications power supply, *Fernmelde-Ing.*, **1972**, Nos 9 and 10 (1972).

Krakowski, H.: Uninterrupted power supply for EDP systems, *etz-b*, **27**, No. 2, 39–41 (1975).

Krakowski, H., and Schott, H.: Power supply – an important factor for the reliable operation of world-wide communications links, *telcom report*, **3**, 185–189 (1980).

Neugebauer, K., and Schwarz, R.: Uninterrupted power supply for EDP systems; increasing the reliability and availability of EDP systems by means of an uninterrupted power supply, *etz*, **100**, No. 12, 606–610 (1979).

Rambold, K.: Power supply equipment for communications engineering – special features and applied principles, *telcom report*, **1**, 404–409 (1978).

Specifications and Standards

VDE 0100 Conditions for the equipping of power plants with rated voltage up to 1000 V.

VDE 0160 Part 1: Conditions for fitting out power plants with electronic equipment; Part 1: Installations with electronic equipment for information processing in power plants.

VDE 0160 Part 1b: Conditions for fitting out power plants with electronic equipment; Part 1: Installations with electronic equipment for information processing in power plants.

VDE 0556 Conditions for polycrystalline rectifiers.

VDE 0557 Conditions for single-crystal rectifiers.

VDE 0800	Part 1: Conditions for equipping and operating telecommunications systems, including information processing systems; Part 1: General conditions.
VDE 0800	Part 1c: Conditions for equipping and operating telecommunications systems, including information processing systems; Part 1: General conditions.
VDE 0800	Part 3: Conditions for equipping and operating telecommunications systems, including information processing systems; Part 3: Special conditions for systems with remote power supply.
VDE 0804	Conditions for telecommunications equipment, including information processing equipment.
DIN 19226	Automatic control engineering; definitions and names.
DIN 40108	Electrical power engineering; current systems, definitions, magnitudes, symbols.
DIN 40110	Alternating quantities.
DIN 40700	Part 8: Graphical symbols; semiconductor components.
DIN 40700	Part 14: Graphical symbols; digital information processing.
DIN 40700	Part 22: Graphical symbols; digital information processing, memory logic elements.
DIN 40706	Graphical symbols; power converters.
DIN 40729	Voltaic secondary cells (accumulators); definitions.
DIN 40736	Part 1: Lead storage batteries; fixed cells with positive box-type plates, cells in plastic containers, capacities, main dimensions, weights.
DIN 40736	Part 2: Lead storage batteries; fixed cells with positive box-type plates, cells in hard rubber containers, capacities, main dimensions, weights.
DIN 40737	Part 1: Lead storage batteries; fixed batteries with positive box-type plates, batteries in hard rubber monobloc containers, capacities, main dimensions, weights.
DIN 40737	Part 2: Lead storage batteries; fixed batteries with positive box-type plates, batteries in plastic monobloc containers, capacities, main dimensions, weights.
DIN 40738	Lead storage batteries; fixed batteries with positive large-area plates, close-fitting, capacities, main dimensions, weights.
DIN 41740	Part 1: Selenium diodes; definitions.
DIN 41740	Part 2: Selenium diodes; marking and general recommendations for particulars in data sheets.
DIN 41740	Part 3: Selenium diodes; measuring and testing methods.
DIN 41745	Stabilized power supply equipment; definitions.
DIN 41750	Part 1: Power converters; definitions for semiconductor converters, construction.
DIN 41750	Part 2: Power converters; definitions for converters, types and designations.
DIN 41750	Part 3: Power converters; definitions for converters, commutating, drive, electrical quantities.
DIN 41750	Part 4: Power converters; definitions for converters, mains operated converters for rectifying and inverting.

DIN 41750 Part 4, Supplement: Power converters; definitions for converters, calculation instructions for mains-operated converters for rectifying and inverting.

DIN 41750 Part 5: Power converters; definitions for converters, self-operated converters.

DIN 41750 Part 6: Power converters; definitions for converters, load-operated converters.

DIN 41750 Part 7: Power converters; definitions for converters, trigger sets.

DIN 41751 Power converters; semiconductor converter stacks and converter equipment, types of cooling.

DIN 41752 Power converters; semiconductor converter equipment, performance characteristics.

DIN 41755 Part 1: Superpositions on a direct voltage, periodic superpositions, definitions, measurement method.

DIN 41756 Part 1: Power converters; loading of converters, modes, load groups.

DIN 41756 Part 2: Power converters; loading of converters, types of load with direct current.

DIN 41756 Part 3: Power converters; loading of converters, types of load with alternating current.

DIN 41760 Power converters; polycrystalline rectifier plates and stacks; definitions and rated values.

DIN 41761 Power converters; converter circuits, designations and identification characters.

DIN 41761 Supplement: Power converters; converter circuits, designations and identification characters; examples.

DIN 41762 Part 1: Power converters; power characteristics for semiconductor converter stacks; polycrystalline rectifier stacks.

DIN 41762 Part 2: Power converters; power characteristics for semiconductor converter stacks; single-crystal converter stacks.

DIN 41772 Power converters; semiconductor rectifiers, shapes and abbreviations of characteristics.

DIN 41772 Supplement 1: Power converters; semiconductor rectifiers, examples of characteristic for chargers.

DIN 41772 Supplement 2: Power converters; semiconductor rectifiers, examples of characteristic for equipment in parallel mode with batteries.

DIN 41774 Power converters; semiconductor rectifiers with W characteristic for the charging of lead storage batteries, recommendations.

DIN 41777 Part 1: Power converters; semiconductor rectifiers and systems, basic requirements for the maintenance charging of lead storage batteries.

DIN 41777 Part 2: Power converters; semiconductor rectifiers and systems, controlled and regulated equipment with constant voltage characteristic for the maintenance charging of lead storage batteries.

DIN 41777 Part 3: Power converters; semiconductor rectifiers and systems, unregulated equipment with W characteristic for the maintenance charging of lead storage batteries.

DIN 41781 Rectifier diodes for power electronics; definitions.

DIN 41782 Rectifier diodes; recommendations for data sheet particulars.

DIN 41783 Single-crystal rectifier cells; measuring methods.

DIN 41784 Part 1: Thyristors; measuring and test methods.

DIN 41785 Part 1: Semiconductor components; abbreviations for use in data sheets, construction of the abbreviations.

DIN 41785 Part 2: Semiconductor components; abbreviations for use in data sheets, abbreviations for semiconductor components in communications engineering.

DIN 41785 Part 3: Semiconductor components; abbreviations for use in data sheets, abbreviations for semiconductor components in power electronics.

DIN 41785 Part 4: Semiconductor components; abbreviations for use in data sheets, abbreviations for digital binary microcircuits.

DIN 41785 Part 5: Semiconductor components; abbreviations for use in data sheets, abbreviations for linear integrated amplifiers.

DIN 41786 Thyristors; definitions.

DIN 41787 Thyristors; recommendations for data sheet particulars.

DIN 41789 Marking of rectifier diodes and thyristors.

DIN 41791 Part 1: Semiconductor components for communications engineering; particulars in data sheets, general.

DIN 41791 Part 5: Semiconductor components for communications engineering; particulars in data sheets, power transistors.

DIN 41791 Part 6: Semiconductor components for communications engineering; particulars in data sheets, switching transistors.

DIN 41794 Part 9: Reliability details for semiconductor components; thyristors.

DIN 41852 Semiconductor technology; definitions.

DIN 41855 Semiconductor components and integrated circuits; types and general definitions.

DIN 57105 Part 1/VDE 0105 Part 1: VDE regulation for the operation of power plants; general conditions.

DIN 57160 Part 2/VDE 0160 Part 2: VDE regulation for the equipping of power plants with electronic equipment; installations with power electronics equipment in power plants.

DIN 57411 Part 1/VDE 0411 Part 1: VDE regulation for electronic measuring instruments and controllers; Protective measures for electronic measuring equipment.

DIN 57411 Part 1a/VDE 0411 Part 1a: VDE regulation for electronic measuring equipment and controllers, part amendment a [VDE Regulation].

DIN 57510/VDE 0510: VDE regulation for lead storage batteries and battery installations.

DIN 57558 Part 1/VDE 0558 Part 1: VDE regulation for semiconductor power converters; general conditions and special conditions for mains-operated converters.

DIN 57558 Part 1a/VDE 0558 Part 1a: VDE regulation for semiconductor power converters; general conditions and special conditions for mains-operated converters, part amendment a [VDE Regulation].

DIN 57558 Part 2/VDE 0558 Part 2: VDE regulation for semiconductor power converters; special conditions for self-operated converters.

DIN 57558 Part 3/VDE 0558 Part 3: VDE Regulation for semiconductor power converters; special conditions for d.c. chopper controllers.

DIN 57800 Part 1d/VDE 0800 Part 1d: Regulations for setting up and operating telecommunications systems, including information processing systems; Part 1: General conditions, part amendment d.

DIN 57800 Part 2/VDE 0800 Part 2: Telecommunications engineering; earthing and potential equalization [VDE Regulation].

DIN 57804/VDE 0804: Telecommunications engineering; manufacture and testing of equipment.

DIN 57871/VDE 0871: Radio noise suppression of high-frequency equipment for industrial, scientific, medical and similar purposes [VDE Regulation].

DIN 57874/VDE 0874: VDE guidelines for measures for radio noise suppression.

DIN 57875/VDE 0875: VDE regulation for the radio noise suppression of electrical equipment and systems.

Index